The Saucer Fleet

Jack Hagerty
and
Jon Rogers

An Apogee Books Publication

The Saucer Fleet

Copyrights

The Saucer Fleet (1st Edition)

Copyright © 2008 Apogee Books/Jon Rogers/Jack Hagerty

ISBN 978-1894959-70-4

All rights reserved. No part of this book may be used or reproduced in any manner whatsoever

without written permission except in the case of brief quotations embodied in critical

articles and reviews.

We acknowledge the financial support of the Government of Canada through the Book

Publishing Industry Development Program for our publishing activities. Printed and bound

in Canada. Apogee Books is an imprint of Collector's Guide Publishing Inc. Box 62034, Burlington,

Ontario, Canada, L7R 4K2

www.apogeebooks.com

Jack's Dedication

This book is dedicated to Jonathan Harris, Les Tremayne, Al Nozaki, Robert Wise, Bebe Barron and especially Jack Hagerty, Sr., all of whom (except the last) played significant roles in creating the productions covered in this book, and all of whom passed away while it was being written. The last is, of course, my dad (see page 228), who at least got to see that photo published in an excerpted version of the Disneyland Flying Saucer chapter in *Filmfax* before he died.

Jon's Dedication

This book is dedicated to the men who created the original Space Heroes of the '30s-'50s, Ed Kemmer, Frankie Thomas, Alden McWilliams and all the others. Veterans of a World at War, they personified in reality, the fictional heroes they created for our entertainment. They gave us real Heroes to aspire to and helped us lift our dreams into the Starry Heavens.

The Saucer Fleet

Table of Contents

Table of Contents

Acknowledgments

Okay, this is the part of the book that no one ever reads unless you're a contributor, and even then you just check your own entry to see if I spelled your name right. Well, I think that it might be worth a few seconds to scan down the list and say a little "thank-you" to all of those that made this book possible.*

Jack's List

Here we go (in alphabetical order):

Jean-Noel Bassior – For guidance with the ins and outs of dealing with a real publisher.

Bob Burns – For telling me where the bodies (props) are buried!

Mike Dunn – For supplying many of the rare images for several of the chapters.

Bob Gurr – For reviewing the *Disneyland's Flying Saucers* chapter (twice!) and helping lay to rest the "Jupiter Missile" myth surrounding them.

Ray Harryhausen – For reviewing the *Earth vs. the Flying Saucers* chapter (twice!) and making sure I kept it in line with the others. It's a thrill working with one of my childhood heroes!

Frederick Hodges – For his thorough reading and comments on the *Lost in Space* chapter.

Justin Humprheys (George Pal's biographer) – For giving the *War of the Worlds* chapter the once-over and providing many insights.

Mike Jensen – For lending his Shakespearian expertise to the *Forbidden Planet* chapter.

Mike Kickham – Webmaster of the *Jupiter 2 Flightdeck* website, for copying his personal set of Fox studio blueprints for the 4 foot *J2* miniature, and then going beyond the call of duty and posting questions I had in some *Lost in Space* chat rooms (where the *real* fanatics are!).

Arnold Kunert (Harryhausen's friend and agent) – For your uncanny ability to spot typos and punctuation errors, and general eagerness to help (are you sure you're a movie producer?).

Henri Nier – For doing a yeoman's job in proofreading the raw chapter text with an eagle eye.

Steve Payne – For so enthusiastically sharing photos of his magnificent *Jupiter 2* model.

David Peer – For graciously emptying his substantial personal archive on *Forbidden Planet* material, photocopying it and shipping it to me without so much as a postage charge!

Robert Rowe – For straightening out the production history on *Lost in Space*.

Bob Stephens – For letting me condense down his entire article on *Day the Earth Stood Still* for the end of that chapter.

Michael Stine and James Wilson – the publisher and editor (respectively) of *Filmfax* magazine for giving early versions of most of the chapters an initial public airing (sometimes the cover story!) which is a great way to find mistakes and omissions.

Elliott Swanson – For the wonderful 3-D shot of the Disneyland Flying Saucer ride (and a whole bunch more that I can't come up with a reason to put in the book, as much as I'd like to).

Bob Welch (my father-in-law) – For helping me identify the military hardware in *War of the Worlds* and *Earth vs. the Flying Saucers*.

Robert Wise (yes, **the** Robert Wise) – For answering, however briefly, my question on the origin of Klaatu's language in *The Day the Earth Stood Still*. Sorry you didn't make it to see the book in print.

The entire staff at Terumo Medical Corporation, Fremont, for their tolerance in the amount of time putting this book together (especially at the end) kept me away from helping push back the frontiers of medicine

…and to Mark Deger for no particular reason!

---

* No prizes for anyone who recognizes that this is the exact same acknowledgement introduction as used in *Spaceship Handbook*!

Acknowledgments

Jon's List

One of the things one discovers about writing a lengthy nonfiction book is that learning is a community affair. The people listed below have contributed mightily to my understanding of the subject and thus helped me make this a knowledgeable and in-depth book on famous fictional flying saucers. My heartfelt thanks goes out to all who contributed.

Elisa A. Rogers – Every writer should have the good fortune to have the loving support that I have had from my wife. She also helped in proofreading and catching flaws that escaped my notice. Thank you, Lisa.

Alden McWilliams – The chapter on *Twin Earths* is gratefully dedicated to Alden "Mac" McWilliams who produced some of the most clear, precise, and elegant drawings the newspaper adventure strips have ever seen.

Roger Steffens – The chapter on *Twin Earths* would not have happened at all if it had not been for the kind help from Roger Steffens. He took time out from his busy schedule of writing, acting and passionate promotion of Reggae music to put me in contact with the people who made it happen. I can't thank you enough, Roger.

Mark Martucci – Mark is probably the most knowledgeable and dedicated collector of *Twin Earths* on the planet. It was only through his kind sharing of his private collection that I was able to put together an adequate presentation of Alden McWilliams' artwork. I owe that and more to Mark and I thank you very much for all your help.

Chris McWilliams – I would also like to thank Chris, Rick, and the entire McWilliams family for their help, understanding and permission to reprint Alden McWilliams classic adventure series *Twin Earths*.

Mark Phillips – I have to admit that the chapter on *The Invaders* would not have had anything close to the information and detail that it has, had it not been for the research, diligence and cooperation of longtime fan, Mark Phillips. Mark's research and help in uncovering behind the scenes details was invaluable to my being able to retell the story of this very significant TV series. Thanks, Mark.

Richard Ingram – Thanks also go to Richard Ingram for his assistance in helping me uncover the history and details of *The Invaders* model saucer.

Bill Hedges – The Lost in Space's *Jupiter 2* would have been far more difficult if it hadn't been for Bill Hedges. His discovering and sharing of original studio blueprints of the Props used in the Jupiter II production sets made my drawings much more accurate. Thanks Bill. You can find him at http://tv.groups.yahoo.com/group/LostinSpaceProps/

Ron Goss – My analysis of the *Jupiter 2* from Lost in Space also benefited from Ron's helpful assistance. Many thanks, Ron.

Christopher Krieg – Chris deserves a tip of the hat for his support in my producing a report on the Gemini and Jupiter spaceships. His technical review of Lost in Space can be found at http://www.jupitertwo.com/

Sometimes a lot of work goes into material that doesn't make the final cut. Rory Coker and Griffith Ingram deserve a lot of credit for helping me on research for Chapter 1, some of which unfortunately didn't get used. Griffith especially helped with the research on the Lee-Evans Flying Saucer. Also My thanks go out to Jim Klotz and his website on UFOs in Popular Culture (http://www.ufopop.org/) for his advice and assistance in presenting evidence of the pervasiveness on flying saucers in the 1950s culture.

I must also mention Jean Noel Bassior's help in many of the more general aspects of putting this book together. Thanks Jean, your help and encouragement was a blessing.

In addition, I'm sure there are several other people who helped me in my research on subject material for this book but due to the length of time and space it took to get it out, I'm afraid I've misplaced their names. Nonetheless, I am grateful for their help and assistance and I apologize for my lapse of memory.

The Saucer Fleet

Foreword

By Dr. Phil "The Bad Astronomer" Plait

The first time I saw *Earth vs. the Flying Saucers*, I was with 700 of my closest friends.

OK, I was actually at a screening of the movie at a science fiction convention in Baltimore, and I only knew three of the people in the auditorium, my buddies who were attending the con with me. But by the end of the movie, the other 697 and I had bonded.

Earth is not the world's best movie. By today's standards it's clunky, slow, and lacks any sort of kludged-on subtext. But we *loved* it. At first the audience participation was reserved; the occasional chuckle after a whispered snarky comment, or a loud moan when a particularly clichéd line was uttered.

But then came the scene where the saucer aliens drain the mind of a military officer, and when they were through with him, they callously chucked him out the airlock to fall to the ground.

I remember there being silence for a moment, and then we all roared. Laughing, none of us (who hadn't seen the movie previously) could believe how cruelly the aliens were depicted. At that point we were unrestrained, yelling out comments, making awful puns, and generally cheering on the aliens in their attack. When the scene turns to Washington DC, with the humans gaining the upper hand, a saucer is seen hovering over the Capitol Building. I started yelling out, "Hit the dome! Hit the dome!" And as if on cue, the wavering saucer crashes right into the building, no doubt advancing the cause of government twenty years. The audience erupted.

I love that movie. It's rarely on TV any more, but if I happen to stumble on it while channel surfing, it's a sure fire time for the remote control to catch its breath.

Other must-see movies for me? *War of the Worlds* (the 1953 George Pal version; I heard a remake was done in 2005, and some ridiculous rumor that Tom Cruise was in it, but obviously these must be completely wrong because THE 1953 VERSION WAS THE ONLY ONE EVER MADE! Are we clear on that?), *Forbidden Planet*, *The Day the Earth Stood Still*, and *This Island Earth*[*].

When these movies are on TV, I *must* watch them. It's a compulsion (my wife calls it a sickness, but she thinks Gort looks "cheesy" so her opinion is suspect). Why? Well, there are grand themes to them, challenging ideas, wonderful visions…

Oh, who am I kidding. They have aliens! And flying saucers!

Face it: flying saucers are *cool*. That may surprise you, since I am a hard-headed skeptic, given to lecture at the drop of a hat how aliens are *not* visiting us, abducting us, and eating our cows' backsides, and most certainly they aren't using interstellar dinnerware as their transport.

But man, I sure do love flying saucers in movies. They're awesome! They zip around, make nifty noises, and there is just something compelling about the shape. I'm not surprised that Kenneth Arnold's initial sighting in 1947 started a craze. And as silly as I think the UFO culture is, I'm kinda glad it spawned such wonderful movies. They shaped my childhood.

And here we have my friend Jack Hagerty, whose knowledge of rockets both real and imagined is almost frightening in its depth and completeness. His *Spaceship Handbook* has an honored place on my bookshelf, and hardly a week passes without me opening it up at random and simply taking in the simple passion of someone who loves rockets.

Now he has taken on the mantle of saucer expert. He plumbs the depths of saucers in movies, including the ones I mention above. But he doesn't just leave us there with a mere dictionary description of the aliens' ships; he goes into loving detail of the plots, the background of the movies, the history of the people involved. This is more like a biography of these movies, written by someone who clearly loves them as much as I do.

And that makes me wonder… where was Jack in 1981? Did he happen to be sitting in the back row of the movie room at Balticon all those years ago, hooting and hollering along with the rest of us? Maybe he was, but I can't help thinking that if he did go to that con, and he did see *Earth* with the rest of us, he was actually sitting in the front row, taking notes.

So, for me and on behalf of the other 699 of my closest friends, as well as the reader who is holding this book in their lucky hands right now: thanks.

---

[*] This Island Earth has the distinction of having the worst line ever said in a movie, ever. At one point, a character introduces the laboratory's pet cat saying, "That's our cat, Neutron. We named him that because he's always so positive!"

Preface

Hundreds of books have been written about Flying Saucers. Why write another one?

In an effort to preserve part of the history of the development of the spaceship, in 2001 we published *Spaceship Handbook*. It contained in-depth background information and scaled outline drawings of many theoretical, fictional, and anticipated spaceships of the 20th Century. The entries were selected because they were, in some way, historically important. While the book was widely acclaimed by readers, critics, and scholars, some readers noted that we had completely omitted one class of spaceships: Flying Saucers.

We acknowledged the omission and defended it on the basis that it had been necessary to produce the work in the time allowed. We also promised that there would be a follow on book that would treat flying saucers in the same manner as the vehicles in *Spaceship Handbook*.

The problem with doing a book on saucers in the style of *Spaceship Handbook* is that flying saucers have no real connection with the history of space flight. Perhaps they will someday, but currently there are no flying saucer type spacecraft in any space program, or even under development. This means we couldn't use the same selection criteria we had used in *Spaceship Handbook*.

Then original meaning of flying saucer is very different from the one it has today. For nearly a hundred years the term, "flying saucer" referred to a type of aircraft. Human beings designed them, built them, and flew them. During all that time, there was no mention of flying saucers as spaceships in popular culture. There were no flying saucer books, movies, songs, or drawings.

All that changed in the summer of 1947 when a fleet of UFO's suddenly appeared in the skies over the U.S. Newspaper headline writers named the UFOs, "Flying Saucers" after misquoting Kenneth Arnold. It was sometime later that "Flying Saucer" acquired the "extraterrestrial spaceship" connotation it still carries. As a result, today the majority of society believes that flying saucers are spaceships from outer space.

Whether or not UFOs actually are flying saucers from outer space, being the craft for extraterrestrials is what all of the other hundreds of flying saucer books are about. That, however, is not the focus of this book; this one is about something different.

We intended this to be a book on historically important flying saucers similar to *Spaceship Handbook*. But what would that be? Since flying saucers haven't contributed to actual human space flight, we decided it would be about flying saucers and how they contributed to the history of the spaceship.

Further, to create an engineering data drawing of a particular flying saucer, similar to the ones in *Spaceship Handbook*, there has to be sufficient information known about it. You can't create a detailed engineering drawing of a fuzzy light in the sky (well, you can, but it would be meaningless). This means excluding UFOs and all the first person eyewitness accounts about them. That information is completely subjective and not independently verifiable. Are there any flying saucers left that are both historically important and have sufficient, documented, physical evidence to do an engineering drawing of them? As it turns out, there are.

Flying saucers that contributed to the history of spaceships appeared in various media. With these we can go back and see what was originally presented to the public that helped establish the public's belief in flying saucers as spaceships. Among the first places flying saucers appeared, besides advertising promotions and hoaxes, were in magazine articles and books. These two sources were very influential in presenting the theory that UFOs were flying saucers from another planet. However, all of the articles and books were about UFO sightings and only peripherally about saucers. There wasn't enough information given in any of these sources that allowed documentation of any specific version.

The first places where we could get sufficient information about specific examples of flying saucers were in the drawn art of the day, mostly the various comic books of the era. But, with one major exception, the drawn art of comics were not very influential. The exception was the syndicated adventure strip *Twin Planets*. For over ten years it appeared in hundreds of daily newspapers worldwide carrying images of flying saucers to vast audiences.

But the media that were the most important in spreading the idea of flying saucers were the movies and, eventually, television. Here the public could and did see believable flying saucer spaceships. Of the two, the movies were by far the most influential in persuading people that flying saucers were alien spaceships. For years, the most popular Sci-Fi films featured flying saucers. The continuing presence of flying saucer spaceships in science fiction movies added much support to the general belief that this is what real flying saucers (UFOs) were.

Preface

So this flying saucer book is different. It is not about Area 51, Roswell or almond-eyed Grays. *The Saucer Fleet* is a detailed in-depth look at real flying saucers created right here on Earth. What you will find here is a detailed and meticulously researched reference about some of the most historically important flying saucers as shown in art, television, and some of the most popular science fiction films of all time. By examining the social history of flying saucers in this way, we think it will help you understand the UFO phenomenon in a different light.

We will omit conjecture and will stick to the facts. We include the history, the context, and many generally unknown facts about them from insiders. There are some simple, strange, and fantastic examples of real flying saucers within these pages. Millions of people have seen these saucers, including you. But, there are many things about them you probably didn't know.

And, if, after the reading of this book, you come to a different understanding of the flying saucer/UFO phenomena than you had before, wonderful. But consider it a side benefit. Our main mission here is to help you discover some new facts about real flying saucers and how their message influences our society. We want to give you a new appreciation for the tremendous human effort that went into creating these popular media events.

What's In This Book

Each entry in this book follows a similar pattern. First, an introductory section gives the background material putting the design in historical context: who did it, what they did, why they did it and what else was going on while they did it that might have influenced them. This is followed by a section summarizing the story, after which there is a section describing the vehicle itself. In most chapters, following the Vehicle section is an "Archeologists Report" by Jon Rogers describing how he researched the physical characteristics of each saucer for the Data Drawings and why the drawings are the way they are. The report naturally includes the actual Data Drawing(s). New to the drawings in this book are sections through the saucers to show what's inside, as best we can determine. For those few chapters without a separate Archeologists Report, the Data Drawing is included in the Vehicle section. Finally, any personal opinions, analysis or other peripheral information on the subject is placed at the end of the section in an "Epilog."

For those not familiar with them, Data Drawings present the dimensional and color data necessary to build an accurate model, if you're so inclined. Note that this is data on the "real" vehicle, not any sort of model design. Turning these data into a model is completely up to you. However, this is not just a book for modelers. The larger audience of science fiction fans and spaceflight enthusiasts couldn't care less about building models. So to not bog down those good people with the extraneous materials, after the vehicle description text there is a Modelers' Note describing any kits that have been produced over the years (if any) and, if we're lucky, examples of builds done by some amazing modelers. Note that this brief section (set off by a grey background to make it easy to skip if you're not interested) is not exhaustive; it's more like a sampler. For an authoritative reference on spacecraft model kits, we can suggest *Creating Space* by Mat Irvine, available through Apogee.

Along with the Modelers' Note, there's something called a "Quickspec" This boils the design down to its briefest possible summary. While we used the same format as for the Quickspec in *Spaceship Handbook*, generally only one body morphology is used (saucer, of course!). There is one exception, though. See if you can find it.

Conventions

As with any engineering text, even civilian grade like this one, consistency and clarity of technical information is a major big deal here. We have tried to be as clear and consistent as possible when presenting all of the facts and figures. All dates are given in the unambiguous day-month-year format (with the month spelled out) preferred by the military and most European nations, even though it looks a bit jarring to U.S. civilians. All measurements are specified in English units first with the metric equivalents following in parentheses. In the Data Drawings, the dimensions are given in the units of the original design. Also, the symbol "Φ" (Greek letter "phi") on a dimension means "diameter." Likewise, a capital "R" in front of a dimension means "radius."

Jon Rogers	**Jack Hagerty**
The Mainland	Livermore, CA
28 June 2007	10 September 2008

Introduction

By Jon Rogers

L et's start by defining the concepts central to the theme of this book.

UFO...An Unidentified Flying Object.

Unidentified Flying Object.................A flying or apparently flying object of an unknown nature.

Flying Saucer...............................Any of various unidentified flying objects of presumed extraterrestrial origin, typically described as luminous moving disks.[1]

The term "Flying Saucer," as we know it today, came into being before "UFO." When it was first popularized (more about that in Chapter 1) it meant only "An Unidentified Flying Object." At that time, (1947) the majority of Americans agreed that they did not have any idea what the "Flying Saucers" in the sky were. By the time the word "UFO" came into popular use, people had already been using the term "Flying Saucer" for over a year. The new, more literally correct term "UFO" had to compete with the more established term for acceptance. While this was occurring, the first proposition that UFOs were "extraterrestrial spacecraft piloted by aliens" was introduced. Over time, the two concepts merged giving us the definition we have today: "A Flying Saucer is a UFO presumed to be an extraterrestrial spacecraft piloted by aliens."

For the purpose of this book, we accept the above definition of "UFO." However, we propose, and throughout this book, we will use, the following definition for the term "Flying Saucer."

Flying Saucer – 1) An aircraft or spacecraft with planar-circular wings that resembles a disc. 2) Any artwork or hardware made to indicate an aircraft or spacecraft that is shaped like a disk.

By this new definition, flying saucers are real. They are also flying discs. They can also be extremely advanced spacecraft from outer space, if the designer or artist says so. What they cannot be is "Unidentified" which also means, "Undefined." UFOs are unidentified. Flying saucers are real and the historically important flying saucers that helped create the "Flying Saucer Spaceships" of today are the subject we are examining in *The Saucer Fleet*.

Today, every "UFOlogist" knows that, in the summer of 1947, a fleet of UFOs suddenly appeared in the skies over the U.S. The saucer fleet had arrived, but not landed. Or had it?

Because they were so sensational, flying saucers (representing the UFOs) started appearing everywhere, overnight. They popped up in advertising promotions, magazine articles and books. They showed up in art and on radio. And, most convincingly, they appeared in movies whose plots revolved around their extraterrestrial origin. In this respect, the saucer fleet *had* landed. Today, flying saucers are everywhere in our society.

The science fiction movies of the '50s that featured flying saucers were largely responsible for this infusion. They were sensational. And they looked convincingly real! To achieve this, producers used the "state of the art" special effects of their time. They built models and full sized sets of saucers. They used real flying saucers (by our new definition) in their productions. They were as impressive to moviegoers of their day as our ultra hi-tech special effects *Star Wars* movies are for us now.

Also, in order to enjoy the movie, moviegoers then, as today, suspended their "faculty of healthy disbelief" upon entering the movie theater. Because of this, they uncritically assimilated the sub-theme or hidden message within the movies' plots (i.e. flying saucers/UFOs are alien spaceships[2]). Over time these movies, and to some degree, the art of the day, redefined what a flying saucer was for most people in society. As a result, that definition is still with us. Today, a large segment of the population truly believes that alien flying saucer spaceships have visited, and even landed on Earth.

The Saucer Fleet is not a study of the belief in extraterrestrial visitations or abductions. We will leave that to the UFOlogists. Also, it is not about the flying saucer in popular culture. For a more in-depth treatment of that subject, I recommend Eric and Leif Nesheim's book, *Saucer Attack!*[3] It shows how pervasive flying saucers were in all forms of media during the

---

[1] All definitions: American Heritage® Dictionary of the English Language, Fourth Edition. 2000.

[2] At first considered wildly ridiculous, this idea followed Nazi Propaganda Minister Joseph Goebbels' principle, "If you tell a lie big enough and keep repeating it, people will eventually come to believe it." The idea that flying saucers were spaceships was BIG, promoted convincingly and became "fact."

[3] *Saucer Attack!* ©1997, Eric Nesheim, Kitchen Sink Press/General Publishing Group, Inc. Los Angeles, ISBN 1-57544-066-0

1950s, the "Golden Age of Flying Saucers."

What we will examine is the major cause of the flying saucer social phenomena—the great science fiction flying saucer movies of the '50's. Since their impact on society still exists today, we feel they deserve close scrutiny. We wish to help the reader better understand the method and magic behind them.

We will look at the world of comic art that was also instrumental in introducing the flying saucer as a spaceship to the general public. That was where the flying saucer spaceship became an "icon," and was spread to millions of readers world-wide.

We will examine the few, but important TV series that featured flying saucers. They also had a large influence on our society. They kept the public entertained and continued to promote the new concept with stories of flying saucers that reached across the galaxy.

Unfortunately, although flying saucer stories on Radio in the 50s also had an impact on society, we cannot examine them like we can the movies and TV, as they left no visual evidence.

Finally, we will also reveal some little known information about the great UFO scare of 1947. Perhaps, in reading this factual presentation of the history of UFOs and real flying saucers, some people will realize that they have willfully suspended their "faculty of healthy disbelief" far too long. However, this is an adjunct background story. It is told so the reader can see how the UFO scare prepared the way for the development of our beliefs today.

That UFOs are flying saucer spaceships is accepted as fact by many. What few realize is that the cause of that "accepted fact" is a fascinating, behind the scenes story, of how real flying saucers were used to persuade the public that some unknown light in the sky was "in fact," a spaceship from another planet.

It is the story of these real flying saucers that we wish to present in *The Saucer Fleet.*

Jon Rogers

The Mainland

June 28, 2007

The Saucer Fleet

The Coming of the Fleet

By Jon Rogers

"This is not anymore a war than there's a war between men and maggots!" [1]

It wasn't a war. The extraterrestrial tripod killers were moving casually among the fleeing, screaming, panicked hoards of humanity annihilating whoever and whatever they pleased with total impunity. They were like gigantic butchers walking among us and we were like flocks of trapped chickens. We were helpless. All mankind, all civilization was being destroyed before my very eyes. It was a horrible scene to watch.

Finally, the image flickered and the screen dimmed. Steven Spielberg's recreation of H. G. Wells' *War of the Worlds* was over. As I turned off the video, I had to admit that Spielberg had really brought the old classic horrifyingly to life. I was thankful I didn't live in that world!

It also reminded me what a powerful medium movies are. Their ability to create reality—to evoke emotional response in the viewer is greater than any other art form. When you watch them, the real world vanishes and you are in another world. Movies can make you believe almost anything.

I stepped back into my office and sat back down at my keyboard. I looked up at the screen. I had been writing a book about the early flying saucer phenomena and its relationship with movies of the 1950's when I had stopped for dinner and had tried—unsuccessfully—to relax with a movie. There on the monitor were the words I'd been writing:

> The true story of flying saucers did not begin in 1947. It did not begin on a cave wall. The true story of flying saucers began with the earliest, struggling aviation pioneers. The birth of aviation was the birth of flying saucers.

I had researched their history and could prove that fly-

ing saucers had been invented, developed, and flown by humans, not aliens. Stemming from the earliest times, they had been a unique type of aircraft (Figure 1) and would have, given time, been developed into a unique type of spacecraft.

But then they fell victim to the UFO panic of the summer of '47. Ironically, they had their name stolen by the very thing they would have become…in time. Now, no one hears of real flying saucers, just alien flying saucer spaceships.

I looked at my notes:

> Why had the UFO sightings caused such a mass panic? There had been UFO sightings for hundreds of years, why was 1947 any different?

I'd thought about this earlier. Of course there had been UFO sightings that year. Just like there were before, like there would be afterward, and like there are today. Why would mankind, especially Americans, view them so differently? Why were UFOs such a threat in 1947?

The nature of UFOs hadn't changed. It must have been the nature of the people seeing them. Humans had changed somehow. And not just a few humans, large numbers of them had changed. How?

I thought of Spielberg's *War of the Worlds*. The obvious answer was a single, ugly word. *Fear!*

An old fear. A fear of unknown things in the sky!

Humans change individually as a result of their own decisions. But they also respond to their environment in order to survive. If a large number of people change significantly at the same time, then the probable cause is a significant change in their environment.

But what was there to fear in 1947? World War II had been over for almost two years. There was peace in the world wasn't there? Then I remembered a scene in *Independence Day* in the lab at Area 51 where the alien invader has his tentacles around Dr. Okun's throat, his face slammed up against the glass, and is using his voice to hiss at us,

Peace?   Nooo Peeaacceee!!!

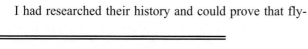

Figure 1
The 1913 Lee-Richards Flying Saucer Monoplane

---

[1] Harlan Ogilvy (Tim Robbins) in *War of the Worlds*, Paramount Pictures, 2005

The Saucer Fleet

The place to start seemed obvious. Check the environment just prior to the summer of 1947, starting with our own, self-made *War of the World*.

1938-1945: The World at War

In my research I discovered that a major cause for the post war UFO hysteria came out of WWII. The cause was an idea. Before the war began, Italian general Giulio Douhet had written an influential military treatise calling for the unrestricted use of bombers in the next war. In order to end a war quickly, he advocated putting civilian populations on the front line and terrorizing them with aerial bombings. This was a major change from the previous methods of warfare. His theory, that it would break the enemy's will to fight, would be disproved by events of the war. However, this did not stop the governments and military from pursuing this policy until the end of the war. Throughout WWII, people would hear stories on the radio or read them in the newspaper about what different military forces were doing to civilian men, women, and children:

Table 1 - Selected WWII Civilian Casualties		
How many killed	**Where**	**When**
German Military		
1,000 civilians	Guernica, Spain	26 April 1937
20,000 civilians	Poland	August to September 1939
20,000 civilians	London (the Blitz)	September 1940 to May 1941
40,000 civilians	Stalingrad	23 August 1942
Japanese Military		
260,000 civilians	Nanjing	December to January 1938
25,000 civilians	Singapore	February to March 1942
100,000 civilians	Manila	February 1945
British and/or American Military		
50,000 civilians	Hamburg	July to August 1943
3,000 civilians	Berlin	23 November 1943
25,000 civilians	Dresden	14 February 1945
84,000 civilians	Tokyo	9 March 1945
78,000 civilians	Hiroshima	6 August-1945
70,000 civilians	Nagasaki	9 August-1945

September 1945: – the War is over!

On 3 Sept 1945, the Japanese formally surrendered, ending World War II.

After hearing news stories like those outlined in Table 1 for almost eight years straight, wouldn't you be glad you didn't live in that part of the world? Wouldn't you be happy to still be alive? Up to the end of the war, you might have been worried that you might be next, but not now, right?

However, actions against civilians during the war told anyone living in 1945 that governments not only had more destructive power than had ever existed before, but that they were quite willing to use it against every living man, woman, and child.

Still, now that the evil governments are gone, we should have some peace at last. At least, that was everyone's feeling in September 1945.

That fleeting sense of security rapidly vanished as people began reading these headlines during the year that followed:

29 October 1945, China: A full-scale civil war has started.

29 November 1945: Yugoslavia has gone Communist.

9 February 1946, USSR: Stalin gives speech saying communism and capitalism cannot coexist. Washington analysts call this a "Declaration of WWIII."

24 April 1946, Austria: Soviet fighters shoot at a US transport plane.

24 May 1946, Sweden: Two night watchmen sight what seems to be a V1 "buzz bomb."

14 June 1946, San Francisco: Bernard Baruch on giving atomic control to the UN "We are here to make a choice between the quick and the dead. We must elect world peace or world destruction."

12 July 1946, London Daily Telegraph: "Ghost rockets" have been reported [flying] over Sweden...a new kind of radio-controlled V-weapon...is in the air today.

21 July 1946, NY Times: *Air Forces to Show JD-2 Rocket Bomb.*

25 July 1946, Bikini Atoll: Fifth US nuclear detonation.

19 August 1946, NY Times: *Russians Unwrap Jet Fighter Plane - 2 Rocket-Propelled Planes.*

26 August 1946, Newsweek: *Russia: A Warning in the Rocket's Glare?* This was the Russian reply to Bikini, the flight of self-propelled missiles over Sweden.

12 September 1946, Madison Square Garden: Commerce Secretary Henry Wallace, "He who trusts in the atom bomb will sooner or later perish by the atom bomb!"

<u>October 1946</u>, Gaedheim, US Zone Germany: Rocket-like object streaks from Soviet Zone, explodes near the town.

There it is in plain English. Peace or perish. It has only been one year since the end of WW II and there has already been confrontation in the skies over Europe, missile testing over neutral countries, development of even newer planes and missiles, and now a government official tells us that our own super weapon, the atom bomb will not keep us safe!

I keep seeing that alien invader from *Independence Day* hissing at us,

Peace?   Nooo Peeaacceee!!!

1947: In the final days before the Fleet

As we moved into 1947, the scary headlines continued:

<u>24 January 1947</u>, Poland: US report on election shows terror and intimidation by Communists. No secret ballot.

<u>14 February 1947</u>, San Francisco: Soviet minister Gromyko demands that US destroy all its A-bombs and not wait for a system of atomic control to be agreed upon.

<u>22 March 1947</u>, NY Times: *"Phantom Bomb" returns to Skies over Sweden.* The missile turned at a certain point and flew back eastward.

<u>23 April 1947</u>, Eastern Evening News (England): *V1 Reported Dropped on French Soil.* A flying bomb exploded yesterday afternoon…about 60 miles south-east of Paris.

<u>May 1947</u>, Gaedheim, Germany: Second time a rocket from Soviet Zone exploded near town.

<u>1 May 1947</u>, Mechanix Illustrated has a cover article on the US Navy's secret Flying Saucer project. (Figure 2)

By June 1947 Americans had either been involved in, or heard about "The War" for eight years, a war in which millions of innocent civilians had died violently, killed by one government or another. And it had only stopped after both sides had acquired and used newer, more horrible weapons than had ever been seen before. By war's end, the Russians had acquired the German V1 and V2 missiles, and America had developed the atomic bomb.

From the end of WWII there had been broken promises, government takeovers and shooting incidents. Europe, coming out of the worst winter on record, is starving. Britain is on food and coal rations. China is in a civil war. Someone has been conducting guided missile tests over peaceful, neutral European countries. Missiles have even exploded in friendly European countries and they all appear to originate from behind the Iron Curtain.

With those A-bomb tests in the Pacific and mystery rockets flying over Europe, wouldn't any normal person be afraid of strange things they might see in the sky? It could

be those rockets from some foreign country (probably Russia) coming to bomb us, and there's nothing we could do about it.

Why, it could happen any day.

Figure 2
The XV-143 Flying Saucer prototype

June-August 1947: Look! Up in the sky! It's a Fleet of Flying Saucers!

<u>26 June 1947</u>, AP wire story: *Flyer Reports 9 Objects Whizzing Over [Mountain] Range.*

Once you have all the ingredients in place, it only takes one small match to set off a big explosion. In this case the match was not the sighting of some UFO's, but the National News Services (AP, UP) reaction and treatment of it.

The sighting happened at 3 PM on Tuesday, 24 June as Kenneth Arnold, a private pilot, was flying over the mountains of Washington. After seeing the UFOs, Arnold continued on to Yakima, Washington where he told a friend, Al Baxter, about it. Baxter didn't believe him, so Arnold left immediately for Pendleton, Oregon. He stayed overnight there and appeared in the office of the *East Oregonian* newspaper at about noon on Wednesday, 25 June.

The Saucer Fleet

There he had "about a five minute meeting" where he discussed the event with two reporters, Nolan Skiff and Bill Bequette. According to Pierre Lagrange's 1988 interview of Bill Bequette,[2] "Nolan jotted down a few notes, then wrote a short story, which I squeezed into the bottom of page one [of the *East Oregonian*]. Then I punched an even shorter (as I recall) version into the AP wire." With their deadline met, Arnold left and Bequette went to lunch.

Flyer Reports 9 Objects Whizzing Over Range

PENDLETON, Ore., June 26.—(AP)—A tale of nine mysterious objects —big as airplanes—whizzing over western Washington at 1,200 miles an hour got skepticism today from the army and air experts.

The man who reported the objects, Kenneth Arnold, a flying Boise, Idaho, businessman, clung, however, to his story of the shiny, flat objects, each as big as a DC-4 passenger plane, racing over Washington's Cascade mountains with a peculiar weaving motion "like the tail of a kite."

An army spokesman in Washington, D. C., commented, "As far as we know, nothing flies that fast except a V-2 rocket, which travels at about 3,500 miles on hour—and that's too fast to be seen."

The spokesman added that the V-2 rockets would not resemble the objects reported by Arnold, and no high-speed experimental tests were being made in the area where Arnold said the objects were.

A Civil Aeronautics Administration inspector in Portland, Ore., added, "I rather doubt that anything would be travelling that fast."

LIKE PIE PAN

Arnold described the objects as "flat like a pie pan," and so shiny that they reflected the sun like a mirror.

He said he was flying east at 2:59 p. m. two days ago toward Mt. Rainier when they appeared directly in front of him 25-30 miles away at 10,000 feet altitude.

By his plane's clock he timed them at 1:42 minutes for the 47 miles from Mt. Rainier to Mt. Adams, Arnold said, adding that he later figured by triangulation that their speed was 1,200 miles an hour.

BIG AS AIRLINER

He said at first he thought they were geese, but quickly saw they were too big—as big as a DC-4 that was about 20 miles away, he said. The DC-4 pilot reported nothing unusual sighted. Then Arnold said he thought of jet planes and started to clock them, "but their motion was wrong for jet jobs."

"I guess I don't know what they were—unless they were guided missiles," said Arnold, who continued here on a business trip.

"Everyone says I'm nuts," he added ruefully, "and I guess I'd say it too if someone else reported those things. But I saw them and watched them closely. It seems impossible, but there it is."

Arnold said he was 25-30 miles west of Mt. Rainier, en route from Chehalis to Yakima, when he sighted the objects. He explained he had decided to look for a marine corps plane, missing since last January, while he was in the area.

BRIGHT FLASHES

He told a reporter the planes remained visible by the bright, "almost blinding" flashes of reflected sunshine as far as 50 miles away.

The DC-4 was closer than the objects, but at 14,000 feet and somewhat north of him, he said, adding that he could estimate the distance of the objects better because an intervening peak once blocked his view of them. He found the peak was 25 miles away, Arnold related.

He also said they flew on the west sides of Rainier and Adams, adding he believed this would make it more difficult for them to be seen from the ground.

The Boise man said at first he thought the window of his plane might be causing the reflections, but that he still saw the objects after rolling it down.

He also described the objects as "saucer-like" and their motion "like a fish flipping in the sun."

Mostly, he said, he was surprised at the way they twisted just above the higher peaks, almost appearing to be threading their way through the mountains.

Figure 3

Original AP Wire Story by Bill Bequette

[2] Pierre Lagrange is a sociologist who lives in France and has written several works on the UFO phenomena. This interview available on the web at http://brumac.8k.com/KARNOLD/KARNOLD.html

Over that one-hour lunch break it began. During the same interview, Bequette later related:

> When I returned to the office after lunch, the receptionist's eyes were as big as saucers - the kind we use under coffee cups! She said newspapers from all around the country and Canada had been calling. They wanted more details on the "flying saucers." I spent the next two hours with Mr. Arnold in his hotel room. From that interview I wrote a story about 40 column inches long. The story was telephoned to the AP Bureau in Portland. Next morning, almost every newspaper in the country published the story on Page 1.

Figure 3 shows the complete text of the AP story. Kenneth Arnold has said many times since that he did not identify the UFOs as "Flying Saucers" during this interview. Arnold states he told the two reporters they were, "flat like a pie pan and somewhat bat-shaped (Figure 4)" but that, "they flew like a saucer would if you skipped it across the water." Arnold later said, and it has been repeated many times, that the two reporters misquoted him, thus starting the association of "UFOs" with "Flying Saucers."

Figure 4

Kenneth Arnold with an artist's concept of his UFO

What actually occurred appears to be somewhat different than the way Arnold recalled it. Note that there were three articles written on this first day, Nolan Skiff's, Bill Bequette's short article and his later, longer article that he phoned in to the AP Wire services. Apparently Bequette's short article was never printed but it did alert "almost every newspaper in the country." It started them checking for similar news on their own. They then printed his longer story the very next day.

In reviewing the two printed articles, Nolan Skiff's article in the *East Oregonian* states, "He said he sighted nine saucer-like aircraft flying…" Near the middle of Bill Bequette's longer AP story, the one that newspapers repeated all over the country is this paragraph:

LIKE PIE PAN

Arnold described the objects as "flat like a pie pan," and so shiny that they reflected the sun like a mirror.

At the very end of the article is a short, untitled paragraph also quoting Arnold as saying:

He also described the objects as "saucer-like" and their motion "like a fish flipping in the sun".

Due to the fact that the two witnesses who heard his story, wrote independent reports and both included the word "saucer-like" within the text (not the headline) of their articles, it is highly possible that Mr. Arnold's memory of the interview is not accurate. It is more likely that Arnold used the term "saucer-like" at some point in both interviews.

What is important to understand is that each paper that carried Bequette's story that first day, wrote its own headline. "Flyer Reports 9 Objects Whizzing Over Range" is one paper's. A different paper that published a shortened version of the same story headlined it, "Pilot Sticks to Story of Mysterious Flying Objects - Experts Skeptical." On the first day, the newspapers were not calling them Flying Saucers…yet.

However, by the next day, Friday, 27 June, the newspapers begin to run stories titled like, "Bellingham Man Believes He Saw Eerie 'Flying Saucers'" and more conservative papers, ran "Mystery of 'Flying Discs' Getting Deeper; Bellingham Man Declares He Saw 'em Too" for the same story. Note that newspaper headline writers coined the term "Flying Saucers." So by the second day, most newspaper's headline writers were calling the UFOs, Flying Saucers or Flying Discs.[3]

Also, all over the nation, beginning on the 26th, as soon as people read Bequette's story, hundreds of people started calling in UFO sightings. The "panic" was on. There had been UFOs sightings for centuries, including the ones that were still being sighted at that very moment over northern Europe, and seldom had anyone used the term "saucer" to describe them, until now.

Later, saucer skeptics would be quick to claim that, "…the global 1947 Flying Saucer wave can be regarded as a media-generated collective delusion unique to the twentieth century."[4] A more accurate statement would be to identify it as a *war fear generated, media triggered panic* unique to the twentieth century.

[3] Strentz, H.J. 1970, *A Survey of Press Coverage of Unidentified Flying Objects, 1947-1966*. Doctoral Dissertation, Northwestern University, Department of Journalism.

[4] Bartholomew, Robert E. and Goode, Erich, May/June 2000, Skeptical Inquirer magazine, *Mass Delusions and Hysterias, Highlights from the Past Millennium*.

With all the horrible events of the past war years, knowledge of unstoppable, horrifying weapons like the rocket bombs flying around Europe, all you needed was some unknown object in *your* sky—coming *your* way—to set off a belief that it was about to happen to *you*.

The rash of UFO sightings would last from 26 June until about 15 August 1947 (Figure 5).[5] The dates are important. The first sighting and the nationwide news of it occurred exactly one week before the 4th of July holiday weekend. Naturally, an entire nation of people out picnicking on the 4th of July was looking skyward, worried that they might see something dangerous coming. All it took was for you to say you saw a flying saucer and you could get your name in the paper.

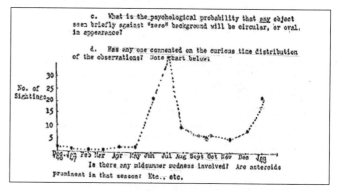

Figure 5
Saucer reports, 1947

<u>29 June 1947</u>, Los Angeles Daily News: *Many Report Seeing Flying Saucers*

<u>5 July 1947</u>, Los Angeles Examiner: *V.F.W. Chief Expects U.S. [Government] To Explain Flying Disks*. Louis E. Starr, National VFW commander-in-chief, told members he is expecting information from Washington regarding the "fleet of flying saucers."

<u>7 July 1947</u>, *Flying Saucers Are Reported Over Seattle, Tacoma, Mt. Vernon, 39 States and Canada*

By Monday, 7 July the newspapers were so full of "flying saucer sightings" that a reward of $3,000 (worth over $30,000 today) was advertised for the first "real saucer" that could be brought in. It was never claimed.

<u>9 July 1947</u>, New York (U.P.): *Hysteria Over "Saucers" Bad Example, Scientists Warn* Three scientists said the hysteria stirred up over the "flying saucers" could well mean that psychological casualties in an atomic or rocket war would far outnumber deaths from atomic bomb explosions. (Oh, **that's** good news! -JR)

[5] US Government, OSI Department Memo, dated 15 March 1949 available at www.foia.cia.gov/browse

The Saucer Fleet

One thing you have to admire about American entrepreneurs, they react very quickly to any new opportunity. The wild public interest in this new phenomenon brought out both the prankster — quick to poke fun at anyone —and the entrepreneur — always on the lookout for a new gimmick.

For the pranksters, the wave of UFO sightings was like the opening day of hunting season. It seemed that each area had its own group who were having a field day creating fake UFOs and getting other people to "sight" them. Some pranksters made up fake photos and then claimed to have seen the UFO themselves. Others made up UFO models and used them to play pranks on their neighbors. In many cases these "jokes" were taken far more seriously that the pranksters had intended.

For example, there was the case of the 11 July 1947 headline that read:

> DOUBLE SAUCER FOUND BY IDAHO HOUSEWIFE.
> An object described officially by the FBI as a "saucer within a saucer" was found today in a Twin Falls yard where a housewife said it landed about 2:45 AM...The gadget is painted gold on one side, silver (either stainless steel, aluminum or tin) on the other....The agent who inspected the object reported to the FBI office in Butte Montana., and to military intelligence at Ft. Douglas, Utah, that he could see three radio tubes inside the plastic dome.

The discovery of this UFO as a fraud can be read in the July 12, clipping (Figure 6). The fact that these sightings were being taken seriously is an indication of the officials' concerns for any foreign object that might be penetrating US airspace during these tense political times.

It wasn't long until all this excitement caught the notice of business people and advertisers.

The UFO wave had just gotten started when the entrepreneurs jumped on the bandwagon. Needless to say, businessmen and advertisers were shameless in their exploitation of this new wave of popularity. There would be many products, hamburgers, ice cream sundaes, car dealers, insurance deals; all embellished with the fancy new terminology all calculated to capture the public's attention. (Figure 7)

However, a couple new products did appear. One item was a new type of toy. The Flying Saucer craze generated the idea that would prompt Fred Morrison to carve his first plastic disc in the shape of what he thought a flying saucer should look like.

In 1948, it was introduced as the "Little Abner Flying Saucer." In 1950 it was renamed the "Pluto Platter," and finally, in 1955 the rights were bought by the Wham-O company who would market it under the now-famous name, "Frisbee" (Figure 8).

Cafe Man Cashes In On Flying Saucers

Tulsan Invents "Flying Saucer Hamburger Lunch"

TULSA, Okla., July 9. (AP)—At least one restaurant owner has decided to capitalize on the "flying saucer" idea and reports considerable success.

Ray Whitehead, owner of a Tulsa coffee shop, read of the mysterious flying discs and quickly invented what he called the "flying saucer hamburger lunch" and advertised it with a large sign in front of his restaurant.

Whitehead said he hadn't seen any of the flying discs and didn't make any claims as to relationship between the discs and his new menu special but added, "as long as these flying saucers in my shop are selling I'll stick to them and let the others search the sky."

Figure 7
Saucer Hamburger Lunch

Imaginative Boys Create Flying Disk That Convinces FBI, Army Officials

TWIN FALLS, Idaho, July 12.—(AP)—Four lads with an imagination that runs to flying disks may or may not be laughing up their sleeves today after their version of a flying saucer had practically the entire local populace, the FBI, the Army intelligence officers and police on the run yesterday.

The boys created and planted in a local yard an object that looked to them, as well as to the Army and civilian officers, just like a flying disk should look.

Their hoax was exposed after Assistant Police Chief L. D. McCracken was tipped one of the boys knew something about the disk.

The creation, which took two days to complete, was made from parts of an old phonograph, burned out radio tubes and other discarded electrical parts. It had a plexiglass dome, radio tubes, burned wires and glistening gold and silver sides.

Since the boys are juveniles their names were withheld. They will not be prosecuted, McCracken said.

The disk, resembling two band cymbals, placed face to face, reposed today at Fort Douglas, Utah, where Sixth Army intelligence officers attempted to decide what to do with it.

The Army's reaction was summed up by Capt. B. B. Zacharias of Fort Douglas who said the disk had no other function than to be "ornamental in a limited manner and to cause considerable expense to federal agencies investigating it."

The object was discovered early in the morning by a Twin Falls woman, and it was not until late evening that the mystery was cleared with the boys' confession. During the period, a plane load of Army officers from Fort Douglas flew here to take over the investigation. Until their arrival police had kept the disk under lock and key.

Figure 6
Typical Saucer Prank

WHAM-O FRISBEE
FLYING SAUCER
Flies Like Crazy!! As Seen On TV
It's Soft - Safe - Unbreakable Round Airfoil with Spinning Gyroaction — gives amazing controlled flights
SKIPS — CURVES
BOOMERANGS
FLIES STRAIGHT
88¢

Figure 8
Wham-O "Frisbee" Flying Saucer toy

Some of the new ideas were just advertising gimmicks. For example the picture of "Waterloo's Flying Saucer" in Figure 9 is of a man holding a paper plate. This is one of 5,000 specially printed paper plates advertising the return of the Waterloo "White Hawks" (a local sports team) for a six day stand. The plates were taken up in a small airplane and dropped over the area where they wanted to advertise. People would look up and, seeing the plates float down, pick one up.

This particular advertising "gimmick", making paper plates advertising some product or company, became quite popular for a while and was used to sell a variety of products from one end of the country to another.[6]

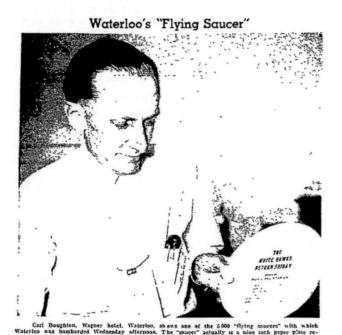

Waterloo's "Flying Saucer"

Carl Boughton, Wagner hotel, Waterloo, shows one of the 5,000 "flying saucers" with which Waterloo was bombarded Wednesday afternoon. The "saucer" actually is a nine inch paper plate reminding that the Waterloo White Hawks return home Friday night to open a six day stand. The plates were dropped from two airplanes.

Figure 9
Flying Saucer Advertising

Entertainers began to learn the value of the Saucer Fleet's unusual appeal. One of the first entertainment icons to embrace the new UFO/Flying Saucer paradigm was that famous Man of Steel, Superman!

In the 31 August 1947 episode, Superman is seen to use the flying saucer advertising gimmick to help a friend advertise his hot dogs (called "teeny weenies"). He took a number of plates, scribed a message on them and flung them over Metropolis so that everyone in the city could see them. And Superman was only one of many nation-wide comic strip characters to use the UFO phenomena to bolster interest.

But, as suddenly as they had appeared, the UFO sightings tapered off. However, even after the summer sighting season was over, the public interest in UFOs and flying saucers did not go away. Mostly this was due to there not being a definitive answer to the question, "What are they?"

The first Gallup Poll sent out in August of 1947 after the rash of sightings had this question, "What do you think these saucers are?" The results came back as:

33% No answer, don't know,
29% Imagination, optical illusions, mirages, etc,
10% Hoax,
15% US secret weapon, part of atomic bomb tests, etc.
3% Weather forecasting devices,
2% Searchlights on airplanes,
1% Russian secret weapon,
9% Other explanations.

It is a human characteristic not to like any significant unknowns in life. Naturally public interest in UFOs continued creating more opportunities to make money.

The entertainment and commercial interests continued to sell the idea long after the rash of sightings was over. Starting in the fall of 1947, there were always new items being introduced to keep interest up. Although they may not have known it, these industries (especially the entertainment industry) were contributing an answer to the "What are UFOs?" question.

By 1948 there were many other magazines on the market featuring flying saucers and stories about them. Most of the early adopters were comic books. Captain Midnight, the Shadow, Flash Gordon, WOW Comics all featured flying saucers on the covers of their issues (Figure 10). Other magazines were also running articles concerning the myste-

Figure 10
Captain Midnight #60, Feb 1948

---

[6] See Appendix A. Also reported in *Saucer Attack*, Eric & Leif Nesheim, 1997, Kitchen Sink Press, pg. 6.

rious objects. Even Hollywood was beginning to warm up to the subject.

The fact that other entertainment media were paying attention to flying saucers is important. In this era before television, the movies were the most influential entertainment medium. However, the cost of making a movie, then as now, slowed the major studios down in their acceptance of radical new trends. They were conservative mostly because of the cost and time needed to produce a film.

Therefore, most of the major movie studios were unable to respond to public interest in UFOs in the way that the newspapers and magazines had. Still, by 1949, there were hints of what was about to come out of Hollywood. And what Hollywood would produce would both reflect and define the public's perception of flying saucers from then

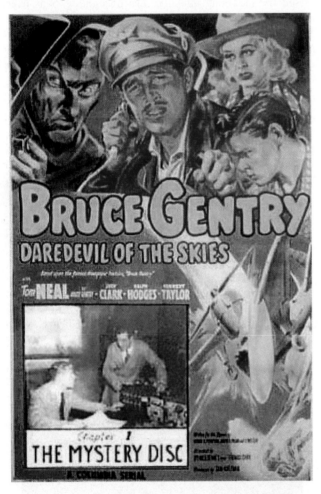

Figure 11
Bruce Gentry, 1st Flying Saucers in movies

on.

In February 1949, Columbia released a new serial named *Bruce Gentry-Daredevil of the Skies* (Figure 11).

The lead character, Bruce Gentry, was the hero of a newspaper aviation adventure strip created by Ray Bailey.

Distributed by the New York Post, the *Bruce Gentry* adventure strip would be published from March 1945 until 1951. Besides the movie serial, Bruce Gentry would also have a little known series of comic books in his name.

Ray Bailey was an excellent artist, who learned much as an assistant to Milton Caniff on his famous *Terry and the Pirates*. *Bruce Gentry* was done in the Caniff style of art and story. His main character and story scenario was similar to Caniff's later *Steve Canyon*. Apparently Bailey's only fault was poor timing. Coming after the aviation heroes of the 1930s but before Caniff's classic series, *Bruce Gentry* was not a big hit. This allowed Columbia to buy the movie rights cheaply and do a serial in the style of the competing Republic Studios.

The 15 episode serial was produced by Sam Katzman (who would later produce Ray Harryhausen's *Earth vs. the Flying Saucers*) and starred Tom Neal as Bruce Gentry. The cast included Forrest Taylor as scientist Andrew Benson, Ralph Hodges as the rancher Frank Farrell, and Judy Clark as his daughter Juanita.

Bruce Gentry-Daredevil of the Skies was a low budget melodrama with a period, 1940's style hero placed in a "save the world" scenario. It received mixed reviews. It was so true to the 1930's-1940's era of black hat villains vs. white hat hero adventures that one had to be able to accept that era's philosophy in order to enjoy it. There were plenty of fistfights and every episode had its "cliff-hanger" ending.

Its claim to fame here is that it had the very first scenes of a flying "disc" ever seen in the movies. In this case, these saucers are remotely controlled flying bombs—a "deadly secret weapon that can be directed against any target." Here too, the attitude of 1940's society is clearly represented. This was a believable explanation for the mysterious UFOs seen recently, i.e. a high tech remote controlled flying bomb.

The discs are controlled by the evil "Recorder" so called because he records his threatening messages and instructions to his henchmen. Bruce must find the evil villain and stop him before he can use his "flying discs" to destroy the Panama Canal and then menace the rest of the world. This was a fifteen episode serial so, if one had managed to see one episode each week, it would have taken to June 1949 to have found out who the evil "Recorder" was and thwart his evil plans.

The special effects that created the saucers on screen were animated drawings done right on the film stock. This was an expensive technique for low budget serials. Due to the expense, there were only two sequences made and they were re-used several times. These sequences were good enough (or available enough) to be recycled in later serials, such as the 1950 *Atom Man vs. Superman*, and 1952's *Blackhawk: Fearless Champion of Freedom*.

The saucers are seen as discs with a central dome. The domes are stationary while the rim spins rapidly around it.

The domes also have rabbit-ear antennas, a porthole on top and a black spot on the bottom.

Of course, in the last episode, the brave Aviator manages to out-fly and destroy those evil flying discs just in time.

The design and purpose of these saucers clearly reflects people's beliefs throughout 1949 and their appearance in this serial did little to change them since the serial's timing was off. Presented in theaters from February to June of 1949, it was just a little too early. The major UFO books that would change public impression about flying saucers would not appear until after it was out of circulation. The next film, though, would not be so unlucky.

The first hint of a new, full length flying saucer feature film appeared in Louella Parsons' syndicated column from 20 September 1949. Parsons had been the major "inside gossip and news" source for Hollywood throughout the 1930's and '40s and had a knack for getting the scoop on any new project that might be coming out.

It was little surprising that she was the first to hear of a small, private studio called "Colonial Pictures" that claimed to have some intriguing film of flying saucers taken in Alaska.

Film to Feature 'Flying Saucers'

By Louella O. Parsons

HOLLYWOOD, Sept. 20—(INS)— So fantastic is the story of the 900 feet of film taken of the "Flying Saucers" in Alaska that at first I thought it was a publicity stunt. The more I investigated, however, the more I believed Mikel Conrad's story.

Over a year ago he was in Alaska filming "Arctic Manhunt," and he heard from the Eskimos of these strange flying discs. He made a trip far into the frozen North to see for himself, and then he let Washington know. The Government sent a man out, asked Conrad for the film he had taken. He turned it over, and it has been in a sealed vault in Los Angeles ever since.

Now the Government is releasing the film to Conrad, who is a producer and director of Colonial Pictures. He will make a picture built around it called "Flying Saucer," which Howard Irving Young is writing. Conrad believes that an official statement will be forthcoming as to the origin of these mysterious discs when the film is returned to him next Tuesday.

Figure 12
News of Flying Saucer Film

Parson's article explained that Mikel Conrad had told the government about the film he had taken over a year ago and they had asked he turn in the film for their review. They had kept it for some time and were returning it soon after the column came out. Conrad said he would be making a film of it shortly titled *Flying Saucer*. (Figure 12)

More interesting is what is not said in the column. The film, according to Conrad's statement, was taken "over a year ago" which puts the date at about late 1947 to early 1948 time frame. This would have been one of the very first movies ever taken of a UFO. It is interesting that it does not appear to be listed as a documented sighting of that period, nor would it appear in the new picture he was producing.

Then, as they say in Hollywood, in December the "plot boils over."

In mid December 1949 the January issue of TRUE magazine was released with an article by Donald Keyhoe. The media's reaction was immediate.

27 December 1949, Los Angeles Times: *Flying Disks Called Spies From Planet*. The author reports that he spent 8 months of intensive investigation. He is convinced that flying saucers are real:

> ...the flying saucers are interplanetary. ...The only other possible explanation is that, the saucers are extremely high-speed, long-range devices developed here on Earth.

What a New Year's present! When people picked up the January issue of *TRUE* magazine, they were astounded to read Donald Keyhoe's conclusions about the nature of the flying saucers. Of course the Air Force immediately released an official announcement that the conclusion of Keyhoe's article was nonsense:

28 December 1949, Los Angeles Daily news: *"Flying Saucer" Myth Blown Sky-High By Air Force Study*. Washington (UP) -- The Air Force has closed "Project Saucer" because the 375 reports that it has made show no verification whatever of flying saucers. It attributes them to "misinterpretation of conventional objects, a mild form of mass hysteria, or hoaxes."

Unfortunately, this announcement did little to quell the rising tide of speculation. In fact it was probably the beginning of the view that possibly the Air Force was biased. The article in *TRUE*, and the on-going sightings of UFOs being continually featured in the nation's papers worked to undermine the Air Force's position.

The "shovel-shaped" and "flying wing" objects that Kenneth Arnold reported less than three years earlier had been renamed "Flying Saucers" by the headline writers of the national press for UFO's of any shape. With this article came the first statement that all UFOs (qua flying saucers) were extraterrestrial spaceships. This completed the conversion of Flying Saucers from actual aircraft to UFOs to Interplanetary Spacecraft, and it took only two and a half years.

1950: The first feature film and pivotal UFO books are released

The first feature length movie to focus on flying saucers was released as "The Timeliest Picture in the Century" (Figure 13) in January 1950. Whatever else can be said about it, the timing was perfect, being released in the middle of the firestorm of controversy that Donald Keyhoe's article in *TRUE* had generated.

11 January 1950, Daily Variety (Review) *"The Flying Saucer" by Films Classics and Colonial Productions*

> Action unfolds at the famed Taku glacier near Juneau. Much of the footage is spectacularly effective. Narrative shows race between the U.S. and Russia to find the saucer. There is fast action in its unfoldment (sic). Scenes purporting to show the lightning-like saucer are thrillingly presented. The producer, director, writer, and star are Mikel Conrad. (Reviewers name not given)

The Saucer Fleet

The smaller, hungrier studios are the quickest to respond to any new opportunity for a sensational film. The article that "leaked" news about it was published in September. From concept to box office in four months is fast for a movie. From what we know, the movie was thrown together from footage Conrad had already taken for another project. The plot had been whipped up to integrate the Alaska scenery footage and the flying saucer theme. They planned to release it very quickly to capitalize on the public interest in flying saucers, but the timely publication of the *TRUE* arti-

Figure 13
The first full length saucer movie

cle must have been pure luck.

The plot centers on an undercover FBI Agent, Mike Trent (played by Conrad), being sent to Alaska to recover a flying saucer before the Russians get their hands on it. To quote viewer, Chris Gaskin from Derby, England:[7]

> Playing a journalist, Mike Trent and his "nurse" Vee Langely (Pat Garrison) are sent to Alaska to investigate strange sightings of flying saucers. His nurse is with him because, as his cover, he is in Alaska "recovering from a nervous breakdown." Not surprisingly, he falls in love with her during the movie. They make a hunting lodge their home during their

> stay but the man, Hans (Hantz von Teuffen), who suppose be helping them to do odd jobs, is actually a Russian spy. He tries to kill the woman a couple of times. He has something to do with the saucer, which appears eventually. The spies are caught out at the end and one of them takes off in the saucer, which then explodes into thousands of little pieces.

More notable than the stars are the bit players in this film, namely Denver Pyle (Turner) who would go on to a lengthy career in TV and films, and Virginia Hewett (Nanette). Hewett would co-Star in ABC's *Space Patrol* TV series for the next 5 years. Also others like Lester Sharpe, Russell Hicks, and Frank Darien were character actors with long careers.

As to the "flying saucer" centerpiece, it is not described as an interplanetary spaceship but rather an invention of an eccentric scientist working on "an entirely new principle." From viewing the picture, it is clear that Mikel did not use any of the "authentic" film of a real UFO he claimed to have shot in Alaska before making this movie. Apparently that was just 'hype' to create interest in the project.

The saucer appears in three flying sequences for a total of about 15 seconds. These sequences show clear evidence of being made with a double exposure technique. There is also a full size saucer shown momentarily in the scientist's laboratory. The images of the saucer in the air are so blurred as to prevent comparison with the 'landed saucer' in the lab. Plus, the scenes of the saucer in the lab are very dimly lit in an apparent attempt to disguise a poorly made prop.

Although this was the first attempt to build a real "full-sized" flying saucer for a movie audience, they failed to convince viewers. The special effects that created the landed and flying versions of the saucer were very low budget. However, they are the first real fictional flying saucers in the movies.

Despite of the review in *Variety*, the film was not very good entertainment. Many of those who have seen it recommend others not waste their time. About the only thing that has been said good about it is that it has some good background scenery shots of Alaska and accurately reflects the culture of the early '50s. In the final analysis it was an opportunistic movie designed to take advantage of the public's interest in UFOs and flying saucers in 1950.

From the time the *TRUE* article came out, the question about the nature of flying saucers was one of the ongoing major topics in America. As sensational as the proposition that UFOs came from another planet was, the majority of Americans were not going to easily believe something so preposterous. The public wanted a better explanation.

Around the end of March, a new theory came to national attention. Since everyone believed that America was the most technically advanced nation on the planet, if it was some kind of a secret, *then it must be our secret.*

Many articles were printed claiming the UFOs were secret developments of the Navy's Flying Saucer project,

---

[7] http://us.imdb.com/title/tt0042469/usercomments

possibly a jet version of the XF5U (Figure 2). This new idea was so logical and backed by what seemed like solid evidence that everyone must have heaved a collective sigh of relief.

This new idea was further strengthened when on 7 April, space flight expert Willey Ley was quoted in the *Los Angeles Daily News* as saying, "There are 3 possibilities: 1) They are a U.S. military secret; 2) They are the secret of some foreign power; or 3) The flying saucers are from another planet. My personal opinion is that the flying saucers are a U.S. military secret." Dr. Ley's opinion was beginning to be shared by many by that time.

22 May 1950, Princeton, NJ: *Most Think Disks Are Secret Weapon.* George Gallup reported that the American Institute of Public Opinion's latest poll (#455 4/28-5/2/50) indicated that the largest number of people believe that flying saucers are new experimental weapons or flying contraptions being tested by the army or navy.

This represents quite a change from 1947 when the largest number labeled them as nothing more than an illusion or a hoax. This time 5 percent thought that they were something from another planet, 6 percent thought that they were some kind of new airplane and only 3 percent thought that they were something new from Russia.

So the American public calmed down a little about the identity of "flying saucers." For the month of April, the number of UFO sightings dropped to one third of what they had been. Then in May, they dropped by half again and stayed there through June.

This was good because bad news was on the way.

25 June 1950, North Korea invades South Korea. Korean War Begins

The United States was at war again, less than five years after the end of World War II. Now flying saucers had to compete for headlines with news from the front, and at first, it was all bad news.

While they were being swept from the front page by the start of the Korean War, the first books on flying saucers were published. These first books were the most important as they established the fundamental beliefs for the UFO culture that would develop; beliefs that still exist in our society today.

The first of the popular and influential books[8] was Frank Scully's *Behind the Flying Saucers*. It was published

in 1950 by Henry Holt & Co. and then reprinted in 1951 in paperback by Popular Library. This book made the national best seller list because it said many incredible things that some people were willing to hear. Many of its themes would be connected to the Roswell incident by later UFO-logists. Its major claims were:

- The Military will deny this.
- They are untrustworthy.
- Don't believe them.
- The Air Force closed Project Saucer and went underground.
- Flying saucers are real.
- Flying saucers are interplanetary.
- Saucers have crash landed on Earth and we have seen them.
- The Air Force has found saucers and is studying them in secret.
- If you talk about crashed saucers the government will try and slander you.

…and much, much more. By the time it was revealed as a total hoax in 1952[9], this one book had already convinced untold numbers of readers and had created many of the UFO myths we still live with today.

The other influential book to come out that year was Donald Keyhoe's *Flying Saucers are Real*. In it, Mr. Keyhoe related his investigation into many of the important UFO sightings that had occurred since 1947. The book was an expansion of his January *TRUE* article, which had started the uproar over the nature of the saucers. In it he further detailed his investigations. More than any other book that year, it was honest in its convictions and convincing in its presentation.

With the release of Keyhoe's book repeating its sensational idea that saucers were interplanetary, (which agreed with a main point in Scully's book) many people began to wonder.

With this shocking new idea going around, the next movie serial Hollywood made about flying saucers just *had* to have a villain that was an alien. So, when Republic released *The Flying Disc Man From Mars*, it starred Walter Reed as hero Kent Fowler, Lois Collier as leading lady Helen Hall, James Craven as scientist Dr. Bryant, and Gregory Gaye as the evil Moto from Mars. Two future stars that were used as unaccredited extras in this low budget serial were Clayton Moore who would become famous as the Lone Ranger, and Guy Teague who would have a long career as a stunt man in westerns.

---

[8] The first book about Flying Saucers was a novel called *The Flying Saucer* by Bernard Newman. The second book published in 1950 was a self published booklet by Kenneth Arnold entitled, *The Flying Saucer as I saw it*. Neither was nearly as influential as Scully's.

---

[9] Cahn, J.P., *The Flying Saucers and the Mysterious Little Men*, *TRUE* magazine September 1952.

Figure 14
The Flying Disc Man from Mars

Stretched over 12 episodes, this serial used a large amount of stock footage to fill in the dialog and action. As Serials go this one was considered below average. The plot was, to quote from Mike Nella's review in the IMDB:[10]

> Mota arrives on Earth from Mars, and takes into confidence manufacturer Bryant (a one time Nazi sympathizer) to launch a wave of destruction that will overwhelm Earth and put the planet under a Martian supreme dictator. Aerial patrolman Kent Fowler sees the theft going around Bryant's plant (by his two henchmen Drake and Ryan) and goes to stop the sabotage despite [the fact] that his employer Bryant is the real criminal. Eventually Mota and Bryant develop their weapons to such a degree that they are able to start their campaign of destruction, unless Kent can stop them.

If you think that this sounds like the usual "invasion from Mars" theme, you are correct. This was not one of the better postwar offerings, and other than the very young, this serial did not influence anyone who saw it.

More important is the fact that the disc in the *Flying Disc Man from Mars* is no saucer at all. In spite of the title and the innuendoes, Mota's ship is a recycled prop from another earlier republic serial, 1942's *King, of the Mounties*. It is a short fuselage, triangular winged, VTOL aircraft with the "Rising Sun" insignia on its tail, came complete from the earlier serial (Figure 14).

Still, interest in flying saucers in Hollywood was on the rise. They were too popular of a phenomenon to ignore for much longer.

1951: Flying Saucers make the big time—the blockbuster movies arrive

17 January 1951, Chinese and North Korean forces capture Seoul.

14 March 1951, For the second time, United Nations troops recapture Seoul.

29 March 1951, Ethel and Julius Rosenberg are convicted of conspiracy to commit espionage. On 5 April they are sentenced to death.

The Korean war (called a "police action" by President Truman) was see-sawing back and forth, the House Un-American Activities Committee (HUAC) was digging for, and finding, Communist spies everywhere, and children were learning how to "duck and cover" in case "The Bomb" was dropped on their community. This was the beginning of the 1950's, "Happy Days" for some, a nightmare to others.

The art of a culture reflects how that culture sees itself and the world it lives in. In April 1951 another small studio, Winchester Pictures Corporation, released a film that perfectly reflected the times and culture within which it was made, the SF-horror film, *THE THING* (Figure 15).

Also known as *The Thing From Another World* this Sci-Fi thriller was a saucer movie in the most fundamental sense. The idea of saucers bringing an Alien from outer space plays a vital role in the movie's premise.

Based on a short story by John W. Campbell, the movie's monster (played by six foot seven inch James Arness, later of TV's *Gunsmoke* fame) *THE THING* was a threat of a type never seen before. And like "the Red Men-

Figure 15
The THING

---

[10] http://us.imdb.com/title/tt0043546/#comment

ace" (world communism) it seemed to be unstoppable.

With Kenneth Tobey as Captain Patrick Hendry, Robert Cornthwaite as scientist Dr. Arthur Carrington, and Margaret Sheridan as Dr Carrington's assistant "Nikki" Nicholson, the movie was well cast. It was also well directed by Christian Nyby. Even the black and white photography helped set the mood. *THE THING* has become a classic of the Sci-Fi horror genre.

The story takes place in an isolated outpost near the North Pole. Manned with both scientists and military (and one woman), the scientists detect an "an impact that could only be made by something composed of twenty-thousand tons of steel, iron or some other heavy element." The scientists and military men fly out to the crash site to investigate.

What the scientists find buried in the ice is clearly a flying saucer. Unfortunately, efforts to remove the saucer, whose edge or fin are all we get to see, (Figure 16) result in the saucer's destruction. However, their Geiger counter indicates another body buried nearby in the ice. After they dig it out of the ice, the realization is made that this creature clearly came from another world in the saucer. Fearlessly, they remove the creature, still in a block of ice from the area and take him back to the lab to examine.

Figure 16
Flying Saucer from The Thing

On the way back in the airplane, one of the crewmembers reading a magazine quotes a statement put out by the Department of Defense that calls all UFO sightings hoaxes. They all laugh at how they have just proven the government wrong. Here again, this is a comment on the cultural feelings of the times. UFOs exist and the government insists on denying it.

They bring the frozen block of ice back to camp and store it in an anteroom. When one of the men is left to guard it, he drapes a blanket over the block; unaware that it is an

electric blanket. The creature gets melted out of his prison and goes on a rampage. What follows is one of the best horror films of the decade as *THE THING* proves smarter, more resilient, and hungrier than they expect.

Note: I am not going to do a synopsis of the last part of the plot and spoil the movie for those who might want to see it. This story was good enough for John Carpenter's 1982 remake to receive three academy nominations.

What is important is the mood that *THE THING* relates to the audience. Although we do not get to see much of the flying saucer he arrives in, saucers are clearly the method of arrival of the monster. This shows the association of fear of the unknown with flying saucers. The point is clear. Flying saucers may be bringing something far more dangerous and frightening than mankind can possibly imagine. This too is a reflection of the times when *THE THING* was produced, and there is still an undeniable element of truth to the idea.

The movie makes one final point as it approaches its final fade out. In the last scene, having barely escaped with their lives from an alien mentally equal and physically superior to humans, the radioman is finally able to get the radio working so they can talk to the outside world. He makes contact with Anchorage where reporters are waiting anxiously for details.

But before telling of their adventure, he gives everyone a warning. In doing so, he looks into the camera clearly indicating that what he is about to say is not just for the characters in the story, it is meant for the audience and the world in general. It is a warning about flying saucers and the danger they represent.

Tensely he speaks: "Every one of you listening to my voice, tell the world. Tell this to everybody; wherever they are…Watch the skies!   Everywhere! Keep looking! Keep watching the skies!"

This final dialog made everyone leaving the theater in 1950 glance nervously up at the sky overhead. And you can guess what they were looking for. Flying Saucers. Alien Monsters. Danger. Death. War. They speak of a frightening time and a worried public.

Peace?   Nooo Peeaacceee!!!

Thus, because of a politically charged environment, the Saucer Fleet had changed the world. And the major movies had finally stepped in to carry their image – and their message – to people everywhere.

Thus, with their "faculty of healthy disbelief" suspended, people were ready for the next big Hollywood picture with a message.

They didn't know it, but they were ready.

The Day The Earth Stood Still

Klaatu...Barada...Nikto. After more than fifty years, it is still the most recognized alien phrase in all of film history. True, it's been eclipsed in recent years by the utterances of Mr. Spock (e.g. "Live long and prosper" and "The needs of the many outweigh the needs of the few"), but these, you have to admit, are paraphrased in English rather than delivered in the original Vulcan.

It was spoken by Klaatu, the alien visitor in the 1951 film *The Day the Earth Stood Still*, an unchallenged classic of the Science Fiction genre, which is all the more amazing when you realize there are very few standard Sci-Fi trappings in the film. Yes, there is a flying saucer and a robot with a heat ray, but repeated viewings of the film (and the author has had **many** repeated viewings!) reveals that the vast majority of the screen time is spent away from these and deals instead with the Earth-bound story.

The film is very loosely based on the Harry Bates story *Farewell to the Master* that first appeared in the October 1940 issue of *Astounding Science Fiction*. In that version, Klaatu and his giant robot Gnut simply appear on the Mall in Washington D.C. (they do not descend from the sky). Their craft is not a flying saucer, but a "time-space ship" capable of moving in all four dimensions, thus able to "appear" without "arriving." Klaatu is shot dead as soon as he steps out of the craft, and from that point the story diverges considerably from the way it was portrayed in the movie.

Astounding Magazine
October 1940

In 1950, 20th Century Fox producer Julian Blaustein was becoming increasingly frustrated with the growing restrictions on the movie industry due to the investigations of Senator Joseph McCarthy and his House Un-American Ac-

tivities Committee, which led to many talented actors and writers being "blacklisted," thus prevented from working. The antics of "tail gunner Joe" were bad enough, but he was further annoyed by what he perceived as media bias (nothing ever changes) on the part of his own industry. Not only did newspapers, radio and newsreels accept the HUAAC hearings uncritically, they were actually exploiting them, using the "Red Scare" to sell newspapers and theater tickets. Blaustein wanted to protest with a strong message film, but didn't want to run the risk of being blacklisted himself. He thought that he could make his political statements "safely" under the guise of a Science Fiction film.[1]

Actually, even if he didn't have a political bone to pick, Blaustein would have made some sort of Science Fiction film in this period anyway. Like any good producer, he was watching movie trends and noticed that Sci-Fi was becoming a more mature and profitable genre. As early as 1949, Blaustein asked story editor Maurice Hanline to start looking for Science Fiction properties suitable for filming:

> Blaustein lectured Hanline for over an hour, making it clear that the story editor was to confine his search to stories of an earthbound nature. There would be no money for an imaginative outer space fantasy. He could only hope to sell the project if he could keep the setting to present day Earth.[2]

Blaustein read a large number of the "pulp" magazines that resulted from Hanline's search. In addition to looking for an earthbound story, he

Julian Blaustein

---

[1] Exactly the same reason that Gene Roddenberry said that he created *Star Trek*.

[2] Steve Rubin, *Cinefantastique*, V4 N4, 1976, pg. 6.

Edmond North

wanted one that didn't have too much in the way of the standard genre clichés to avoid the derogatory perception of Sci-Fi by the general public at the time. He came across the Bates story and thought it was ideal. "There were many parts of the story that I didn't care for at all," he said in a 1995 interview,[3] "but there were certain aspects that really got my wheels turning." What his wheels were turning about was the fact that in the story, the alien was the peaceful, benevolent one and the government was imperious and destructive. In this one story he found everything he was looking for; a solid science fiction tale from a respected source that takes place completely on Earth in the present day, and in which the alien is the sympathetic character fighting against an overbearing government.

He hired future Academy Award winning screenwriter Edmund H. North[4] to create a much-modified version of the story that made the political statement he was after. He then enlisted the support of studio head Darryl Zanuck. Zanuck's help was crucial since Blaustein was planning to make an anti-war, anti-military film just as the United States was becoming involved in the Korean War. Zanuck approved the project based on its Science Fiction aspect since the timing was perfect. Both *Rocketship XM* and *Destination Moon*[5] had been back-to-back box office successes earlier that year (1950), and Fox had no other Sci-Fi projects in the works.

Darryl Zanuck

Zanuck recommended 36 year-old Robert Wise as director. Wise, who had started out as a film editor on *Citizen Kane*, was thrilled at the assignment and welcomed the opportunity to make political and social statements similar to Blaustein's. Zanuck also recommended Spencer Tracy, who was at the peak of his career, to play Klaatu. This violated

Robert Wise

Zanuck's own self-imposed rule of never interfering creatively with a producer.[6] Blaustein hated the idea saying that Tracy would have "no credibility" as his spokesman since he was such a well-known actor, and that's all audiences would see. Wise and North suggested the English actor Claude Rains who, while still well known (*Casablanca*, *Now Voyager*, etc.), was not as familiar to American audiences. Rains, though, was tied up with a Broadway play and unavailable. Incidentally, Blaustein had talked Zanuck out of using Tracy before Robert Wise was even involved in the picture. That explains a 1988 interview where Wise had no idea what the interviewer was talking about when asked why he didn't use Spencer Tracy for the part.[7]

Violating his own rule a second time, Zanuck made another suggestion to play the alien, this time a new hire he had just made. While in London, he had seen the English stage actor, Michael Rennie, in a play and immediately signed him to a contract.[8] With his chiseled features, imposing stature and smooth, cultured delivery, he was the perfect spokesman for Blaustein's message. The fact that he was a complete unknown to film audiences was a

Michael Rennie

[6] *Making the Earth Stand Still*, bonus feature documentary on DVD release of the movie.

[7] *The Science Fiction Film Reader*, Limelight Editions, 2004, pg. 53.

[8] While it's true that Zanuck discovered Rennie on the London stage, he had acted in nearly 30 British films by that point, mostly in very small bit parts.

[3] AMC Backstory: *The Day The Earth Stood Still*, American Movie Classics cable channel, 2001.

[4] The Oscar® came quite a bit later, in 1970, for *Patton*.

[5] See *Spaceship Handbook* for entries on both these films.

bonus.[9]

Patricia Neal was cast as Helen Benson, the war widow whose son, Bobby (Billy Gray) befriends the alien visitor. Wise had just finished working with Neal in *Three Secrets* the previous year and they both had tremendous respect for each others' talents. Neal said that she thought the script was "the funniest thing I'd ever read. I couldn't believe that people would take it seriously."[10]

©20th Century Fox Film Corporation
Billy Gray and Patricia Neal

For the "Judas" role of Tom Stevens, Helen's boyfriend, Wise hired Hugh Marlowe, one of Fox's contract players. Marlowe had recently played Col. Ben Gately in the Oscar® winning *Twelve O'clock High*. A few years later

©20th Century Fox Film Corporation
Hugh Marlowe

he would have the lead in Ray Harryhausen's *Earth vs. the Flying Saucers* (see page 199).

The most controversial casting was for the Einstein-like professor Jacob Barnhardt. Blaustein wanted to use the renowned character actor Sam Jaffe whom he knew from a similar role in *Gentleman's Agreement* (1947) in which Jaffe played the Einstein-like professor Fred Lieberman. Blaustein, however, immediately ran into problems with middle management. Jaffe's liberal political views had gotten him blacklisted[11] and the casting director pleaded with Blaustein to choose someone else and keep the studio out of the controversy.[12] Seeing that this was exactly the sort of thing the film was protesting, Blaustein went all the way up to the head of production (Darryl Zanuck) to get the problem resolved and found a sympathetic ear. Even though Zanuck had been the target of much criticism over *Gentleman's Agreement* and its theme of anti-Semitism, he went out on a limb and allowed Jaffe to be hired for the project despite his non-grata status. Ironically, Albert Einstein himself, being a pacifist, had been watched since WW II by various U.S. government agencies as being a potential security risk.

Sam Jaffe

Since one of the major points the producers wanted to make was the overbearing nature of political and military force, they needed a significant amount of military hardware to surround the ship and for the climactic chase scene through Washington DC. But after reading the script, the Army, understandably, refused to support the film. Undaunted, Wise sent his second-unit director, Bert Leeds, across the Potomac where the Virginia National Guard was only too happy to lend them some men and a few trucks. These were used for the "Martial Law" scenes where people are prevented from leaving the city, and the climactic nighttime chase through Washington D.C. The soldiers, tanks and artillery used to surround the ship were supplied by the California National Guard since all of the scenes with the full size saucer set were actually shot on the Fox backlot in Southern California.

The spaceship exterior and interiors were created by Oscar® winning production designer Lyle Wheeler and art director Addison Hehr. Wheeler was one of the most accomplished art directors ever to work in Hollywood. Over his extraordinarily long career he worked on nearly 400 movies and television series. He was nominated 29 times

---

[9] Ironically, both Rains and Rennie starred together in Irwin Allen's *Lost World* nine years later.

[10] *AMC Backstory.*

---

[11] Jaffe was a rather benign liberal, and while sympathetic to the communist movement, there is no evidence that he ever joined the American Communist Party. His views were public enough that his name was published in an informal industry pamphlet called "Red Channels" listing over 150 "Known Communists and their Sympathizers" in the entertainment business.

[12] *Making the Earth Stand Still.*

Lyle Wheeler with model of saucer interior set

for the Academy Award and won five of them. Prior to *The Day the Earth Stood Still*, Wheeler had won the Oscar® for his work on *Gone With the Wind* (1939) and *Anna and the King of Siam* (1947). After this film, he won further Academy Awards for *The Robe* (1953), *The King and I* (1957) and *The Diary of Anne Frank* (1959). The only genre film that he was nominated for was *Journey to the Center of the Earth* (1959). Strangely, he was not even nominated for *The Day the Earth Stood Still*, although he was nominated for four (!) other films that he worked on in 1951: *Fourteen Hours*, *House on Telegraph Hill* (another Robert Wise film), *On the Riviera*, and *David and Bathsheba*. Conversely, this was Addison Hehr's first film as art director. We will discuss the functionality (in the context of the story) of this brilliant minimalist design a bit further on in "The Vehicle" section.

Three versions of the outside of the ship were built. There was the full size set, of course, with the operating ramp and rotating dome port.[13] The set, however, went only slightly more than halfway around, leaving a wedge-shaped open area in the back for the stage crew to operate the mov-

©20th Century Fox Film Corporation
Seven-foot model in a "hanging miniature" shot.

ing parts. To show the ship from the side, a "hanging miniature" had to be used. A seven foot (2.1 m) diameter fiberglass model was built for this purpose but used only in the long shot where Klaatu descends the ramp seen from the side. It was also used for the "takeoff" scene at the end.

There was also a two foot (60 cm) model built for the landing scene. The image of the saucer and its shadow were filmed against a white background and rotoscoped onto the high altitude shot of The Ellipse where it set down.[14] The miniatures were built from Lyle Wheeler's design by special effects supervisor Fred Sersen and his team of Ray Kellogg, L. B. Abbott and Emil Kosa.

©20th Century Fox Film Corporation
Two-foot model rotoscoped into landing shot.

Aiding Wheeler and Hehr in the design of the ship, and other scientific aspects of the film, was Dr. Samuel Herrick, a professor from UCLA, who was hired as the film's technical advisor. The 39-year-old Herrick was one of the world's foremost authorities on celestial mechanics at the time, and he personally wrote the equations seen on Professor Barnhardt's blackboard in the film. Sam Jaffe had been a mathematics instructor in New York before he took up acting, so he was able to be quite convincing when making chalk dust, but Michael Rennie had to be coached when putting the "corrections" on the board.[15] It was Herrick who suggested the use of non-contact controls in the spaceship, after which Blaustein forbade the use of anything that

Samuel Herrick coaches Michael Rennie

---

[13] Blueprints of this set are included in the bonus feature section of both the DVD and Laserdisc versions of the movie.

[14] *Cinefantastique*, pg. 21.

[15] *Cinefantastique*, pg. 18.

looked like a conventional switch or knob.

To create Klaatu's space suit, Wise went outside of Fox's wardrobe department and hired noted New York fashion designer Perkins Bailey. The design remains one of the most elegant and effective alien costumes ever filmed. Bailey also designed Gort, that is to say the smooth, featureless look of the giant robot. The wardrobe department had to figure out how to render the drawings into something wearable and filmable. The biggest problem was the featureless perfection of Bai-

ley's design. The only surface details on the robot were some "breaks" on the wrists, ankles and waist (combined with a couple of lines around the top of the legs forming his "underwear" – modesty in a robot?), plus, of course, his helmet visor and its side hinges. With the rest of the costume completely smooth and seamless, how was the actor to

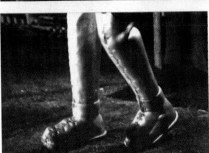

Production stills of the laces on the front of the "rear-shot" suit

get in and out?

The answer was to create two suits, one for front shots that opened in the back, and one for rear shots that opened in the front. Despite many references in later articles to "zippers" for the costume, it was actually laced with cord as can be seen in these production stills. The material was sponge rubber, like a diver's wetsuit, painted silver with some internal fiberglass forms to help keep its shape.

Now they had a costume, but they were at a loss as to who to put in it. You just can't call central casting for someone eight feet (250 cm) tall! Wise remembered that the doorman at the famous Grauman's Chinese Theater was the extraordinarily tall Lock Martin. At 7' 7" (230 cm), he was a good "fit" for the part, but still not tall enough for Blaustein and Wise. To boost him up in the costume six

inch (10 cm) platforms were built into the feet.[16] But even at that Martin's eye line barely came up to the bottom of the visor slit. Air holes added to the underside of the chin also helped him see where he was walking. For scenes shot from the sides or back, a clear visor replaced the normal opaque one. In a later interview,[17] Blaustein mentions a prism assembly in the helmet (essentially a periscope) to allow Martin to see out and down from his vantage point in the suit, but we have found no supporting evidence of this. From this production photo, it's clear that there is nothing in front of his face visible anywhere in the helmet, although it's possible that there could be a mirror in the helmet's chin to help him see out the air holes (which are visible in the photo at left).

Although giant in stature, Martin was physically quite frail. He could only work in the heavy Gort costume for half an hour at a time plus, as you will recall, he still

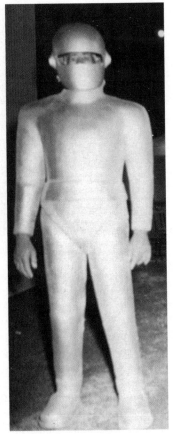

Lock Martin in Gort costume (note missing gloves as well as visor)

wasn't quite tall enough for the producers. To make the robot even more imposing when standing guard over the ship for several days, the prop department built a rigid fiberglass version. This was not a suit intended to be worn by anyone, it was just to tower over the army engineers as they tried to make sense of the ship and robot. And tower it did,

Gort statue under construction

[16] *Making the Earth Stand Still.*

[17] Ibid.

being made a full foot taller at nine feet (2.8 m) than the costume version.[18]

©20th Century Fox Film Corporation
Gort mechanized helmet

The final Gort prop was a mechanized head-and-shoulders casting of the helmet with the moving visor. There were lights behind the visor that moved back and forth as a guide for the optical effects department when they added the warm-up and discharge of the heat ray.

Martin's frailty meant it was impossible for him to carry, even less pick up, Patricia Neal. For the scene of Gort carrying Helen Benson past the camera, a body board was used to support the actress. The board was suspended from thin wires hung from a moving crane just above camera range. Since this shot starts from the side and moves to the back of Gort as he passes the camera, they used the "backside" version of the costume, and some of the laces can briefly be seen on the front as they emerge from behind the fence. For the rear shot of him walking up the ramp, a lightweight mannequin was substituted for Miss Neal. Wise later said that of everything in the film, this "stunt" was the technical detail he was proudest of.[19] Ironically, Neal's biggest fear in her scenes with Gort had nothing to do with

©20th Century Fox Film Corporation
Gort "carrying" Helen Benson. Note the support wires in this contrast-enhanced frame. Note, too, the laces on the front of the Gort costume.

being carried; she was afraid that she would burst out laughing when trying to recite the line "Klaatu, Barada, Nikto."[20]

Speaking of, this is a good place for a diversion into the alien language spoken by Klaatu at several points in the film since there seems to be considerable confusion over it among the movie's fans.

There is a "final shooting script" provided as a bonus feature on the laserdisc and DVD versions of the movie, which is dated "February 21, 1951." It does not contain any of the alien phrases, with the exception of the very first command ("Gort! Deglet ovrosco!") used near the beginning. The dialog in all of the other scenes where Klaatu speaks in his native language is handled by stage direction. For example, in the scene where he "phones home" to report, the script simply reads: "Klaatu starts speaking into a built-in microphone in his own strange language." Well, if it's not in the script, then where did it come from?

In an e-mail exchange with Robert Wise,[21] he said that scriptwriter Edmund North created all of Klaatu's language. Michael Rennie has supported this in several interviews where he said the alien words were there in the script for him.

To resolve this conundrum of the script not containing the lines that both the film's director and the actor that spoke them says were there, we turned to a professional in the industry. Warren Chaney has had a long career as both a producer and director of film and television (and is a distant cousin of silent film great Lon Chaney). He's also a huge fan of this and other films in the genre. When asked to comment on this apparent discrepancy he said:

> The fact that the dialog is not in the script on the DVD probably means that it's a "commercial" rather than an actual "shooting" script. An awful lot happens to a script in pre-production. A writer just hopes that it's not a disaster since it's usually others doing these updates rather than him. However, Robert Wise was well known for keeping the original writers involved right up to the point of principal photography.[22]

Dr. Chaney's comments are supported by the fact that the script is dated 21 February, but the film didn't start principle photography until 9 April. A lot of additions and modifications were probably made in those six weeks.

As a final note on the language issue, the first issue of *Fantastic Films* magazine[23] had an article by Tauna Le Marbe, a professional linguist, purporting to have translated the phrases. Ms. Le Marbe, however, provided no corroborating statements by anyone involved in the production supporting the validity of her efforts.

---

[18] It's easy to tell which version you're looking at, even without some sort of scale reference. The fingertips of the suit with the actor in it come down to mid-thigh, while the statue's arms are shorter, ending right around the groin.

[19] *The Science Fiction Film Reader*, pp. 56-57.

[20] *AMC Backstory*.

[21] 3 April 2003.

[22] Interview, 4 April 2003.

[23] *Fantastic Films* V1, N1, April 1978, pg. 46.

The Saucer Fleet

The final element that assured the film's classic status was the brilliant music score by the legendary Bernard Herrmann. Wise had worked with him twice previously on the Orson Welles' productions, *Citizen Kane* (1941) and *The Magnificent Ambersons* (1942). At the time, Herrmann considered himself an east coast concert and radio composer (he conducted the famous CBS radio orchestra) who took on the occasional film composing job. However, the disbanding of that orchestra in 1949 forced him into a career decision. He moved to California in 1950 to work full times in movies, and *The Day the Earth Stood Still* was his first film after arriving.

Bernard Herrmann in 1951

For this film Herrmann used a minimal orchestra, not much more than an ensemble, really, of piano, harp, brass and tympani. He also used nearly every electronic instrument available at the time: organ, electric violin, electric guitar, electric bass (string bass, that is; not bass guitar) and, of course, two theremins. There were no woodwinds. "My goal was to characterize a man from another world, and the music had to reflect an unearthly feeling of outer space without relying on gimmicks. The result seems to have been successful and most certainly predicted the shape of things to come for electronic scoring."[24]

Hermann biographer Steven Smith, a composer in his own right, noted that the score contained more than just general predictions of trends in Science Fiction scores.[25] In an almost eerie bit of cultural foreshadowing, Herrmann incorporated a phrase in the main title consisting of three ascending notes, played by muted trumpets, which are perfect fifths. In this score they are a heraldic call, announcing the arrival of someone important. However, to any Sci-Fi fan after 1968, they are instantly recognizable as the opening notes of Richard Strauss's *Also Sprach Zarathustra*, which Stanley Kubrick made famous as the title theme to *2001: A Space Odyssey*.

His use of the theremin was influenced by Ferde Grofé's score for *Rocketship X-M* a year earlier. Dimitri Tiomkin also put the instrument front-and-center in his score for *The Thing* the same year (1951). Being used in three popular Science Fiction films in less than two years pretty much sealed the fate of the instrument as a genre

cliché from that point on. Almost as important as the theremin was the electronically amplified vibraphone that created the eerie oscillating bell-like tones.

This haunting and evocative score, using only a handful of musicians, has, in this author's opinion, never been surpassed, or even equaled, in doing so much with so little.

After 42 days of shooting, the film wrapped primary photography on 22 May 1951 and started post-production. This is when Bernard Herrmann recorded the music and Fred Sersen's crew went to work on the optical effects. At the same time, Harry Leonard and Arthur Kirbach in the sound department were creating the sound effects ranging from the exotic (Gort's heat ray) to the mundane (traffic sounds). Incidentally, while the "whoop-zap" of the heat ray was created in the sound department, much of the overall effect actually comes from the music. The pulsing rhythm of the beam as it melts the weapons is actually orchestrated.

As the September release date drew near there was one last detail that Blaustein had to deal with: the title. There wasn't one. The working title on North's script was *Journey to the World*, but they needed something more commercial. Blaustein invited members of the marketing department to his house for a brain storming session. He said it was the head of marketing who observed, "Since your spaceman makes everything stop all over the world, how about 'The Day the World Stopped'?"[26] Blaustein admits in the cited interview as to being fuzzy on the details, but thinks it was himself that suggested that "Earth" sounded better than "World", and that "Stood Still" flowed better than "Stopped." The only problem with this recollection is the timing. According to Steve Rubin in the referenced *Cinefantastique* article, the title was changed in late January towards the end of preproduction. Blaustein's shaky memory not withstanding, this seems more likely. It is supported by the "shooting script" included in the DVD which, as previously mentioned, is dated February 1951 and it already contains the final title throughout.

The usual practice at the time was to preview the film for test audiences prior to general release to gauge public opinion and decide if any final "tweaking" was necessary. When the preview print was ready, they screened it for Zanuck to get his permission to release it for previewing. Both the producer and director were taken aback at the enthusiasm of Zanuck's reaction: "That's it!" he exclaimed at the end of the viewing. "Don't touch it! Don't do anything to it! Don't preview it!"[27]

Blaustein, though, was still unsure about how audiences would react to many of the details in the film. He waited until Zanuck was out of town on business, then arranged a "private screening" at a small theater in Inglewood, California in mid-August. At first he thought his fears would be

---

[24] Bernard Herrmann, liner notes to *The Fantasy Film World of Bernard Herrmann*, London Records SP 44207, 1974.

[25] *A Heart at Fire's Center - The Life and Music of Bernard Herrmann*; Steven C. Smith, University of California Press 1991.

[26] *Making the Earth Stand Still.*

[27] *Cinefantastique*, pg. 22.

confirmed when the audience snickered at the scene of the tanks skidding when making the sharp turn in their hurry to get to the spaceship. He was afraid that they would laugh outright when Gort appeared. He feared for nothing as everyone loved the film and the review cards were overwhelmingly positive. Not quite convinced, he did two more preview screenings and got the same result. They didn't make any changes in the film before release.

The film went into general release in September and was a huge hit with both audiences and critics, making back more than twice its $1.2 million production costs in initial release. Critics were universal in their praise of its message of peace and non-violence. It was also immediately recognized as a classic by Sci-Fi fans. At the 9th World Science Fiction Convention in New Orleans in September 1951, they made a big public presentation of a "Certificate of Merit" to a phony Klaatu in front of the newsreel cameras.[28]

Robert Wise went on to become one of America's most eclectic directors. Working in a wide variety of genres his films ranged from musicals like *West Side Story* (1961) and *The Sound of Music* (1965) (both of which won the Academy Award for Best Picture, and earned Wise a Best Director Oscar®), to the supernatural with *The Haunting* (1963) and *Audrey Rose* (1977), to more war/anti-war films like *Run Silent, Run Deep* (1958) and *The Sand Pebbles* (1966). He revisited science fiction twice more with *The Andromeda Strain* (1971) and, of course, *Star Trek, the Motion Picture* (1979). He was awarded the American Film Institute's Lifetime Achievement Award in 1998. He died in September 2005, just a few days after his 91st birthday.

Wise directs Spock, Kirk and McCoy while Gene Roddenberry hovers.

After the acclaim accorded to Michael Rennie as Klaatu, you would think that he would have gone on to a career of leading roles, but that was not to be. While his

---

[28] *Movietone Newsreel*, September 1951.

career was long and steady, it was not especially distinguished.

Although the very next year he did play Jean Valjean in Lewis Milestone's *Les Miserables*, most of his roles were what could be called "substantial supporting roles", such as the Apostle Peter in *The Robe* (1953). That same year he worked with Robert Wise one more time, doing the voice-over narration to *The Desert Rats*. Also that year he did the uncredited narration to the Fox version of *Titanic*. Whenever a director needed a scrupulous, aristocratic and somewhat mysterious British character, Rennie usually got the call, such as for Disney's *Third Man on the Mountain* (1959) or Irwin Allen's *The Lost World* (1960). After the mid '50s, though, most of his work was on television. He did over 100 guest appearances on series as wide ranging as *Shirley Temple's Storybook* to *Route 66*, and *Wagon Train* to *Alfred Hitchcock Presents*. He even made appearances in the two TV series featured in this book, namely *Lost in Space* and *The Invaders*. In the former he played "the Keeper" in the first season two-part episode of the same name. In *The Invaders* he did two episodes: the first season's "The Innocent" and the two-part second season "Summit Meeting." A lifelong heavy smoker, he died of emphysema in 1971 at the age of 61.

For Sam Jaffe, the reprieve he got from Darryl Zanuck was only temporary. After *The Day the Earth Stood Still*, he did not work again for an American studio until 1958 with John Huston's *The Barbarian and the Geisha*. Things took off again for him the following year when William Wyler cast him in a supporting role in *Ben Hur*. Like many others, he did most of his work after that in television. He appeared in nearly 100 TV episodes in the 1960's, '70s and '80s in shows ranging from the serious (*Playhouse 90*) to the silly (*Batman*). He's probably best known for his recurring role as Dr. Zorba on *Ben Casey*. He only did a handful of theatrical films, including Disney's *Bedknobs and Broomsticks* (1971), and continued working right up to the time of his death in 1984 at the age of 93.

Patricia Neal's life was as dramatic and tragic as any of the roles she has played over the years. At the time she played Helen Benson, Neal was involved in an affair with actor Gary Cooper (who was married…but not to her). The following year she appeared in *The Breaking Point*, and *Operation Pacific;* the latter with John Wayne. Between the stress of working and the crumbling affair with Cooper, Neal suffered a nervous breakdown. While recovering, she returned to New York to act on the stage in a revival of *The Children's Hour*, and in 1953 she married author Roald Dahl.

She continued working on both stage and film. In 1959 she played Helen Keller's mother in the Broadway production of *The Miracle Worker*. Two years later she co-starred with Audrey Hepburn in *Breakfast at Tiffany's*.

In 1961 her daughter Olivia died from measles and the very next year her son Theo's carriage was hit by a taxi when he was just four months old. Despite grieving these terrible losses, in 1963, she won the Best Actress Oscar® for her performance in *Hud*, with Paul Newman.

As if the death of one child and the severe injury of another wasn't enough, in 1965 while pregnant with her daughter, Lucy, Neal suffered a triple stroke from three burst cerebral aneurysms and was in a coma for three weeks. Her husband was in charge of her rehabilitation and she had to completely relearn to walk and talk. (Incidentally, Lucy was born healthy in August 1965.)

She was offered the role of "Mrs. Robinson" in *The Graduate* (1967), but turned it down, feeling it was too soon after her stroke. Later, back in television, she played Olivia Walton in the pilot episode for *The Waltons* (1971). Her performance earned her a Golden Globe, but she was not asked to play the role in the television series.

In 1981, Glenda Jackson played her in the television movie, *The Patricia Neal Story*. She, along with Billy Gray who played her son, are the only cast members still alive as of this writing.

Even Lock Martin's career got a boost from this film. While there aren't that many roles for giants, he did appear in the genre classics *Invaders from Mars* (1953) (he played one of the mutants that carries the octopus-head leader) and the title character in *The Snow Creature* (1954). His part in *The Incredible Shrinking Man* (1957) was edited out before the film was released, and in the mid '50s, he hosted his own children's show called *The Gentile Giant* for a local TV station. Like most people of truly giant stature, his health was always an issue and he died in 1959 when he was only 42.

The only props from the film still extant are the 9-foot (275 cm) Gort statue, restored and still owned by über-fan Bill Malone (we'll describe how it became his in a bit), and the special effects Gort helmet (which had the motorized visor and lights for the heat ray "warm-up") belonging to Bob Burns in Burbank, California. Bob also owns the seven-foot (212 cm) diameter saucer miniature used for the "takeoff" scene at the end.

Bob Burns with the seven-foot model

The whereabouts of the two foot model used in the landing scene are unknown. There was a different seven-foot saucer prop on display for a while at the Universal Studios museum in Florida. That one was pretty rough and had a wedge cut out of it as if the door opening in the dome extended all the way out the skirt. Bob Burns has identified this prop as probably a test molding for the version that he has, and was never used in the film.

The Gort costumes worn by Lock Martin no longer exist since they were not designed to last more than a few wearings, and were probably discarded after principle photography was completed. The fiberglass statue, as previously mentioned, now belongs to Bill Malone, although it very nearly didn't.

Gort as "Rotar"

The prop department sold the statue in early 1952 to Larry Harmon, a television producer best known for playing one of the earliest versions of Bozo the Clown on TV. Harmon bought it to use it in a program called *General Universe* that he was developing. While that show never made it on the air, he did use the statue in another called *Commander Comet*. There it was called "Rotar" and the prop was adulterated with fins on its head and speakers in its mouth and stomach.

After *Commander Comet*, the prop was placed in Harmon's garage where it stayed for more than a decade. Producer/ director Ted Bohus learned of it and offered to buy it from Harmon but couldn't afford Harmon's asking price of $500. He let Malone know of it, and though he, too, was a starving producer just starting his career, managed to raise the money and rescue Gort. "I scraped all the junk off that Larry put on, the fins and all that stuff and got him back to the way he was."[29]

©Jon Rogers
A Gort replica guards the entrance to the SF Museum in Seattle

[29] Bill Malone interview by Ted Bohus on *Monsters 411*: http://www.monsters411.com/billmalone.html

Today, full sized Gort replicas are available from Fred Barton Productions in southern California, one of which stands guard at the entrance to the Science Fiction Museum in Seattle, Washington.

The film has enjoyed continuous favor for half a century as not just a Science Fiction classic, but a genuinely great message film. There are some, though, that question the message, as we shall see later in the "Epilog" section.

The Story

Right from the very first frame of the title sequence, we are told that this is not your usual Science Fiction film. We are treated to views of galaxies, nebulae and, finally, the Earth as we pass the moon on our way in; but never do we see a spaceship. It eventually dawns on us that we are seeing this from the point-of-view of the ship itself. It is a new perspective. In this moment we are the ship; we are the aliens.

As we enter the atmosphere over the Pacific, the ship is

©20th Century Fox Film Corporation
First Contact over the Pacific

picked up on radar; first by an American unit on a Pacific island, then by a British station in Hong Kong. It is tracked around the globe as it passes over India, Eastern Europe, France and England before finally approaching the East Coast of the U.S. To show us the fear and concern of the whole world in this short sequence, Wise uses the same technique that Orson Welles used to scare half the country 13 years earlier. He presents the ship's flight as snippets of real radio broadcasts from each of the countries that it overflies showing both the announcers at the microphones and the worried citizens gathered around their receivers. To achieve greater realism, he used well-known radio personalities of the time, like Elmer Davis and H.V. Kaltenborn, playing themselves (although the impact is largely lost on viewers today).

Throughout all of this, we have yet to see a spaceship. That moment finally comes as tourists on the Mall in Washington D.C. start looking up and pointing to the sky. A glowing dot appears over the Capitol building that grows larger as it proceeds to circle the Mall over the Smithsonian Institution, the Washington Monument and other landmarks before finally revealing itself to be a large, smoothly curving saucer just as it touches down on The Ellipse behind the White House.

©20th Century Fox Film Corporation

©20th Century Fox Film Corporation

Davis (top) and Kaltenborn at their massive microphones

A man runs through the streets shouting "They're here! They've landed!" as police cars and motorcycles pour out of the station and head to the landing site. The Tank Corps from Fort Meyer in Virginia race out the gate, skidding in a most un-tank like fashion in their hurry to leave.

To showcase the alien's emergence, Wise re-uses his "reality" technique, but upgrades it to the brand-new medium of television. It's two hours after landing and we see the best-known TV commentator of the day, Drew Pearson, in his signature fedora, in the studio. He is reassuring viewers that rumors of invading armies are "absolutely false." He describes the scene

©20th Century Fox Film Corporation
Drew Pearson

around the saucer (looking into the monitor) when suddenly his eyes go wide as he exclaims, "Just a **minute** ladies and gentlemen, I think **something** is happening!"

The Herrmann score provides tremendous tension as, from the formerly featureless side of the saucer, a ramp extends. Simultaneously, the dome at the top of the saucer opens to create a wedge shaped door from which a completely human-looking figure strides forth. It's wear-

The Saucer Fleet

©20th Century Fox Film Corporation
"We come in peace"

ing a sparkly blue spacesuit[30] with a translucent helmet. There is a transparent band in front of the eyes and a grid in front of the mouth, which has a built in amplifier, as he speaks in a booming voice, "We have come to visit you in peace and with good will."

As he strides off the ramp onto the grass he reaches into the front of the suit and extracts a complex looking device. With every eye and gun trained on him, he advances towards the military cordon and snaps the device open. An over-anxious soldier shoots it out of his hand, the bullet lodging in the alien's shoulder. As the nearest soldiers rush forward to help him, an enormous robot emerges from the ship and advances down the ramp. The civilian viewers retreat in panic as he raises the visor on his helmet-head to expose a pulsating light. As the soldiers continue to back away slowly, a heat ray issues forth from the robot's visor and strikes their weapons. The GIs drop their M1 carbines as the guns begin to glow and melt. The robot turns his attention to more small arms, then a tank and finally an artillery piece, which are all melted to slag.

By this time the alien has recovered enough to issue a cancellation order, "Gort! Deglet ovrosco!"[31] after which he is approached by the lieutenant in charge of the squad. "It was a gift for your President," the alien says, picking up the remains of the device that was shot from his hand. "With this he could have studied life on the other planets." The colonel in charge of the operation arrives in a jeep and orders the alien taken to Walter Reed Hospital.

At the hospital the President's personal secretary, Mr. Harley, is admitted to see the alien who says that his name is Klaatu. He has been traveling five Earth months and over 250 million miles (400 million Km) to bring an important message to all the nations of the world, but he won't deliver it to any one nation; all must be present together. Harley says that such a meeting is impossible due to the "present

©20th Century Fox Film Corporation
Klaatu and Harley

international situation." Klaatu says that he's not here "to solve your petty squabbles" to which Harley replies, "Believe me, they won't sit down at the same table." Klaatu starts to get irritated and says that he doesn't want to resort to threats, but "that seems to be the only thing your people understand." Harley says that he'll do what he can.

Back at the ship, Army engineers are trying to do a materials analysis on both the saucer and the robot, but aren't making much progress. A sergeant is blasting away at the ship's hull with a torch. An officer

©20th Century Fox Film Corporation

Out-of-this-world materials

---

[30] The film is in black and white, of course, but nearly all of the color promotional materials, such as the headshot of Rennie on page 15, show the suit as being a deep blue. An unfortunate exception is the poster shown on page 14. Posters are often commissioned before a film is finished, and the colors, especially on black and white films, were left up to the individual artists. There were dozens of different lobby posters customized for local markets around the world, so we wind up with green spacesuits, blue robots and blonde damsels wearing red cocktail dresses where Patricia Neal is a brunette and wore only conservative black (at least very dark) outfits.

[31] All of the phonetic transcriptions of Klaatu's native language are from a very complete web page on this film, www.dreamerwww.com/tdtess.htm, which is, unfortunately, now defunct. While this particular phrase is actually in the "shooting script" provided on the laserdisc and DVD versions of the movie, the rest are not and have been transcribed by fans. As you can imagine, there are many variations in the spelling of the words in different sources.

The Day the Earth Stood Still

asks if he's having any luck. "No, sir," he replies. "I saw the ramp come through the side of the ship right here, and now I can't even find a crack!" Carlson, a civilian materials expert, supports this as he reports, "We've used everything from a blowtorch to a diamond drill. For hardness and strength it's out of this world." The officer replies, "I can tell you, officially, that's where it came from!"

We return to the hospital for one of the most inadvertently funny scenes for modern audiences. Two doctors are discussing the results of the medical tests they've run on Klaatu. Even though he looks to be about 38, he's actually 78, and that the life expectancy on his planet is 130. "I don't believe it!" exclaims one of the doctors while lighting a cigarette. "How does he explain that?" "He says that their medicine is just that much more advanced." says the other, while also lighting up. "He was very nice about it, but he made me feel like a third class witch doctor!" They continue talking in a cloud of smoke when a third doctor joins them, having just left Klaatu's room. "I removed a bullet from that man's arm yesterday," he says dejectedly, "and now it's completely healed! I don't know whether I'll just get drunk or give up the practice of medicine."

How do the aliens live so long?

Mr. Harley returns with the bad news that all of the invitations to Klaatu's meeting had been rejected. Klaatu gets even more irritated. He says that he is "impatient with stupidity" and that "my people have learned to live without it." He suggests that maybe he should get out among Earth people and see what the basis is for these "unreasoning attitudes." Mr. Harley says that he can't be allowed to leave, but Klaatu just smiles as the guard locks the door after Harley leaves. In the next scene, a nurse walks up with a tray of food, but when the guard goes to open the door, Klaatu is gone! As officials go into a panic, we cut to a tall man carrying a satchel calmly walking down a residential street while shrill radio and TV announcers are heard coming from the open front doors, inciting fear at the "escaped monster." A close-up reveals that it's Klaatu. He's wearing the clothes of a Major Carpenter, which he presumably has stolen from the Army post's dry cleaners.

He enters a boarding house to rent a room where he startles the residents (who are watching Drew Pearson, still in his hat, on the TV). They are much relieved as they turn on a light and see he is just an ordinary man. He is introduced to the other residents including Helen Benson, a war widow, and her son Bobby.

Mrs. Barley (Frances "Aunt Bee" Bavier) weighs in on where she thinks the alien *really* comes from

At breakfast two days later, the residents are discussing (what else?) the "spaceman." They're listening to the radio, as another real commentator, Gabriel Heatter (pronounced "heater"), delivers a particularly inflammatory invective. Klaatu notes this while reading the paper, where he sees a story on a local savant, Professor Barnhardt, who is calling a meeting of scientists to study the spaceship.

Helen's boyfriend, Tom Stephens, arrives to take her out for the day, but he is distressed that they have to take Bobby with them. Klaatu steps in and offers to watch him for the day. Together, he and Bobby visit Arlington National Cemetery (and Bobby's father's grave) where Klaatu expresses his dismay over the loss of life. He says that they don't have wars where he comes from. "No wars? That's a good idea!" says Bobby, mouthing words directly from Blaustein. They next stop at the Lincoln Memorial where Klaatu laments, "That's the kind of man I'd like to talk to."

Klaatu and Bobby admire Daniel Chester French's imposing statue of Abraham Lincoln

The Saucer Fleet

As he promised, he takes Bobby to see the spaceship where they are interviewed by a man-on-the-street radio crew wanting to hear how frightened everyone is. The interview is cut short when Klaatu doesn't conform to the predetermined opinions the interviewer wanted (another Blaustein point). After the crew moves on, Klaatu asks Bobby who the greatest thinker in the world is. Without hesitation he answers "Professor Barnhardt." Remembering the newspaper story, Klaatu says, "Let's go see him."

They arrive at Barnhardt's residence, but no one is home. They look into his study through the glass doors, but while Bobby finds them locked, Klaatu casually opens them and walks in. He examines some very complex mathematics on the blackboard,[32] and declares, "He doesn't have the answer." He starts to add some equations of his own at the bottom

©20th Century Fox Film Corporation
Helping Barnhardt with his "arithmetic"

when Barnhardt's housekeeper returns and insists that they leave. He writes his name ("Carpenter") and the address of the boarding house on a note pad before complying.

That evening, a police detective named Brady shows up to take him to the Professor.[33] The savant is busily at work at the blackboard when Klaatu is brought to him. Behind closed doors, he tells Barnhardt his true identity, after which Barnhardt dismisses the guard.[34] Klaatu reveals the nature of his mission, that the Earth faces total destruction if we don't change our violent ways as we move out into space

("the planet Earth will have to be…eliminated"), but still withholds the specifics. Stunned, Barnhardt asks if Klaatu could give a "little demonstration" of the power behind the threat to convince people of the seriousness of the situation. "Something dramatic, but not destructive. An interesting problem," Klaatu muses.

Back at the boarding house, Klaatu borrows a flashlight from Bobby, who, being curious, follows him as he leaves the house. He walks straight to the spaceship, which is, oddly enough, being guarded by only two sentries.[35] Klaatu sneaks behind them and signals Gort with the flashlight. Gort lumbers silently over and knocks out the guards after which Klaatu orders the ship opened ("Gort, barringa!"). He strides in and fires up the communications system to phone home: "Emray Klaatu narruwak. Micro pu val barata luke dinsal inkaplis. Yabu tari axel bugettio barengi degas…"

Bobby witnesses everything up to Klaatu entering the ship. He flees home in terror and waits up for his mother to get home. He tries to tell her what happened, but naturally she doesn't believe him and tries to convince him it was all a dream. She tells Tom to go up and get Mr. Carpenter to prove to Bobby it was all in his head. Naturally, he's not there, but instead Tom finds one of Klaatu's diamonds (which he uses as money). Combined with Bobby's story, Tom starts getting suspicious. This is reinforced when they see Bobby's shoes, which are soaking wet from the grass on the Mall.

Klaatu meets Helen the next day just as she's leaving for lunch. He says he must talk with her and they take a private elevator. No sooner do they start their descent, than

---

[32] The equations, incidentally, are a real problem in celestial mechanics known as the "three body problem," which describes the motion of a "free body" (a spaceship) traveling between two planets (the other two bodies). While crucial for interplanetary travel, it has no "closed form" solution and can only be approximated. At the time this film was made, the best way to resolve the three body problem was being hotly debated among scientists.

[33] Many people erroneously think that the detective is an FBI agent. This is understandable from Bobby's exclamatory "Mr. Brady is a government agent!" even though Brady himself says nothing of the sort.

[34] The sharp-eyed viewer will notice that although Klaatu left with a plain-clothed police detective, he arrives at Barnhardt's residence escorted by an Army captain. The reason is that there's a scene deleted at this point. As originally shot, Klaatu was not taken to Barnhardt's, but to police headquarters for interrogation. Stills of

this scene are included on the laserdisc and DVD versions of the movie, and the action/dialog can be read from the included script. In the scene, Klaatu sees first hand the effects of the intense police dragnet looking for "the spaceman" and the mob mentality gripping the city. Just as the police are about to send him back to the hospital (where the doctors can identify him) an Army captain arrives with orders to take him directly to Professor Barnhardt. Apparently Robert Wise thought that this little detour interrupted the flow of the main story too much and removed the scene.

[35] You'd think that something this monumental would be surrounded by a whole battalion of Marines!

the car jerks to a stop and the lights go out. Klaatu tells her that they will be there for some time since "the electricity has been neutralized all over the Earth." We are treated to a wonderful montage of people and machines all over the world, immobilized due to the lack of electricity.[36] We see Tom getting an appraisal on the diamond (as the jeweler tries to find enough light in the darkened store). "There are no diamonds like this anywhere in the world that I know of!" the jeweler remarks, very excited. Back at the elevator, Klaatu finishes up telling Helen the whole story just as the electricity comes back on. "It must be 12:30," he says. "Yes, just exactly" she replies, her mechanical watch apparently functioning perfectly through the shutdown.

The military, having discovered that Gort had moved (and conked out two guards), encase him in "KL 9-3, a new plastic material stronger than steel." General Cutler, convinced that the robot is under control turns his attention to the spaceman. "Up till now we have agreed on the desirability of capturing this man alive. We can no longer afford to

©20th Century Fox Film Corporation
Tom betrays Klaatu's where-abouts to the Army

be so particular. We'll get him. Alive if possible, but we **must** get him!" He places the city under martial law and no one is allowed to leave. Tom is now convinced as to "Mr. Carpenter's" true identity. Helen pleads with him to stop as he calls General Cutler to tell him that Mr. Carpenter is really the spaceman and where to find him. She races back to the boarding house to warn Klaatu, one step ahead of the Army.

They hop into a cab and head towards Barnhardt's residence where he plans to hide until the meeting of scientists at the spaceship that evening. They are soon spotted by the Army's dragnet, however, and Klaatu worries what Gort might do if anything happens to him. "But he's a robot," says Helen. "Without you what could he do?" "There's no limit to what he can do," replies Klaatu ominously. "He can destroy the Earth!" He then gives Helen the famous instructions, "If anything should happen to me, you must go to Gort. You must say these words: 'Klaatu, Barada, Nikto.'" Helen repeats them once, and then mouths them silently as the cab is caught in an Army roadblock. Klaatu makes a run for it, but is brought down by a brief burst from a jeep-

mounted .50 caliber machine gun. As he lies dying, he tells Helen, "Get that message to Gort!"

We next see why Klaatu was so concerned. Gort is freeing himself from the block of KL 9-3 by burning it away from the top down. As soon as his head is free, he raises his visor and disintegrates the two guards (note that prior to this, Gort had only attacked the weapons, or disabled the guards temporarily. With Klaatu dead, his priorities have obviously been changed and he wastes no time in executing his new program by killing the guards quickly and efficiently). Helen arrives just as he finishes melting the plastic. She tries to approach him, but is stopped by her own fear. He advances on her, backing her into the fence. He raises his visor, exposing the heat ray, but before it's fully charged, she delivers the famous message. This causes Gort to close the visor, pick her up, and carry her into the spaceship where he also phones home (silently) for instructions. Leaving her locked in the control cabin, He lumbers out of the ship.

©20th Century Fox Film Corporation
Helen delivers the strange alien phrase

We next see the Colonel reporting to General Cutler from a local jail that they got the alien and "he's dead, all right," and that they've put the body in a cell. Gort appears outside the cell wall, which he proceeds to melt with his heat ray. He carries the body back to the ship[37] and places it in a strange-looking machine.

Outside the ship, Barnhardt is telling the assembled scientists that the Army has ordered them to leave since "the robot is on the loose." Back inside, we see Gort operating the controls that bring the machine with Klaatu's body to an ear splitting crescendo as Klaatu starts breathing again. The machine abruptly quits with the "bzzzt" of a high-current contactor opening, and Klaatu's eyes flutter open. "I thought you were…" Helen stammers. "I was," returns Klaatu. The exchange that follows was added at the insistence of the Breen

---

[36] Just playing devil's advocate, why would diesel engines stop? True they need electricity to spin their starter motors, but once running, they are completely mechanical.

[37] This sequence really stretches credibility. Moving as slowly as he does, how did Gort make it all the way from the Spaceship to the jail and back (carrying a dead body!) through the massive Army dragnet without being seen? On the other hand, there was more to this sequence in the original script where Gort **is** spotted, but simply lays waste to all of the Army forces a lá the Martians in *War of the Worlds*. This scenario has its own problems in that Gort could then no longer return to the ship unobserved, and the scientists would not be calmly gathering for the meeting while Gort is inside resurrecting Klaatu.

Office (an industry "watchdog" organization) over Blaustein's strenuous objection.[38] "You mean he has the power of life and death?" asks Helen, nodding towards Gort. "No," replies Klaatu. "That power is reserved to the almighty

©20th Century Fox Film Corporation
Gort begins the resurrection process at the ship's biocenter

spirit. This technique, in some cases, can restore life for a limited period."

Back outside, the scientists are startled when the ship opens up and Gort emerges followed by Klaatu (who is back in his sparkly spacesuit, but without the helmet) with Helen at his side. She continues down the ramp to a few raised eyebrows as Klaatu steps forward to finally give us his message. It's delivered in full I've-had-it-up-to-here-with-you-people mode as he speaks bluntly. To avoid war, the people of the other planets have entered into what we would consider a Faustian bargain with technology. They have created a police force of robots, and have given them the irrevocable power to destroy any aggressor. Even if we don't join them in this pact, the robots will still destroy us if we start threatening other planets. Note the logic here. We have the choice of joining them and voluntarily submitting ourselves to this

©20th Century Fox Film Corporation
Klaatu delivers his terse message

robotic tyranny, or we don't join them, and the robots will destroy us anyway at the first sign of "bad behavior." As we shall explore later in the "Epilog" this is no choice at all, and is 100% dependent on having completely infallible machines. Let's hope that the robots' algorithms that interpret our monitored behavior have been well tested!

Klaatu marches back into the ship, calling Gort after him ("Gort, barringa!" the only alien phrase in the film to be repeated). The ship immediately, and without warning starts humming for liftoff, causing the scientists to run for their lives. It takes off smoothly and recedes into the distance. Just before it disappears completely, the dot of light rushes back to declare "The End."

©20th Century Fox Film Corporation

The Vehicle

As with the music in this film, the design of Klaatu's spaceship is a masterful use of minimalism. There is far, far more implied here than is actually shown on the screen.

There is one scene in the film where we almost get to learn the workings of the ship, and from Klaatu himself! When viewing the spaceship from the visitors' area, Bobby asks him "what do you think makes it go?" to which Klaatu answers, "A highly developed form of atomic power, I should imagine. There are several ways of reducing landing speed. The basic problem is to overcome the inertia, and…" At this point he becomes aware of a couple of buffoons listening in as they burst out "Keep goin', mister, he was fallin' for it!" As the cretins stagger off, laughing in that smug, superior way that only true ignoramuses can, we lose our chance to hear how the ship works.[39]

---

[38] The censors at the time refused to allow any sort of purely technological device restore life to a corpse. They were afraid of a religious backlash at this implicit "proof" of the lack of the need of divine intervention. This exchange pushes the dialog in the exact opposite direction, thus avoiding a potential "Frankenstein" scenario.

[39] By the way, if Klaatu is such a hotshot celestial navigator, why did he approach the Earth from the wrong direction? By flying east-to-west the way he did, he had to not only neutralize all of his ship's velocity from space, he had to "back up" about 850 MPH (1,370 Km/hr) to match the rotational speed of the Earth at Washington, D.C.'s latitude. If he'd come around west-to-east he could have saved himself 1,700 MPH (2,740 Km/hr) in "delta vee."

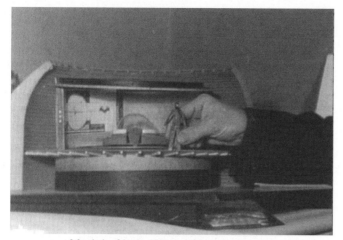

Model of Lyle Wheeler's interior set

The technology represented by the ship and Gort are the perfect embodiment of Arthur C. Clarke's famous observation that "any sufficiently advanced technology is indistinguishable from magic," and more than a decade before he said it, to boot.[40] Designer Lyle Wheeler's deceptively simple sets exemplify the principle that as technology gets more advanced, the machinery (or at least the part that you see) gets simpler. We have a perfect example of that in nearly every home today, the personal computer. At the time this movie was made, computers were monstrous things requiring three staffs of professionals to run each one: a staff of mathematicians and "information scientists" to write the program, a staff of engineers and programmers to enter the program, and a maintenance staff to keep the thing running as the thousands of vacuum tubes inside expired at random times. Today, even the most humble home computer is a million times more powerful, but your average pre-schooler has no problem switching one on and playing games on it.

The locals have a laugh at Klaatu's expense

That's because the technology has advanced to the point that the complexity (and they are orders of magnitude more complex than the old spaghetti-wired, tube busters of yore) is hidden. We don't see the hundreds of thousands of engineers and technicians in the semiconductor industry making the chips, or the millions of people in the software industry sweating over the code that makes the PC work. All you see, and indeed need to see, is a keyboard, a screen and a mouse.

Klaatu's ship embodies this same principle. The controls are simple in appearance, but amazingly complex in operation. First off, they are all non-contact. While this was jaw dropping to audiences in 1951, today we think nothing of having a supermarket doors swing open or lights turn on as we approach them, or unlocking a door at work by simply waiving our badge in front of it. In a like manner, Klaatu (or Gort) simply waves a hand over a bank of con-

The ship's communication center

trols and the communication system springs to life.[41] The controls are also intelligent, being somehow aware of who is operating them. When either of the legitimate operators passes a hand over the door, the row of lights either comes on (open), or goes off (locked) depending on the wishes of the operator. But when Helen Benson rushes to the door after Gort leaves (and the lights go out, locking her in), the door ignores her.

As magical as the controls appeared, there had to be some common touchstones for the audience so that they could understand what it was they were looking at, even if they didn't know how it worked. For example, on either side of the main communications screen is a large "tuning eye." On the upscale radios of the time, this device measured signal strength to aid in tuning the radio to the desired station. It was a small phosphorescent screen about an inch (2.5 cm) in diameter. When there

---

[40] This oft-misquoted phrase is also known as "Clarke's Third Law." It first appeared in *Profiles of the Future* (1962).

[41] This is also, incidentally, exactly the way a theremin is played. Thanks to Kris Harber (the artist who did our wonderful cover art) for recognizing this probably inadvertent bit of self-reference between the movie and its own music.

was no signal, only half of the circle was lit up, but as the signal got stronger, the edges of the lit area rotated towards each other forming a "V." The narrower the "V," the stronger the signal. In the communications scene, you can see these in operation as they "home in" on the headquarters signal.

The large, central screen was patterned after radar screens of the period with its circular calibrations superimposed on the round screen. There is a square grid on a wide rectangular band hovering just off the screen's surface that extends to the left and the right. Somewhat curious to our modern ears, for an electronic device, is the whine of a small, high-speed electric motor when the communication system is in operation. A possible origin for this comes from television. Even though TV was a commercial reality in 1950, it was still black & white. Color television was still highly experimental, and one process under development was a mechanical system by CBS. It used a small, high-speed motor to spin a colored filter wheel in front of a regular black & white picture tube. There was a similar filter wheel being spun in front of the TV camera in the studio. If (and it's a big "if") the two wheels were in sync with the red, blue and green filters passing in front of both the camera and picture tube in the same sequence at the same time at both ends, you'd get color TV using only a black & white camera and receiver.[42] The downside is, of course, that your TV makes an annoying high-pitched whine and the picture flickers constantly. So what do we observe as Klaatu and Gort phone home? A high-pitched whine and a flickering light across their faces. Obviously they are using the most advanced communication devices on the planet!

But, what about the rest of the ship? The short answer is that we just don't know. The underlying strength of Lyle Wheeler's set design is that everything is implied but almost nothing is shown. In the best tradition of Greek theater, the parts we fill in with our imagination are far more vivid and "real" than anything that could be shown on the screen. However, that hasn't stopped fans from doing some pretty inventive speculating.

In the center of the command area, presumably the dead center of the ship, is what must be the pilot's station. It is a round kiosk with a clear half dome over it and a circular console with some simple controls lining it. There is a narrow break in the kiosk to allow access to the control area. We never see anyone sitting there. In fact, without an obvious chair we're not sure if the controls are operated from a sitting position or standing up in some sort of well recessed into the floor. It could even be a hatchway to the engineering areas below, and the half dome might be a sealing hatch that just happens to be rolled back into the open position.

An article by M. Darktower[43] proposes that the annular

space around the central command area, which serves no apparent purpose other than as a corridor where no corridor is needed (why not just have the outer door open up directly into the command area?), could be part of the ship's radiation shielding. Once the ship is in space and that area is in vacuum, magnets lining the walls of the corridor would turn it into a "storage ring" similar to those in nuclear particle accelerators, thus trapping energetic particles before they can get to the central crew area.

The article further speculates that the large outer diameter of the saucer contains a Tokamak fusion generator[44] that is the power source for the main drive. Power from the generator is fed directly into the drive coils, which consist of both magnetic (to allow easy maneuvering on planets with magnetic fields, like Earth), and gravitational field coils. Of course, you'd think magnetic coils strong enough to lift a ship this big on the Earth's weak magnetic field would be sucking in cars and trucks from the surrounding parking lots when they lifted off...

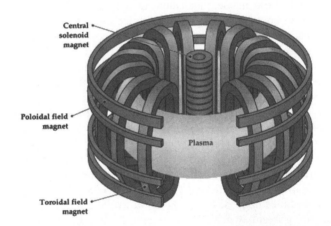

Central solenoid magnet

Poloidal field magnet

Plasma

Toroidal field magnet

Courtesy Lawrence Livermore National Laboratory

Tokamak fusion reactor schematic (above) and in operation showing the yellow plasma ring edge-on.

That leads to the enigma of the boarding ramp. If the outer rim of the saucer contains a Tokamak generator ring, then where is the ramp stored? Darktower's wonderful lateral thinking suggestion is that it wasn't stored anywhere. The ramp was made of "memory material" that "remembers" its ramp shape. When not needed as a ramp, the material is simply part of the wall of the saucer skirt. When the ramp is required, the correct electric/

---

[42] This is how the color TV cameras carried on the Apollo missions worked, although the signal was processed to be compatible with standard NTSC color TV receivers.

[43] *Fantastic Films*, April 1978, pg. 49.

[44] A Tokamak is a type of magnetic mirror fusion generator whose main chamber is in the shape of a torus (doughnut). It was invented about 15 years after this movie was made.

magnetic/thermal/whatever stimulus is applied to make the material remember its ramp shape, and the ramp just grows from the side of the ship. This would explain why the Army engineers "can't even find a crack" since there would be no crack to find. Of course, that isn't really what we see in the film. The ramp extends in two parts, an upper part, which is part of the saucer skirt when closed, and a hidden lower part that pushes out linearly. As the lower part comes out, it pushes the skirt part up out of the way, which then falls neatly behind forming the upper part of the ramp. There is, incidentally, a very pronounced gap visible between the fixed part of the skirt and the ramp where it emerges.

Next we have Jon's Archeological Report on Klaatu's saucer, after which we'll be back to discuss this movie's impact over the years and how its message might not be quite as benign as it first appears.

◆ ◆ ◆ ◆ ◆ ◆ ◆ ◆ ◆ ◆ ◆ ◆ ◆ ◆ ◆ ◆ ◆ ◆

Archeological Report on Klaatu's Saucer
By Jon Rogers

What you see on the Silver Screen is really nothing more than flashing lights-- it is all just an Illusion. If there ever were an example to prove that statement, it's Klaatu's saucer from *The Day the Earth Stood Still*. The most amazing thing about this saucer is that when you see it in the movie, it looks so believable. As stated in the main "Vehicle" section,

> The underlying strength of Lyle Wheeler's set design is that everything is implied but almost nothing is shown.

That's an understatement! What you've seen—or think you've seen—you really haven't seen at all. In the finest tradition of the saucer sightings of the '50s, what seems obvious at first glance, becomes quizzical on a second, then mysterious, and then finally a complete enigma. And what begins as mysterious, stays mysterious.

Art Director Lyle Wheeler

Figure 1

At first Klaatu's saucer seems so simple and obvious, however, when you carefully examine different aspects about it, you discover that one fact contradicts another until in the end, you discover that you really know almost nothing about it at all. And that is what turned Klaatu's saucer into one of the most challenging and intriguing archeological "digs" this engineer has ever tackled.

Let's begin by reviewing our source of material for our knowledge of Klaatu's saucer. The main source is, of course, the film itself. Other primary sources are still pictures taken during the filming of the movie and the (poorly copied) plans for the construction of the set that have come down to us over the years. Of special interest is a picture of Lyle Wheeler showing an architectural model of the set used for the interior of the Spaceship (Figure 1). There is also the 84 inch (213 cm)[1] diameter saucer miniature (see photo page 22) which was used in the long shots of it landing and taking off. Taken together we should be able to get a good understanding of what the saucer looked like and what it was, or rather was supposed to be.

So what was the Flying saucer seen in *The Day the Earth Stood Still*? We have two views of it from the film, externally and internally. The External view is more defined and consistent so we will deal with that first.

Exterior

We first see it flying over Washington DC, past many monuments only to land in the outfield of a park, right between two baseball diamonds. It hums, it glows, it settles to earth and is still. From this point forward the saucer is largely an immobile backdrop until the last scene in the movie where once again, it glows, it hums, and rises up to the heavens and fades off into infinity.

---

[1] In a 1995 documentary Bob Burns displays the original miniature (see page 22) and states that it is "7 ft across, made of fiber-glass." He also points out that the model's originally flat bottom was later modified when it was reused in "Voyage to the Bottom of the Sea" by adding an extended base with a ring of lights to it.

Klaatu's Saucer

1/300 scale
Dimensions in inches
© 2003 by Jon C. Rogers
Sheet 1, Exterior Views
Sources:
The Day the Earth Stood Still,
20th Century Fox, (1951)

Notes:
1. Saucer color is silver. It glows white during operation.
2. Saucer sections are seamless.
3. For details of Saucer Interior see sheet 2 and 3.
4. Bottom base (shown in black) is not a part of the Saucer. It was used to show the Saucer "floating" where it was parked. Since it is shown in the original set drawings, it is included here for reference.
5. Door is shown at maximum opening. Only the right side moved. The grey area is the ramp down to the walkway.
6. The Interior room is visible through the door. Centered in the opening is an interior door that is never used and *may* be located directly behind the control console.

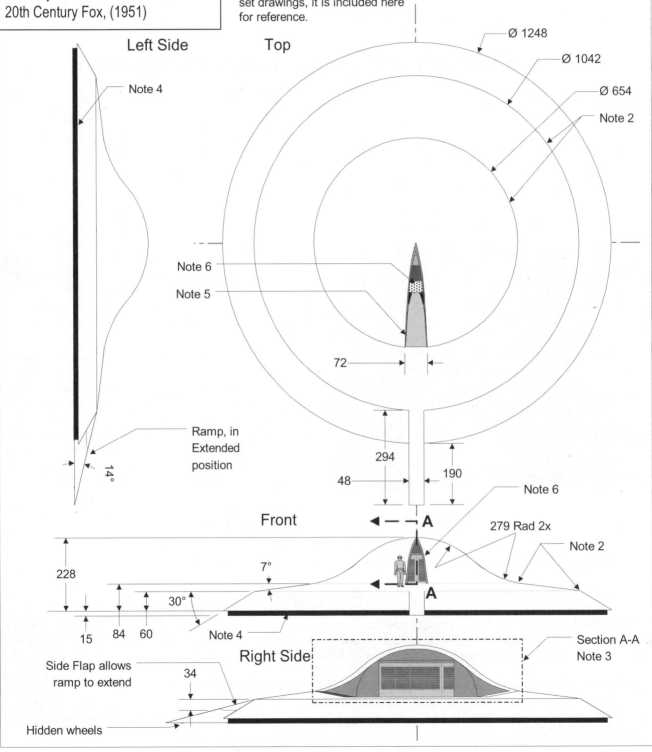

Left Side

Top

Note 4

Ø 1248
Ø 1042
Ø 654
Note 2

Note 6
Note 5

72

Ramp, in Extended position

14°

294

48

190

Front

Note 6

279 Rad 2x

Note 2

228

7°

30°

15   84   60

Note 4

A

A

Section A-A
Note 3

Side Flap allows ramp to extend

34

Right Side

Hidden wheels

Klaatu's Saucer

1/108 scale
Dimensions in inches
© 2003 by Jon C. Rogers
Sheet 2
Interior Plan and Side View

Notes: (See Archeological report)
1. Mysterious light above control room.
2. Opening in ceiling of control room.
3. Ramp to decend to control room level which is below entrance port.
4. Two aft doors in control room.
5. Two forward doors in control room.
6. Door seen through ship's entry way. Not used. Possible maintenance access to control console.
7. Bench, approximately 18 x 72
8. Shape of interior set walls.
9. Main door opening (ref)

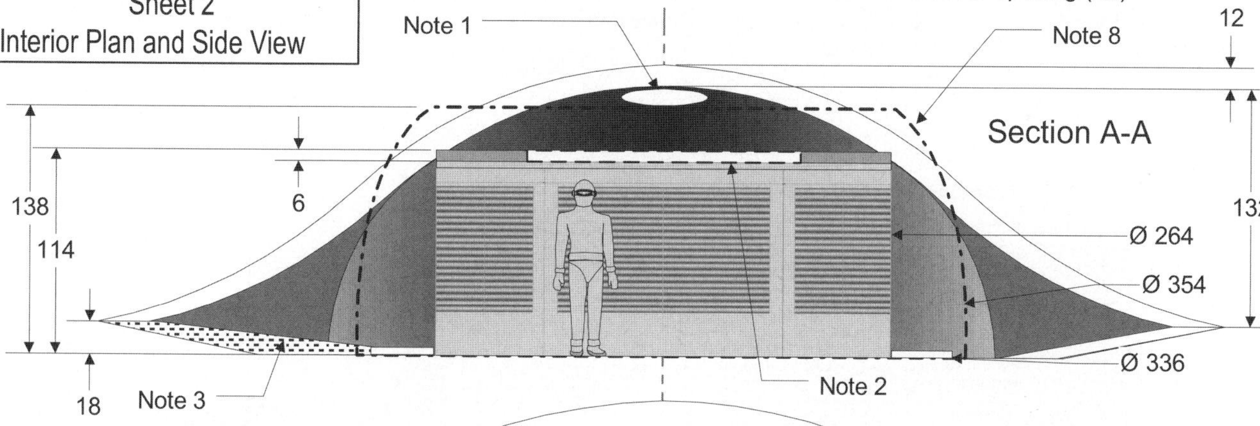

Section A-A

Note 1
Note 8
12

138
114
6
132

Ø 264
Ø 354
Ø 336

18
Note 3
Note 2

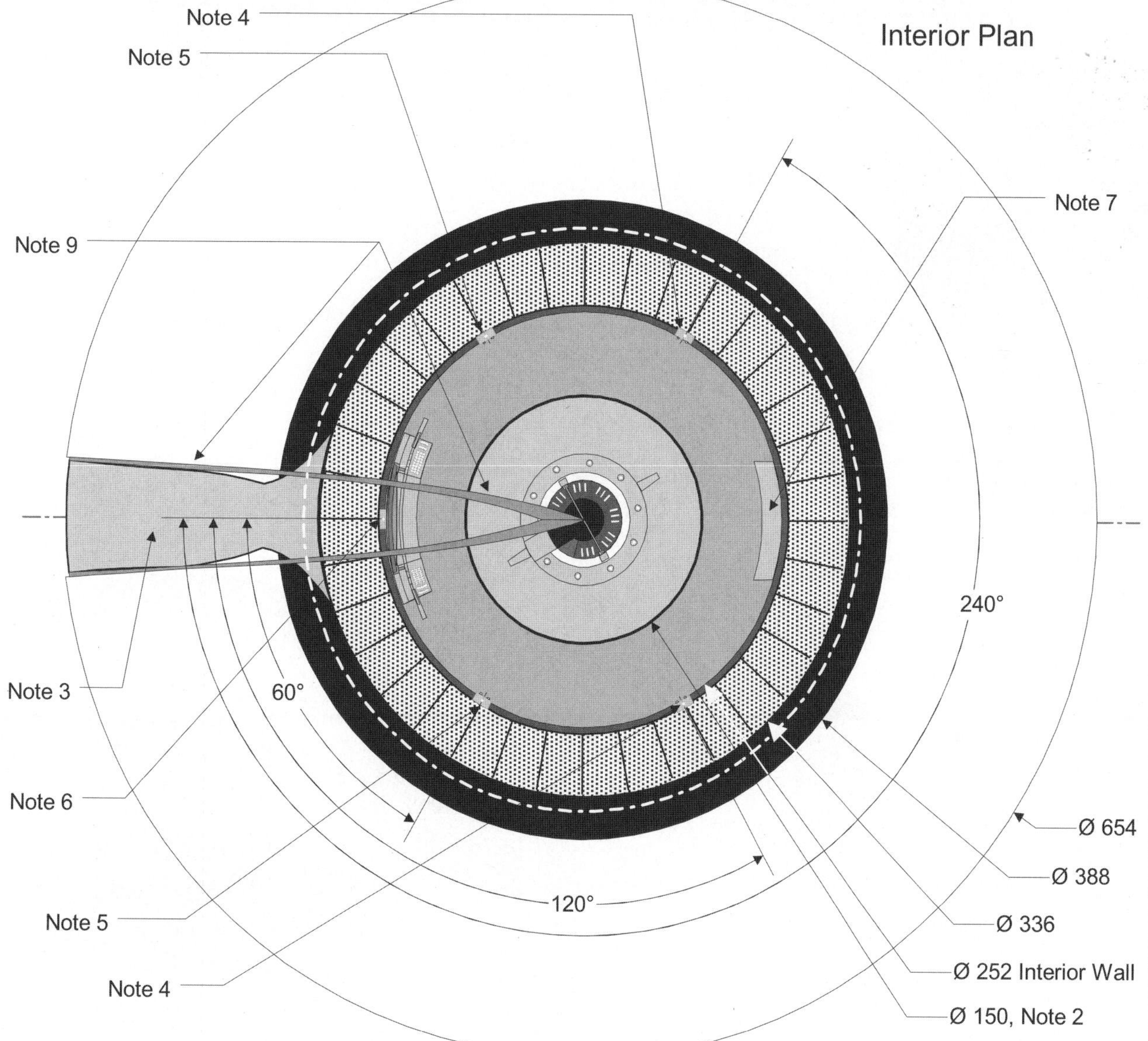

Interior Plan

Note 4
Note 5
Note 9
Note 3
Note 6
Note 5
Note 4

Note 7

240°
60°
120°

Ø 654
Ø 388
Ø 336
Ø 252 Interior Wall
Ø 150, Note 2

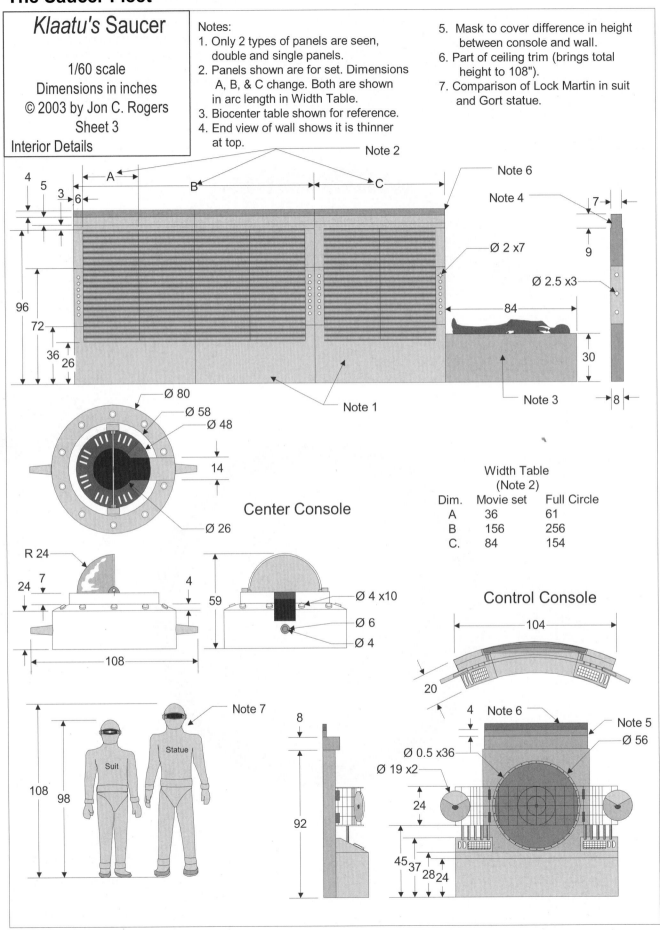

Klaatu's Saucer

1/60 scale
Dimensions in inches
© 2003 by Jon C. Rogers
Sheet 3
Interior Details

Notes:
1. Only 2 types of panels are seen, double and single panels.
2. Panels shown are for set. Dimensions A, B, & C change. Both are shown in arc length in Width Table.
3. Biocenter table shown for reference.
4. End view of wall shows it is thinner at top.
5. Mask to cover difference in height between console and wall.
6. Part of ceiling trim (brings total height to 108").
7. Comparison of Lock Martin in suit and Gort statue.

Center Console

Control Console

Width Table
(Note 2)

Dim.	Movie set	Full Circle
A	36	61
B	156	256
C.	84	154

Can there be any doubt about what we see? The saucer shape is clear and evident when it flies overhead and lands. The nearby baseball diamond gives us an excellent comparison to judge its relative size. The set construction plans give even better data, so that its size is closely, and accurately determined. We see many close-ups of it during the movie so its features, or rather lack of features, are well understood.

When we start comparing some of the sources we have of the saucer, we discover a couple interesting anomalies.

©20th Century Fox Film Corporation
Figure 2

- In the plans for the set, the left side appears shorter than the right side. Also when it is seen on the ground surrounded by tanks, workers, and light poles, it also is shorter on the left than on the right. However in other shots and when it lands and takes off, it is clearly symmetrical.

- When we see it on the ground, it appears to have a black base recessed within the skirt of the saucer but when it takes off, there is no base. The bottom of the saucer is entirely flat.

- When it opens up for Klaatu or Gort to enter, only one side of the dome moves, the other is stationary.

- Finally, when this "entrance port" is open, we can see the interior room slightly. In the middle of the entrance port is what we learn later to be a door to the inner room of the ship. However, it is clearly avoided by Klaatu when he enters the ship. Why have doors you don't use? (Figure. 2)

Some of these enigmas are cleared up easily.

- The question of the saucer's symmetry comes from the fact that the "landed saucer" set was designed, not as a full saucer but as only half a saucer with the rear always open. This was necessary for stagehands to manually move the "mysteriously emerging ramp" out on cue. When this set was still photographed from an acute left angle, (never seen on screen) the truncated left edge of the set looks shorter than the right. The plans of the actual set show how it was made to look round when it was actually "one half pie shaped." In addition, the use of the miniature saucer in takeoff and landing scenes clearly shows that the saucer was supposed to be fully round with a hump on top and flat on the bottom.

- The question of the black base it seems to have when sitting on the ground is not quite as clear. The base shows up clearly in the plans for the set. However, almost all scenes of it in the movie are shot to mask any details of it. Almost like the director wanted it to disappear. Furthermore, as stated above, when the saucer takes off it clearly has a flat bottom. So was it there or wasn't it?

  In this case, we have to take into account the state of the visual effects art in 1951. As with the ramp and entrance port, effects which could be easily "morphed" in films fifty years later, at the time they had to be built, out of real materials, and made to *seem* like they worked as though they were mysteriously evolving from the ship. Much work with tape and plaster was put into the saucer set to make it appear seamless. All movements were planned to hide any mechanism to make it appear more mysterious. With this in mind, it becomes clearer that the purpose of the black base under the saucer was to give it the appearance that it was floating in the air a few inches off the ground, like it was weightless. It was simply the best method that special effects could do at that time. So as we view the film now we see something that the audiences of the day would have not seen (the black base) and we do not see what they would have seen and accepted (the spaceship floats a few inches off the ground when parked).

- The port in the dome opening to one side only is likewise a product of the state of the special effects art of the day, and probably the film's budget. It was much simpler to make one side of the entrance port move than to co-ordinate two groups of workers pulling on two doors to make them open smoothly and mysteriously. Any bobble or mis-syncronization and the illusion would be lost.

It is this forth item that doesn't seem to go away. We clearly see the door Klaatu doesn't use, but why have it?

After thoroughly reviewing the interior layout of the ship, one of the features seems to give us an answer. It becomes apparent that when anyone uses either of the two visible doors in the interior (in any one scene), that the entrance port was not seen from any of them. Furthermore, while the interior walls are translucent they do not show any shadow or indication of an outer door opening. Since we know the ship has a port opening that we never see from the inside, it must be obscured. The only thing big enough inside to obscure a door is the main communications console. But why have a door blocked by the main screen? The answer fairly jumps out: for construction and maintenance. On commercial aircraft, there are always additional doors used by "authorized personnel only" for repair and flight preparation. This answer explains why we see this door from the outside, but never from the inside. But remember, this is our logical interpretation of the facts, not necessarily the truth. In fact we do not know what the door in Figure 2 is for, or where it leads.

Interior

This brings us to the subject of the interior of the saucer. Here the mystery deepens. It is not just the existence of the mysterious, unused door, but the interior itself.

We see the interior of the saucer in a number of scenes. Specifically:

- When Klaatu enters to "phone home" and arrange to stop the world's electricity for half an hour.
- When Gort carries Helen inside and locks her in while he too calls for instructions and then leaves to get Klaatu's body.
- When Gort returns with Klaatu's body. This scene continues while Klaatu revives, talks to Helen, and then leaves.

Within these three scenes we have a number of shots where the action and the interior are seen from a specific camera angle. We must always realize that each shot must be set up and the whole set adjusted so the camera can see what the director wants the audience to see. With each shot is the possibility that the set might not be consistent and the audience might get disoriented. Art directors work hard to insure this doesn't happen.

When one watches these three scenes in the movie, the first impression is that all shots were taken in and of the same set, that is, all shots look consistent, and the action flows smoothly in a familiar space. However, if we review these scenes with the intention of re-creating the interior of the saucer, problems begin to crop up. Lets begin with Interior Scene 1[2]

Scene 1, shot 1 (Figure 3), taken from inside the ship but outside of the walkway, shows Klaatu walking briskly counter-clockwise in a hallway around the outside of control room. He then opens a door on the far right to go inside. We notice that apart from one break in the control room wall on the far left (top picture in Figure 3) and the door Klaatu enters on the right, the control room wall is made of continuous corrugated paneling. We notice the door he opens slides to his right only (left side stationary). This all seems straightforward.

Scene 1, shot 2 (Figure 4) taken inside the room, we see Klaatu enter the control room on the right side, turn on the lights, cross over to our left behind the control kiosk,[3] and walk toward us on the left of the kiosk.

©20th Century Fox Film Corporation
Figure 3 – Scene 1, Shot 1

©20th Century Fox Film Corporation
Figure 4 – Scene 1, Shot 2

---

[2] In the following discussion, when describing position in the room, "front" refers to the direction towards the communications panel, but "left" and "right" have to be taken in context. When describing motion or position, "left" and "right" refer to the sides of the movie frame relative to the viewer. Other times they are used to describe position relative to a character's left or right.

[3] "Control kiosk" means the strange object in the center of the room with a transparent half dome. Its purpose as a control center is strictly inferred from its shape and features, as it is never seen being used. Its true purpose is unknown.

We see there is another door on the left rear of the room like the one Klaatu enters, and the entire back wall is made of four continuous corrugated paneling. The two doors look about 90° apart. A simple shot, and all looks consistent with the first shot because you mentally assume that the right wall (most of which we cannot see) is what Klaatu passed in the hall in order to enter the room by the door on the right rear of the room. (A sharp viewer might catch the fact that the door he enters in shot 2, closes from Klaatu's left, not right, opposite from the way it opened.)

We know he didn't pass the back wall, to come in the right door in the first shot (because he would have been shown walking clockwise with the wall on his right side), so it must have been the door to our right as seen from inside the room through which he entered. Therefore the wall on the right in shot 2 must be just like the back wall—long continuous corrugations with no door. The only doors shown are two in the back of the room. This gives us an understanding of almost ¾ of the room's wall construction. Or does it?

Scene 1 shot 3 (Figure 5) continues Klaatu's walk around the kiosk as seen from the back of the room. We see Klaatu approaching the large communications console on the front wall that must have been behind us in shot 2. We see two corrugated panels on each side of the console ending at two doors with some paneling beyond. We see the back side of the kiosk in the center of the room. Normal movie goers would assume these are the same two doors we saw in the last shot. Now we have seen the entire 360° of the interior.

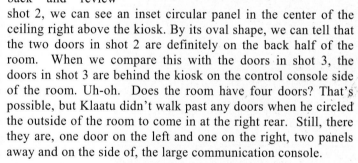

Figure 5 – Scene 1, Shot 3

However, if we go back and review shot 2, we can see an inset circular panel in the center of the ceiling right above the kiosk. By its oval shape, we can tell that the two doors in shot 2 are definitely on the back half of the room. When we compare this with the doors in shot 3, the doors in shot 3 are behind the kiosk on the control console side of the room. Uh-oh. Does the room have four doors? That's possible, but Klaatu didn't walk past any doors when he circled the outside of the room to come in at the right rear. Still, there they are, one door on the left and one on the right, two panels away and on the side of, the large communication console.

Scene 1 has several more shots of Klaatu using the control panel (Figure 6); one medium close and then progressively closer ones. Sure enough, we see the right "front" door in the background. They help us understand

Figure 6 – Scene 1 completion

the panel, but they do not reveal any further information about the room. We are left to wonder how the room is laid out, especially when we thought we knew on first glance.

Scene 2 shot 1 (Figure 7) is seen from the floor level in the outer walkway, Gort walks counter clockwise around the room and we see a shadow of the door opening as he brings Helen into the control room. We do not know which door this is. The scene fades into shot 2 and we don't actually see him putting her down.

Shot 2 (not reproduced here) is from inside close-up of Helen shows Gort turning away while Helen cringes, sitting down on a bench under the back four panel corrugated wall.

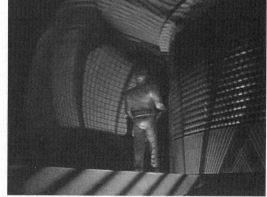

Figure 7 – Scene 2 Shot 1

The Saucer Fleet

Figure 8 – Scene 2 Shot 3

Figure 10 – Scene 2 Shot 6

Shot 3 (Figure 8) is from close up behind and over Helen's shoulder showing Gort going to the communications panel (Note: there's more discussion later on where this shot was taken). It shows one door on the left half near the front of the room. The kiosk now appears turned backwards with the entrance of the kiosk to the rear of the room and the bubble to the front (more on this later).

Shots 4 and 5 (Figure 9) are done from high over Gort's shoulder. They show Helen sitting on a bench on the back wall while Gort concentrates on his task. This is repeated with close ups of Helen for effect.

Shot 6 (Figure 10), again from high over Gort's shoulder shows him finally finishing his task and turning to leave the room. We see that there is one door on the left side, back half of the room, behind Gort. If we look closely at the wall, we can see a ruffled skirt at the bottom of the wall instead of a clean edge. This is not a feature of a solid wall but of a movable partition.

Shot 7 (Figure 11) from close up behind and over Helen's shoulder again, shows Gort leaving the control panel and going out the door on the left-front of the room near the console. Note: this time, the left side wall moves. Also at this point we can see that the door is barely 4 inches (10 cm) taller than Gort. (This is also

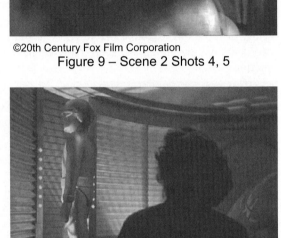

Figure 9 – Scene 2 Shots 4, 5

Figure 11 – Scene 2 Shot 7

the shot where the Gort costume's transparent visor is visible momentarily.) Also, over Gort's head we can see, two track lines in the ceiling trim on the inside and outside the open doors leading around the wall. These tracks are further evidence that the walls move.

Shot 8 (Figure 12) from inside and fairly close we see Helen jump up and make an attempt at escaping out the door as it is closing behind Gort's back. The door closes as she futilely tries to stop it. This close up gives us a clear estimation of the scale of the door molding and the corrugated wall panel.

At first, all of Scene 2 appears consistent with what we saw before. Then we remember that the kiosk was turned completely around from where

Figure 12 – Scene 2 Shot 8

it was in Scene 1. Before the kiosk had an opening pointing to the left of the control panel, (Figure 5) now it is pointed toward the door on the right rear of the room (Figure 8). Did the kiosk move? Did the wall move? What happened? We're not sure. The bench Helen sits on, did not seem to be there in Scene 1 but it could have been hidden by the kiosk. Also from shots 4 through 8 in Scene 2 we are pretty sure there is only one door on each side of the room. But they seem to be on the left rear and right front when viewed like in Figure 9, with the communication panel behind us.

After considerable study we realize that in Scene 2, shots 3 and 7 were taken from far enough behind Helen that the camera had to be **outside** of the room. There has to be an opening in the wall next to Helen, but shots 2, 4, and 6 say there isn't one. Closer examination shows there is indeed a small break between two panels above the paneling in the back of the room. Even then, to split those panels for that camera shot like that would have required moving the whole wall to Helen's right. If the seemingly continuous back wall can split and walls can move aside for a shot like that, then possibly all the walls can move. This could explain why the kiosk is backwards in this scene. If it didn't move, perhaps all the walls and the communication panel did!

Scene 3 shot 1 (not reproduced here) begins from outside in the hallway, showing Gort carrying Klaatu's body into the control room. This seems to be a door in the right-rear of the room—the same one Klaatu entered in Scene 1. The door opens to Gort's left (this proves we can open that door from either side). For a fleeting second there is a view past Gort that shows the door on the opposite side of the room already opened, and the biocenter in place, ready for Klaatu's resurrection. To the right we can see Helen standing beside the bench she sat on earlier.

Shot 2 (Figure 13) from inside near the back wall shows Gort passing Helen and approaching the biocenter. At this point, Helen moves back away from Gort. If we count the panels behind her as she moves, we note she crosses three panels, one door, and (almost) two panels before she stands to the left of the communications console. There can only be one door on that side of the room and according to this shot it is on the console (front) side of the room.

Figure 13 – Scene 3 Shot 2

Shot 5 (same setup as Figure 14) shows

In Shot 3 (Figure 14), taken from outside the room, we see Gort putting Klaatu's body onto the biocenter. Past him we can see Helen standing next to the control panel.

Gort operating the biocenter. Helen is in the background. Here we can see now that the kiosk is back in the same orientation as Scene 1. The bubble is again toward the rear and the opening is toward the communications console.

Skipping the shots that happened outside the ship,[4] shot 10 (Figure 15) is the famous view of Klaatu's outstretched body, with Gort standing over him and Helen in the rear near the communications console. We get a good view of the inset center of the ceiling and the insides of the door on both sides of Klaatu's body.

Shot 15 (Figure 16) is a close up of Klaatu

Figure 15 – Scene 3 Shot 10

Figure 14 – Scene 3 Shot 3

Figure 16 – Scene 3 Shot 15

on the table. In the background we see the outer wall of the hallway. Klaatu is lying on a surface that is in the space where the hallway was. Now there is a panel with a space for his head.

---

[4] For conciseness, from this point on, I will count, but not analyze, shots in the sequence that are not pertinent to understanding the room's design.

Figure 17 – Scene 3 Shot 20

Figure 18 – Scene 3 Shot 24

Shot 20 (Figure 17) is taken from high up in the hallway as Gort backs away from the bio-center to stand against the wall next to the door near the communication console. Again, we see only one door on the left. Helen is standing next to the communication console.

In shot 24 (Figure 18) Klaatu starts to rise from the biocenter surface. Again we can see past him through the wall opening into the hallway outside. We can clearly see the outer wall of the ship's hallway.

Shot 30 (Figure 19) is taken from outside the room, over the biocenter. Helen approaches and she and Klaatu speak. When the conversation is done, Klaatu exits the room by going around the kiosk to its right, and out the door between Gort and the communications console. The door opens to the left this time. Again, we clearly see details of the outer hallway wall (with no outer port) and the kiosk.

As Klaatu's departing shadow is seen through the corrugations, this is the last time we see the interior of the ship.

Figure 19 – Scene 3 Shot 30

Conclusions

So if we take all the shots in these scenes together what do we know of the inside of Klaatu's saucer?

It has a round control room with translucent corrugated walls, a bench and a control panel. Does it have four visible doors or two? We're not sure. Is the kiosk immobile and the walls move or vice versa or do both move? We're not sure. Are the doors in the back of the room or toward the front? We're not sure because in one scene they're one way in another they're opposite or missing.

Using the known height of Gort and Michael Rennie, we can get an accurate height of the room and from that we can calculate the room's diameter. Using close observations of the doors, we can get good approximations of the frames lights and the corrugated panels. The problem is, when we get the size of the panels and fit them into a model of the room, there is way too much open wall space to define a room. It's clear that the set was not much more than *half* a room, a stage that the director always showed us, the viewers, in a way to convince us it was a circular room. The rest of the set was always open to the stage crews.

We can understand what the *set* of the interior of Klaatu's saucer was like. In fact, there's a photograph of Lyle Wheeler with a model of it (Figure 1).[5]

In the set, the kiosk was fixed in the center while the walls (which only went halfway around) and communications console were on overhead circular tracks (we can see both the tracks in the ceiling and the skirts on the bottom of the walls in some shots). We saw the doors open both to the left and to the right, but always away from the camera's point of view. Essentially, they did whatever camera setup was necessary for a shot, and then rotated the walls into position. This does not, however, eliminate the probability that the Kiosk rotates too. But if it does, there is no absolute reference to define the placement of objects in the set to where they would be in the saucer.

It is also clear that the set was intentionally arranged "backwards" for Scene 2. I have found production stills of the control room set in both configurations. (See Figure 20) Since production stills are used by the art director to approve the final appearance for each set, it is evidence that the director clearly arranged the room "backwards" intentionally, not accidentally. Here his in-

---

[5] The model in Figure 1 was at first considered to be a preliminary concept model. After analysis, it appears to be the final configuration.

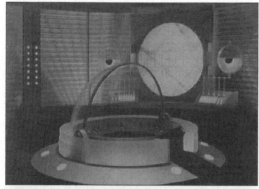

Figure 20 – Production stills of both set arrangements

tention must have been to give the audience something subconsciously unfamiliar to "unsettle" them just a little. Like Gort entering the room going counter clockwise in the hallway and Klaatu by going clockwise. Unsettle, but not to confuse them.

And since it can be shown that probably both the walls and the kiosk rotated in different scenes, we cannot know what their orientation to the front entrance was. It is most likely that there was no static alignment.

So the result is that we can identify the size and dimensions of the *set* of the interior of Klaatu's saucer with accuracy. This we have done in the data drawing. This is what can be proven to be what the set of the spaceship's interior actually was. Also, we have included the size of the room panels as they would have had to be to complete the room the saucer would have had. As you compare the views you will see that, although the movie showed us what seemed to be a complete room, it was far from complete.

This leaves us with an enigma. What was Klaatu's ship "really" like? Our guiding credo when describing these fictional ships is "as seen on the screen" but as I've just shown what we really see of this saucer on the screen is self-contradictory. There is no way to create a realistic version of the interior that is faithful to every scene and shot in the movie.

There are several ways to go here. One is to create a composite version of the interior that represents most of what is "seen on the screen," This approach will create a complete, 4 door interior. The components to do this are seen on Sheet 3 of the data drawing. This is the way the room would be if one were to put *all* of the elements into a room that size.

However, this approach requires the use of a stretched interior panel that was not seen. And while it answers some questions, it doesn't address some of the issues of the peripheral walkway (i.e. some doors should have been visible during Klaatu's initial walk around the interior).

Another way is to admit this is, after all, a movie set and acknowledge the brilliant simplicity of Lyle Wheeler's design. This approach allows the modeler to build an open room, like the set actually was. Here you can use the set sized panels and just leave the rest open, as in the real set.

Or the modeler could also choose to build the 5-door version as there is a lot of evidence that supports the view that that is how the "real" saucer was. It creates a complete room and uses the correct width panels. I've shown this layout on Sheet 2 of the data drawing.

In the end, it's really up to the modeler. The interior of Klaatu's saucer was shown several different ways, and each of them contradicted the others. The challenge then becomes one for the modeler to decide which scene he wishes to depict in his presentation. Make his model match that scene, display his proof of that particular scene and don't worry that its all wrong for a different scene.

But what was the inside of Klaatu's saucer really like? We kind of know, but then we don't. Like all good flying saucer stories, after thorough examination, we are no longer quite sure of what we saw and when we try to explain it to others, we sound even less convincing.

In the end we are left with a mystery. Naturally, we've used our imaginations to fill in the gaps. Still, that does not answer the many questions we have. The mystery remains. And that's the way I leave the analysis of Klaatu's saucer's interior with you. It is a mystery!

Modelers' Note

There were no commercial models of Klaatu's ship contemporary with the film, since the plastic model industry was just coming into existence at the time, and the only vehicles people seemed really interested in were jet fighters and cars.

Fortunately, that isn't the case today. Several of the smaller kit companies produce styrene and resin kits of the saucer, Gort, and even dioramas containing both, such as this one from Skyhook Models.

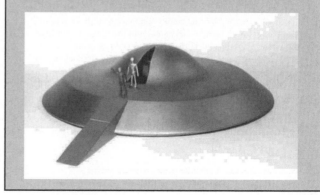

Quickspec: Klaatu's Spaceship

Vehicle Morphology.................Saucer

Year...1951

Medium........................Theatrical film

Designer.....Lyle Wheeler/Addison Hehr

Diameter.....................104 ft (31.7 m)

Height...........................19 ft (5.8 m)

Epilog

We run the risk of overusing the term "brilliant" when describing this movie. Everything from the understated set design, to the minimalist music, to the film-noir direction shows a well-integrated team of creative people at the peak of their powers. Before we delve into the impact the film has had over the years, there are a few loose ends to tie up.

Not everyone shares the view of a completely altruistic intent of this film. Bob Stephens, writing in the special flying saucer edition of *Outré* magazine[45] presents a completely valid, alternate interpretation of the film's message:

> Klaatu and Gort arrive in a flying saucer [to] deliver a warning: If the people of Earth do not choose to live in peace and, instead, venture into space with their weapons, the world will be incinerated by robot police, the products of an unimaginable technology.
>
> Are we not justified in being suspicious of a declaration backed by such ruthlessness? Belief in such a proposal leads to the perverse concept of utopian violence, a twisted type of peace through strength.
>
> Though [Klaatu] utters charitable sentiments before the Lincoln [Memorial], and despairs at the loss of life during a visit to Arlington National Cemetery, there's some slight, but telling, thing that's wrong. Is Klaatu, with his magnificent avian head, the eagle of political ideals, or is he the shrieking predator that murders something small, off in the distance of our restless nights?
>
> His space vehicle has the luminosity of an object of terrible, transcendent awe, or, more hopefully, divine promise. Its glowing descent should be familiar to those who are acquainted with images of religious rapture or the Old Testament dread of a cruel, omnipotent deity. (Such emotions are reinforced by Bernard Herrmann's majestic, yet ominous, score-highlighted by Dr. Samuel Hoffman's eerie theremin performance.)
>
> Finally, there's an undeniably totalitarian strain in Klaatu's declaration of protection. His solution to the threat of apocalyptic warfare, the loss of human liberty, could be worse than the problem. My view is obviously not that of the filmmakers, but their acceptance of dubious assumptions, and lack of skepticism about the use of force to eradicate war, strike me as dangerously naïve. Power exercised in this fashion, and with this purpose, would seem easily corrupted, unless one is a devout believer in the infallibility of intelligent machines. So I am, retrospectively, trying to describe an alternative film that lurks behind the one

whose superficial principles have been accepted without question for half a century.

The Day the Earth Stood Still presents a vision of man stripped of self-determination in exchange for security, watched over eternally by vaguely anthropomorphic, but inhuman, robots. Klaatu says his people have consented to the arrangement as a way to eliminate terrible wars. But how could our little, cosmically inexperienced planet be sure of promises made by someone so remote from us?

One odd thing, for a Sci-Fi film, is that we never really find out where Klaatu comes from. There are oblique references, such as when speaking to Mr. Harley, he says, "I've travelled 250 million miles [to be here]." You'd think that would locate it for us. After all, you can't go 250 million miles (400 million Km) inward from the Earth since we're less than half that distance from the sun. Going outward 250 million miles puts us smack in the middle of the asteroid belt and even in 1950 they knew that the asteroids were just a collection of rocks leftover from the solar system's formation with no possibility of supporting life as we know it. However, if we think of that distance as the total flight length and not just the distance between orbits, then it's just about right for the length of a transfer orbit from Mars to the Earth. Jon Rogers noted the supreme irony in having a man from the Planet of War coming here to tell us to give up our warlike behavior.

Then there is the issue of the "Christ Story" embedded in the film. In order to present the subliminal message that Klaatu was the messiah who has descended from heaven to save us from ourselves, writer Edmund North filled the story with allegorical references to the Biblical story of Jesus. However, Robert Wise has always maintained that the film is only "accidentally representative" of the Christ story. In a 1989 article, for example, Al Taylor and Doug Finch, interviewing Robert Wise, said:

> One of the "messages" in the picture that Wise claims was unintentional on the filmmaker's part (and only brought up by critics and film buffs) was the "religious symbology" that the film contained.
>
> Some saw Klaatu as God's messenger on earth, pointing out eerie parallels in [the film] to Christ and his life, such as Klaatu's "resurrection scene."
>
> "We were not trying to say," insisted Wise, "that this is a version of Christ's return. Over the years I've been asked about this matter of symbolism or parallelism, but none of us had anything like that in mind at all. I guess why this symbolism comes through so strongly to a lot of people is the look of Michael Rennie himself – the tall, thin, ascetic, almost unreal looking visitor from the heavens. You would have had a marvelous Christ figure had you put the 'Christ beard' on him."
>
> So, rather than label the theories as foolishness, Wise seems pleased that this meaning has been read into the film – just another valid interpretation of a multilayered, carefully conceived work.[46]

In a 1976 interview, though, Edmund North revealed his little plot:

---

[45] *Outré* #23, pp. 52-53

[46] *Filmfax* #17, November 1989

It was my private little joke. I never discussed this angle with Blaustein or Wise because I didn't want it expressed. I had originally hoped that the Christ comparison would be subliminal. It gave the film a kind of form, and it was there for anyone who wanted to see it, but I didn't think it was part of the fabric and I didn't want to make the audience aware of it. I wouldn't have been bothered an instant if nobody had ever mentioned it.[47]

It may have been subliminal, but it was also pervasive. With only moderate effort, one can find a large number of allegorical symbols to the Biblical account. Whether or not North put all of the following symbols there deliberately, no one can say, but it's hard to believe that they are **all** coincidences.

Klaatu descends from the heavens and says he is from a vastly powerful (almighty) multi-planet civilization. He is the representative (the Son) of this civilization (the Father), and he travels with a mysterious and powerful, but non-human, companion who emits tongues of fire from his head (the Holy Spirit. Okay, that one is a stretch, but it makes sense further down).

He performs miracles like the healing of the sick (his own bullet wound) and walking through walls (escaping from his upper floor hospital room with the door locked and guarded, and effortlessly opening the locked doors of Barnhardt's study). And, of course, the films title event, which, with our modern understanding of the Solar System, is analogous to stopping the Sun in the sky (yes, we know that this was an Old Testament miracle, but they're the ones that put it in the title).

Even though he is powerful beyond our comprehension, he chooses to walk among us as an ordinary man, taking on the guise of a carpenter (Mr. Carpenter). He surrounds himself with a small group of disciples (the boarding house residents) who are from many differing backgrounds, and eat together around a large table.

When still very young in his dealing with humans, he goes to the temple (Barnhardt's study) and astounds the rabbi (Barnhardt) with his knowledge of scripture (celestial mechanics).

He enters the Holy City (Washington D.C.) to great acclaim (the man running through the streets yelling, "They're here! They've landed!"), and the adoration of the masses (the crowd surrounding the spaceship) to bring us a message of peace and divine sovereignty. But before he can deliver it, in only a week he is betrayed by one of his disciples and murdered by the military occupants of the city (who are not the rightful power of the land). As if to hammer the point home, when Klaatu is shot at the film's climax, actor Michael Rennie assumes a "crucifixion pose" (lying on his back, arms outstretched, one knee drawn up and head rolled to the left) for a brief second or two before Helen reaches him.[48]

The body is placed in a stone tomb (concrete jail cell), but the Holy Spirit (Gort) rolls back the stone (melts the wall with his heat ray) and resurrects him from the dead.

After his resurrection, he is transfigured (glittering spacesuit), and appears only to the faithful (the scientists gathered outside the ship). Since he has only a short time left on Earth, he delivers his ominous message that even though we have free will ("It is no concern of ours how you run your own planet"), our only choice is to follow him ("Join us and live in peace"), or burn for all eternity ("This Earth of yours will be reduced to a burned-out cinder").

After delivering this very terse ultimatum, he promises to return ("We shall be waiting for your answer") then ascends bodily into heaven.

Yeah, there might be a vague similarity there.

$$\alpha \qquad \Omega \qquad \alpha \qquad \Omega \qquad \alpha \qquad \Omega$$

The impact of *The Day the Earth Stood Still* on both the movie industry the general public was profound. While *Destination Moon* was a big hit with the techie crowd, and raised the Sci-Fi consciousness of the public in general, this film was different. Its somber tone and serious message drilled it straight into the mainstream American psyche.

Coincidence? Klaatu Crucified

---

[47] *Cinefantastique*, pg. 18.

[48] Robert Wise's protests not withstanding, there is nothing subliminal or open to interpretation about this. It's blatantly right there on the screen and easily seen in the image above. If Wise did not direct Rennie in this action, then the actor must have been in on the scriptwriter's private joke.

The Saucer Fleet

Even though Klaatu's saucer would soon be eclipsed by more technically sophisticated versions (e.g. the *C57-D* from *Forbidden Planet*, see page 157), the movie remained significant in the mind long after its hardware became obsolete. For decades after, if a filmmaker wanted to add some "class" to his production, all he had to do is use some stock footage, or music, or other aspect of this film's production, and their project was instantly elevated.

Ray Harryhausen's *Earth vs. the Flying Saucers* (see page 199) not only uses footage from *The Day the Earth Stood Still* in wholesale lots, it "borrows" the premise, the major effects setups and even actor Hugh Marlowe. Of course, the aliens aren't very benevolent in that film. After we shoot first, they shoot back and don't stop.

Irwin Allen knew quality when he saw it, and used large amounts of the music (as well as pieces of other Bernard Herrmann scores) in his pilot for *Lost In Space*. It worked well enough to sell the series to CBS, but he replaced nearly all of the music with new compositions by John ("Johnny") Williams by the time the show made it to the air. There's one place, though, near the end of the fourth episode when they are exploring the ruins of an extinct civilization, where we hear the "Spaceship at Night" sequence from this film. One can almost hear Klaatu's footsteps echoing among the ancient stones.

Allen's other big show of the period, *Voyage to the Bottom of the Sea*, reused Klaatu's ship in the first season episode "The Sky is Falling." It even directly uses the sequence showing the saucer over the Smithsonian and other buildings in Washington, before patching in their own new footage. After it dives into the ocean, the *Seaview* encounters it underwater and we see that it's Klaatu's ship, all right, now with a funky pod sporting flashing lights protruding from its formerly all-flat bottom.

Musician Ringo Starr produced a post-Beatles album in 1974 called *Goodnight Vienna*. For the cover he used a parody of the most popular image from this film (that of Klaatu and Gort on the ramp of the ship, with Klaatu raising his hand in the universal gesture of peace) with his own face replacing Klaatu's helmeted head. The film was 23 years old at that point, but still instantly recognizable and, apparently, commercially viable.

The influence of the film continues to this day. In the 2002 film *One Hour Photo*, writer/director Mark Romanek wanted to show how Robin Williams' character, Sy, had become delusional and considered himself an avenger, des-

tined to hold a wrongdoer to a higher, absolute standard. Rather than showing Sy directly in this scene, he instead shows us his TV set playing the last minute or so of Klaatu's speech at the end where we are told that we will be held to an absolute standard of behavior, thus indicating that Sy now felt justified in pursuing his vengeful course. There was no setup or explanation needed, the film's message (whether you consider it good or ill) still comes across strongly.

More recently, mischievous pranksters reprogrammed a highway construction information sign that had been left unattended for several days. The highway department got several call about the sign now reading in a "strange foreign language."

In the mid '70s, Fox tried to get a sequel off the ground, one of many attempts over the years. They even hired Ray Bradbury to write the screen story. As Robert Wise relates:

> My agent was called [by the studio]. They wanted to know if I'd be interested in doing a sequel. I told him, "Hell, no" – I'd done my piece; I didn't want to do that, I don't believe in those things. Ray Bradbury worked on the story for a sequel for some time over there; he turned it in and they didn't like it, and it was just put on the shelf. I've heard that there's talk of *remaking* the picture, which of course I don't like at all.[49]

Naturally, a film this famous couldn't escape the Hollywood remake machine forever. Once Wise had died, all objections to remaking the film were removed. His grave was barely cold before Fox restarted the project, casting Keanu Reeves as Klaatu. At press time, the film had an announced December 2008 release.

©20th Century Fox Film Corporation

The Day the Earth Stood Still has remained a significant and enjoyable film for over half a century, inviting endless repeated viewings. It has done so due to its message, quality of production and probably that it's a bit controversial.

---

[49] *The Science Fiction Film Reader*, pg. 58.

Twin Earths

By Jon Rogers

© Alden McWilliams Estate

Figure 1[1]

If it weren't for Pythagoras, we would not have had the flying saucers of *TWIN EARTHS* in our Daily and Sunday newspapers throughout the '50s. But Pythagoras, that wise Greek of 2,400 years ago, believed there were 10 planets in our solar system at a time when only 5 were known. Furthermore, he said one of them was the same distance as we were from the Sun, directly opposite us. And because it was behind the Sun and orbiting at exactly our rate, we would never see his "Counter-Earth ."

In 1952, writer Oskar Lebeck proposed the same concept for a new, syndicated, daily adventure strip. It would feature flying saucers coming to Earth from our "twin planet" behind the Sun. However, many newspaper editors hearing the proposal were afraid of it. They thought their readers would mistake the story for fact! They didn't want to add to the clamor over flying saucers that was currently rampant throughout the country.

Besides, as one later article about the strip put it:

> By all the rules for adventure-strip writing, *TWIN EARTHS* should have been a bust. It makes no attempt to interest children. It doesn't even try to appeal to all adults—just those who are interested in science. And its fantasy isn't very fantastic.[2]

---

[1] All *Twin Earths* artwork illustrating this chapter is the creation of Alden S. "Mac" McWilliams. It is the copyrighted property of the Alden McWilliams estate, used with permission. It may not be reproduced in any form without the express permission of the McWilliams family.

[2] Popular Science, January 1953, Don Weldon, *Living off Another Planet*

To reassure the editors, Lebeck pointed out that the idea was actually 2,500 years old, and that Pythagoras had thought of it first. It was largely because of his reputation as a great, classical thinker that they thought they might take a chance on it.

The advertising machinery of the United Features Newspaper Syndicate swung into action. They promoted it in all the major city markets. They contributed advertising for many of the smaller, independent papers to use in their launching the new strip (Figures 2 & 6). By mid-1952, almost everyone who read newspapers knew it was coming.

The new daily adventure strip, *TWIN EARTHS* was slated to appear in newspapers, nationwide, on 16 June 1952. As the release date approached, they all held their collective breath.

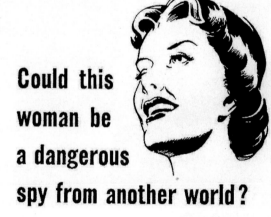

Could this woman be a dangerous spy from another world?

Follow the adventures of VANA, heroine of TWIN EARTHS, exciting space-adventure strip starting in Monday's Chicago Tribune!

© Alden McWilliams Estate

Figure 2

From the very beginning, *TWIN EARTHS* was a hit. It "got off to the fastest start in newspaper-syndicate history"[3] and it skyrocketed to the top of the popularity charts. Soon there were over ten million readers worldwide[4] clamoring for more of the adventures of its most unusual heroine, Vana, from Terra and hero, Garry Verth the FBI agent. The syndicate and editors breathed a sigh of relief.

---

[3] Ibid.

[4] Ibid.

The Saucer Fleet

TWIN EARTHS was not only a success in America; it was popular worldwide. The strip was translated into German, Swedish, French, Spanish and published in many countries. It was avidly read in Europe, South Africa, Australia, and even behind the Iron Curtain. The Spanish version appeared throughout South America. Essentially, *TWIN EARTHS* entertained and influenced people of all ages everywhere on the planet for over a decade .[5]

TWIN EARTHS was one of the biggest winners of the '50s. To produce such a performance takes talent and experience. The men behind *TWIN EARTHS* were two of the best. Both Oskar Lebeck and Alden McWilliams were accomplished as artists *and* writers. Being double talented made them an unusually powerful team.

Figure 3: Oskar Lebeck

Oskar Lebeck (1893-1966) was responsible for the basic concept and also for selling the idea for the strip to the newspaper syndicates. He also wrote the storyline for the first four and a half years.

Lebeck was born in Germany and started his career in Berlin. He worked for the stage and film director, Max Reinhardt as a theatrical set designer during the early '30s. When he saw Hitler and the Third Reich coming to power he decided to leave the country. He and his wife, Ruth, left Germany for the United States. Once in America, Lebeck settled down in New York. There he began work as a set designer for Florenz Ziegfeld and Earl Carroll on Broadway.

After Broadway, he joined Western Printing and Lithography Company. In 1938 *Dell* Publishing Company contracted with Western Printing to produce comic books featuring characters from Disney, Warner Brothers and other studios' cartoons. Lebeck became an art director and editor of Dell's comic books, and helped many aspiring artists and writers along in their career. He also worked as a book illustrator during this time.[6] He illustrated the 1939 version of L. Frank Baum's *Wizard of OZ* which was published concurrently with the MGM movie.

Lebeck also wrote and illustrated a number of his own children's books in the late '30s. Some of his titles included, *The Diary of Terwilliger Jellico* (1935), *Stop Go, the Story of Automobile City* (1936) and *Clementina, the Flying Pig* (1939).

It was during 1941 that he first worked with Alden McWilliams. Together with Gaylord DuBois, Lebeck wrote and Alden drew the art for such children's books as *Stratosphere Jim and His Flying Fortress* and *Rex King of the Deep*. DuBois and Lebeck also wrote other books together like *The Hurricane Kids on the Lost Islands* (Figure 4).

DuBois went on to become a famous author in his own right. He produced thousands of "Big Little Books", comic books, and comic strips. In a later interview, DuBois said of Lebeck:

> I soon met Oskar Lebeck who was on the editorial staff of Whitman and through him I began writing comic book scripts, full time. Oskar was a man of immense drive and had a way of developing the best ability and the fervent loyalty of the artists and writers who worked under him.[7]

Oskar Lebeck was a pioneer in another respect. In the late '30s, the rise of the "Superhero phenomenon" had increased the popularity of comic books to the point where many parents and "authorities" were becoming concerned.

> George Hecht, publisher of the popular *Parents Magazine*, noted that reading comic books absorbed 75% of the leisure

Figure 4

---

[5] In the final strip (25 May 1963) the last panel showed the hero and heroine planning to get married and live on Terra. It was unusual for any strip to have such a clean ending.

[6] Robert Susor, *A Biography of Alden McWilliams,* published in *TWIN EARTHS, Special Edition,* 1993, pg. 25

[7] *Gaylord Dubois Speaks,* an autobiography available at http://www.erbzine.com/mag0/0072.html

time of children nine to fourteen years of age. He reported with some alarm that 125 comic book titles were now featured in some 100,000 newsstands across the country. Some 15 million comic books were sold a month for a total of 180 million copies a year.[8]

Since this large outpouring of Comic books in the late '30s and early '40s was made up largely of Superheroes, fights and mayhem, many thought it was harmful to young readers:

> …Already a concerned observer, George Hecht, the publisher of *Parents Magazine*, decided to start his own line of comic books… [as] an alternative to the garish super-heroes that concerned parents could buy their children…The first of these, *True Comics*, debuted in April 1941.[9]

True Comics was soon followed by other titles. He started the *Parents Magazine* line of comics that emphasized true stories with real heroes and famous people so that these comic books would be inspirational as well as entertaining. And they became moderately successful at it. Another pioneer, Albert Kanter, noticed this and decided to start a competing "wholesome comic book" line of his own. He decided to adapt famous literary classic novels to comic format. His *Classic Comics* would eventually become *Classics Illustrated*, which enjoyed a modest success for decades. As Mark Carlson describes it:

> In the fall of 1941, the first issue of *Classics Comics* appeared. Kanter had 200,000 copies printed up of his first adaptation, *The Three Musketeers*. … The book proved to be a financial success.
>
> *Classic Comics* didn't inspire many imitators at first…[but] Dell Comics was an exception. Dell stuck their toes in the literary adaptation market. They ended up issuing only two issues of their *Famous Stories* in 1942, adapting both *Treasure Island* and *Tom Sawyer*. Editor Oskar Lebeck hoped that a classic novel, "told through a new and vivid medium, will recommend itself to parents and teachers everywhere, as well as to the boys and girls for whom it was created." The [managing] folks at Dell, however, barely gave the new venture a chance at succeeding before they canceled it.[10]

And so Lebeck's attempt to bring famous classic novels to comic book format failed. One wonders what new inspirational titles Dell might have produced if they had allowed him to continue.

Lebeck had a whimsical side too. In 1942, he submitted artwork to Disney studios. It was a clever takeoff of an Academy Award-winning cartoon and song that Disney had produced called, *Der Fuehrer's Face*. It was to be a book and party favor tie in to the Disney cartoon. Based on the German "Schnitzelbank" song, Lebeck's artwork showed Donald Duck ready to throw a tomato at Hitler. Unfortunately it was never produced.[11]

Lebeck worked at Western/Dell throughout World War II. After the war, while he was editor at *Dell* producing comics like *Fairy Tale Parade*, he and Alden McWilliams teamed up again on several new comic book covers including *Space Cadet*, and *Dick Cole*.

By the time the *TWIN EARTHS* strip idea came to him, Lebeck had been in the business for twenty plus years and was approaching 60. When it came to picking an artist to join him in creating his brainchild, he had plenty of contacts. He chose someone he knew to be one of the best, Alden McWilliams.

Alden McWilliams (1916-1993, Figure 5), or "Mac" as he was known to practically everyone, had intended to be a magazine illustrator. He graduated from the NY School of Fine and Applied Arts (later Parson's) in the late '30s. However, he had already been illustrating books and magazines for Street & Smith and Popular Publications, for some time. He had contributed to *Argosy, True, Boys Life,* and various pulps, but, as he put it in one interview:

> …but magazines were starting to have difficulties, and were buying fewer stories. As a result they had command of more good artists, and I found myself bumping into the likes of Norman Rockwell in that end of the business—which was impossible! The pulps were fading too, but comic books were coming in, so I became one of the first half-dozen fellows, I believe, who worked in the field.[12]

As an early pioneering comic book artist, one of his first jobs was assisting Lyman Young on *Tim Tyler's Luck* in 1935. As he said, "I began by cleaning brushes for him and gradually got to doing backgrounds."

© Alden McWilliams Estate
Figure 5: Alden "Mac" McWilliams

---

[8] Mark Carlson, *"Hey! That Ain't Funny!,* A history of Classic and True Story Comics in the Forties" www.nostalgiazone.com/doc/zine/06_V2N1/thataintfunny.htm

[9] Ibid.

[10] Ibid.

[11] http://toonsatwar.blogspot.com/2007/03/der-fuehrers-face-part-5.html

[12] Interview, *Cartoonists Profiles*, September 1972

The Saucer Fleet

From there he drew for Dell and Centaur comics with titles like *Keen Detective Funnies* and *CRACKAJACK Funnies*. He worked for "practically all the publishers" of comics at one time or another. During this pre-war era he even worked with Oskar Lebeck, illustrating several of his children's books. Regrettably much of Mac's earliest work was unsigned making later recognition of his early contributions to American comic art difficult.

When World War II came, Mac showed his other talent, courage. He was one of the men who landed at Omaha Beach during the D-Day invasion. He fought in the St. Lo breakout. He was in the V Corps during the Battle of the Bulge and the relief of Bastogne. He was one of the first men across the Remagen Bridge and he stayed with the US Army until it met the Russian Army at the Elbe River in Germany.

He won a Bronze Star for taking out a German machine gun nest. He also won the French Croix de Guerre for his work with the French resistance in preparing for the invasion of Paris. He earned the European Theater of Operations Ribbon with 5 Battle Stars. After Victory in Europe Day, he was scheduled to be one of the men in the invasion of the Japanese homeland until the atomic bomb ended the war.

Interestingly his winning the French "Croix de Guerre" (Cross of Honor) was related to his doing artwork– sort of. Prior to the liberation of Paris, the US Army's V Corps had no accurate maps of Parisian streets. Mac was chosen because he could draw maps better than anyone in the Corps. So the French resistance secreted Mac behind enemy lines and into Paris by way of the sewer system. There he was able to draw maps of the street layout because there were actually street signs in the sewers.[13] Fortunately, Paris was later declared an open city and the planned street-to-street fighting did not occur.

Once back home, he returned to illustrating comic books for Avon, Quality and others. One of his notable achievements during this time was his creation of the classic *Tom Corbett Space Cadet* comic book cover (he actually drew the entire book).[14]

However, when the war was over, the comic book market had changed. Now it wanted, "Romance" comics, something he hadn't done much of professionally. When one editor suggested that he needed to learn how to draw "girls" he laughed. He had always been able to draw women and in fact, he was one of the best artists of the female form in the medium.

Later in life, Mac would create other famous strips like *Dateline Danger* and *Davy Jones*. He also helped other artists produce their strips. For example he contributed to *Rip Kirby, Mary Perkins, and Dan Flagg*. He also continued to create comics, including *Masters of the Universe, I Spy, The Man from U.N.C.L.E., Captain Marvel, Buck Rogers, Flash Gordon, Star Trek,*[15] *Star Wars* and over a hundred others.

It should be noted that one of his strips, *Dateline Danger* was *the first* integrated newspaper comic strip with a black lead character (Danny Raven). Once again Mac broke new ground in social commentary and art. When asked how he was selected for such a controversial strip, he replied:

> I had done comic book work for Western Printing, and it happened that I had drawn the *I Spy* series, based on the popular TV show, for them.
>
> Publishers-Hall [Syndicate] said, "You've had some experience doing blacks so we'll try you first." I guess they had 3 or 4 possibilities in mind for the job. Fortunately for me, I got the assignment.
>
> I really had to dig in and study black characters when I started drawing them regularly for a daily strip printed in black and white. First of all, I got the magazine, *Ebony* and copied faces religiously from there. Getting the shadows down was really something![16]

The strip, *Dateline Danger* became another of Mac's worldwide popular successes.

Mac was well known in the industry. He was a contemporary and close friend of such famous cartoon artists as Al Williamson, Gil Fox, John Prentice, Dik Brown, Mort Walker, and Hank Ketchum. He was nominated for the National Cartoonists Society award for the Best Comic Book story of the year four times in a row: 1975, 1976, 1977, and 1978. He won the award in 1978. Considering the experience and talent these two men had when they teamed up to produce *TWIN EARTHS,* it becomes less of a mystery why the strip became an instant hit.

After its successful launch, Lebeck wrote the stories and Mac drew *TWIN EARTHS* for the first four years. Then, having been in the field for almost 30 years, Lebeck decided to retire. Although, the strip continued to carry both their names, Mac started writing as well as drawing the strip. From about 1957 on, *TWIN EARTHS* was completely Mac's.

Although, Oskar had created the original concept, Mac showed the readers its vision through his precise, clear and dramatic artwork. Mac's art is classified as an excellent example of the "Alex Raymond" school of art, noted for the most realistic, detailed, using thin lines, and presenting portrait faces. Considered one of the best techniques, it entailed considerable effort and attention to detail. Drawing all of

---

[13] Mac later related that, on one occasion, he peeked out of a storm drain to get his bearings and there was a pair of German Storm Trooper boots 6 inches in front of his nose—so close he could have touched them!

[14] After *Twin Earths*, he continued in comics and produced work for other famous Sci-Fi series like *Buck Rogers, Star Trek, and Star Wars*.

[15] You'll find "Commander Alden McWilliams confirms [James T. Kirk's] honorary appointment to the UFP Academy" on one of *Star Trek*'s creator's website.

[16] *Cartoonists Profiles*, September, 1972

TWIN EARTHS, both the Daily and the Sunday editions, became a tremendous load.

> [Mac worked] literally seven days a week regularly, usually starting about 0900 [9:00 AM] and frequently not leaving the board for the day until 2100/2300 [10/11 PM]. While his hours were interrupted sometimes by various errands, gardening, lunch, dinner, etc, he easily put in 60 hours/week, averaging about 10 hours a day. And from the time he took over *TWIN EARTHS* and during most of the time he did Davy Jones and Dateline [Danger], he did his own lettering and balloons."[17]

To complete the eighteen black and white drawings needed to fill six days of daily strips, he did the initial drawings in pencil and then drew over them with pen and ink. After that, they were sent to an assistant to create the balloon lettering. When the Sunday strip started, Mac said, "It was like adding another two and a half days work to the effort."

Since the Sunday strip was in color, the drawings had to be colored in. When asked how that was done, Mac said:

> The Sunday proof sheet is colored up with crayon or watercolor by the fellow who does the lettering. Then the sheet is sent to the engravers who are supposed to follow the colors when they make color plates which print your Sunday paper. Of course, they never do, so I don't know why we bother in the first place.[18]

Mac enlisted his younger son, Rick, to assist with the coloring, a task he felt slowed down his drawing work.

All this work was just for the drawing needed. Later he began writing the plots too. And what plots they were. Together, his great art and novel plots combined to create a unique adventure series.

This unusual new series had other things going for it too. For one, its timing was perfect. Not only was 1952 the biggest year for flying saucer sightings ever, but in July and August, right when *TWIN EARTHS* came out, people reported more

© Alden McWilliams Estate

Figure 6

UFOs than they had in the previous five years combined. Interest in UFOs and flying saucers was at a peak. This was an ideal time to present a strip that was loaded with "flying saucers." But that's not all it was loaded with.

TWIN EARTHS was also an "Adventure strip of Scientific Progress." Almost immediately readers discovered that the Terrans were more scientifically advanced (naturally) than their cousins on Earth. Developing this angle, Lebeck and McWilliams had the "saucer people" show the readers what wonders they might expect to see in their own future. Among the many wonderful predictions were "Wireless television telephones that fit into a handbag[19]", Automobile-airplanes (called Autoplanes) with safety devices that prevent crashes, "Chemical food factories that actually grow organic foods, such as meats," "indestructible fabrics that repel or attract heat," "houses made of Polaroid glass that turn with the position of the sun" and many others.[20] (Figure 6)

While many of these technical marvels seem wonderful today, imagine their impact on people in 1952. But Lebeck and McWilliams were not hanging *TWIN EARTHS'* success on just the "whizbang" effects of wowing the audience with predictions of a marvelous future. The series was also loaded with action, [more about that later] *and* one of its biggest surprises was the presentation of a controversial change in the social relationship between men and women.

While Science Fiction as a genre had been more "liberal" in its presentations of women as the equal to men in society, Lebeck and McWilliams took the issue to new extremes. On 15 August 1952 they presented readers everywhere to the startling fact that the scientifically superior society of Terra, was *dominated* by women. Men were not allowed to vote or hold

---

[17] Interview Chris McWilliams (Mac's son), 9 May 2007

[18] Roger Steffens, letter to Editor Robert Susor, *"Twin Earths Special Edition"* 1993, pg. 20. Used with permission.

---

[19] Of the many futuristic inventions, this one has become reality. Cell phones, recently capable of sending moving pictures, are even smaller than they predicted.

[20] *TWIN EARTHS* advertisements and original strip preview, 13-14 June 1952

office. As Vana, the former secret agent of Terra explained to Garry Verth, FBI agent,

> VANA: You see Garry, [out] of our entire population, only eight percent are males.
>
> GARRY: Wow!
>
> VANA: So naturally, the females took over leadership and are running everything! Men became a small minority with few rights and live under government protectorate for the preservation of the Human Race!
>
> GARRY: {Ulp}...Excuse me, Vana... I've got some [thing caught in my throat.]...

Vana explains that this was caused by the constant wars their men pursued for centuries. These wars finally resulted in a great plague that almost killed everyone on her planet, including most of the males, and affected the females so that very few males were born afterwards.

Vana continues explaining that now there is little to no social contact between men and women. Later in the evening she rejects Garry's amorous advances toward her because, as she says, women from Terra have lost "romantic sentiment through centuries of living in an all female society…"

Thus Lebeck and McWilliams set up the plot situation where a peaceful, scientific, female dominated society of Terra will interact with the warlike, male dominated society of Earth of the 1950s. This allows for some interesting complications. (Figure 7)

It is amazing that this was published during the period of McCarthyism, the Korean War, the flying saucer-Cold War-ICBM and Nuclear Holocaust scare. While modern feminists may take exception to some of the details of Lebeck and McWilliams' social commentary, it must be remembered that, for the tenor of the times, this was shockingly new. And the strip did have a positive influence on social attitudes towards women. The fact that it did not immediately change society is true, but you have to start people thinking first. By showing a society of strong, independent, largely rational women leading a peaceful society gave the public something to think about...especially in contrast to the tense, paranoid times around them.

But Terra was no Utopia. *TWIN EARTHS* also pointed out, through the character and actions of Col. Zena Alotera and others, that women-in-command would face the same problems that men-in-command had always faced. Women could be just as altruistic, patriotic, and heroic under fire, but they could also be ambitious, jealous, scheming, and more interested in their own good than the welfare of others. In short, Terrans could be just as human as we were.

However, social commentary and experimentation was not the primary purpose of the *TWIN EARTHS* strip, entertainment was, and for that the authors included some good old-fashioned struggle between good and evil. The villains that would appear in *TWIN EARTHS* began with Russian secret agents and escalated to enemies far more potent and dangerous.

Loaded with exciting stories of space travel, spies, futuristic gadgets, mystery, action and beautiful women, *TWIN EARTHS* succeeded in entertaining people in many countries throughout the 1950s, from the time of the great saucer scare to the dawn of the space age.

There would never be another adventure strip quite like this one. None would keep people as amazed, intrigued, and craning their necks, looking to the skies, hoping to catch sight of a flying saucer bringing that next startling vision of the future to Earth from our sister planet, Terra, just like in *TWIN EARTHS*.

Figure 7

The Stories

Movies, in general, have several minor plots but only one main plotline which, when played out, concludes the movie's story. Illustrated adventure strips are different.[21] Adventure strips have many consecutive stories that play out individually, but at the end, the strip itself (and usually its characters) goes on to the next adventure. It is not done for humor; its purpose is primarily serious entertainment.

TWIN EARTHS was an illustrated adventure strip. Like many other adventure strips, it had two separate plotlines that ran in parallel. There were two completely different sets of characters in each series. The daily series of adventures were written for adult audiences and the Sunday's were written for younger readers.

Research has identified 36 distinct stories for the daily strip and 22 stories for the Sunday strip. There are, most likely, many more as we did not have access to the entire strip. Writing a plot synopsis for all its stories would take more space than available. Therefore, the following will give you the feeling of each strip by doing a synopsis of their beginnings and some of their stories.

TWIN EARTHS -Dailies

The first story series begins with a young woman anxiously catching a taxicab and racing to the FBI headquarters in Washington D.C. In desperation she rushes in and asks to see the head of the department—it's an emergency. She is ushered in to see his assistant. His name is Garry Verth (Figure 8).

© Alden McWilliams Estate

Figure 8a

© Alden McWilliams Estate

Figure 8b

What if a young, beautiful woman came into your FBI office and told you she was from another planet, that she was a spy and was willing to give you information in return for asylum? What would you think?

Naturally, Garry thinks she needs a psychiatrist rather than the FBI. However, she pulls out a metal disc from her purse and shows him that it floats in midair. Then she shows Garry a short wave transceiver about the size of a walnut, hidden in a small locket that she is wearing.

Vana says the reason she needs help is because she was almost murdered the night before. Garry investigates and discovers the bodies of two dead girls in her apartment, one of whom looks exactly like Vana. To further prove her story, Vana takes Garry and the chief of the FBI, to an observatory where she directs them to aim the large telescope at a location in low Earth orbit. To the FBI chief's amazement; they see an orbiting space station with a flying saucer floating around it. Later, tests made of Vana's blood reveals a blood type that is totally unknown. Vana tells them that this is the one way her people differ from us.

Garry and his boss are stunned by what they have learned within only a few hours of talking to Vana. Later that evening, the FBI's chief reports their findings to the President of the United States. (Figure 9, next page)

This is how *TWIN EARTHS'* story begins. It offers a solution to the flying saucer mystery that has been making front-page news all across America during the first months of the strip's release. It postulates a whole new planet in the solar system and, it uncovers spies from another planet infiltrating us and "observing our daily lives."[22]

But Vana has more shocking news for the FBI agents, *if* they can keep her alive.

---

[21] They are often confused with "comic strips" because they both appear together in daily or Sunday newspapers. Comic strips are situational drawings done for humor that are complete in themselves. Once popular, Illustrated Adventure strips have almost disappeared from newspapers.

---

[22] The "spies from space" hypothesis appeared in Frank Sculley's best selling book a year earlier. During the strip's run the first contactee, George Adamski, would surface.

The Saucer Fleet

Through Vana, Garry learns that her people gave up conquering space with rocket power because they have found a better means. Developing their flying saucers they first discovered the earth back in 1903. They now use them to regularly travel back and forth between Earth and Terra.

One evening over cocktails, Vana tells Garry why males make up only 8% of the population on Terra. A great plague decimated her people and altered their gene structure so that few males were born afterwards. This is why the women on Terra took over society and have largely lost their romantic impulses.

Back at the FBI office the director instructs the special effects department to create a life like, animated dummy of Vana to smoke out the intended assassins. They set a trap by putting the dummy in the back seat of a convertible and driving it through the local park. Sure enough, a sedan pulls alongside and riddles the dummy with bullets. The FBI springs the trap, chases down the sedan, and, in a final shootout, the sedan they were chasing, explodes. The G-men are amazed to learn that it had been carrying two little old ladies who had been the assassins attempting to kill Vana (Figure 7, page 50).

© Alden McWilliams Estate

Figure 9

Garry and Vana think she is relatively safe now that whoever is trying to kill her must think she's dead. Later that night, after dropping Vana off at home, Garry stops at the observatory. He observes that the Terran space station has had an additional deck added, doubling its size. The chief and Garry wonder what this means.

The next day they discover a small two-way video communicator in Vana's purse. While it is being examined in the lab, a woman's face appears on the miniature screen. At Gary's question she states that she is head of Terra's intelligence program on earth. She tells them that Terra's purpose in observing Earth is to protect itself from our war-like nature. Suddenly, she cuts the contact. The FBI men trace the broadcast and are surprised to find it is coming from within Washington, DC. Greatly concerned, they set up an observation station inside the Chiefs office.

Before they can discover the source of the communicator signal, a Miss Millard arrives at FBI headquarters and says she recognized the Terran woman on the videophone and could lead them to her. Almost simultaneously, a messenger delivers Garry a package. Suspicious, Garry puts it

in another room. Moments later, it explodes. They attempt to capture the messenger who delivered the package but fail when he is accidentally run over by a truck. While searching his body they discover a microfilm, which proves that Russian agents had sent the bomb. Vana says communists have been trying to steal Terra's scientific secrets. In the excitement, they discover Miss Millard has left. As they leave, Garry mentions how he wants to establish a personal contact with Terra's Intelligence Service even if he has to resign to do it.

While taking Vana home that evening they are both surprised by the mysterious Miss Millard who was hiding in the backseat of Garry's car. She orders them to drive to a little traveled road where they park to talk. There, she reveals that she is Colonel Zena Alotera, commanding officer of Terra's Secret Service and the woman they saw on the communicator. She offers Garry a chance to contact Terra's officials, but he may have to leave Earth to do it.

While they're talking, communist agents drive up behind them and try to kidnap them. A running gunfight ensues. The three take refuge in an old abandoned mansion. The communist agents surround the mansion and pepper it with gunfire. Colonel Alotera calls for help on her two-way communicator. Together Zena, Vana, and Garry manage to hold off the communist agents until a flying saucer arrives overhead and ends the fight. The saucer's crew takes the communist agents captive. (Figure 31) Vana is also taken on board and Garry watches with Zena as the saucer flies off into the night.

Later, Garry calls his chief and reports that Terra's secret agents are leaving Earth. Garry tells him that he's leaving for Terra too. He tells the FBI chief where to point his office scanner to see his departure in a flying saucer. The next day a picture of the saucer over Washington DC is in all newspapers.

On board the saucer, Garry is shown a fluoroscopic ray projector that makes the solid wall of the ship appear like a transparent view port. Garry's first stop is the Terran space station that he and his chief have been watching through the telescope. Gently the saucer lands atop the double ringed station.

Garry is shown around the space station. He is amazed

to discover that the station has its own gravity field and atmosphere surrounding it. The girls show Garry how to swim in space. They temporarily adjust their gravitational field to allow them to float outside the space station with nothing around them but open space. Garry is enjoying himself immensely. However, he notices a slight tension arising between Zena and Vana whenever he is around.

At this point, the Terran Supreme Council decides to return the communist agents back to their own country. To demonstrate their technical superiority, the Terran space station lowers itself into Earth's atmosphere and hovers over the Kremlin to release the captured Russian agents. They broadcast their peaceful intentions, but the Russians attack the station from all directions with jets, bombs and rockets. (Figure 10)

The station's powerful magnetic defensive fields easily deflect the Russian attacks. Seeing that the Russians have stopped attacking them, the Terrans use the saucer to take the agents the final distance down onto Red Square. But when they open the saucer's ramp to release the men, Russian tanks open fire on them, killing some and damaging the saucer. The hovering station drives the tanks off with heat rays. Then they make a risky maneuver. They bring the space station below its safe operation limits to hover above the crippled saucer and rescue those on board. Rather than return to Russia, several of the agents decide to stay with the Terrans. Taking their station back skyward, the Terrans detonate the damaged saucer still sitting in Red square. It leaves a *big* crater as they ascend into space.

Figure 10

All artwork © Alden McWilliams Estate

Figure 11

Once back in orbit they receive a message that the huge space liner, *S.L. Galaxy,* will be arriving to take Garry and others back to Terra. Garry boards the *S.L. Galaxy* and is taken on a several week long trip around the sun to Terra.

During the trip, the authors indulge in a little education. When Garry gets bored after weeks in space and asks how long he's been on the ship, he gets a little elementary education about the true size of space.

Finally the *S.L. Galaxy* approaches Terra. The news of the first earthman coming to visit Terra has preceded them. After they land, Garry looks out from the elevator descending the ship and sees thousands of Terran women who have come to give him a "royal" welcome. While the few men in the crowd are amused by his "quaint costume," the ladies

get out of hand. Since it's such a momentous occasion, they turn into a throng of souvenir hunters. They tear most of Garry's clothes off before Colonel Zena and Vana can rescue him. Garry is not amused (Figure 11).

Although this little scene is played for humor, it is an interesting commentary on what can occur when one sex vastly outnumbers the other. Historically, when men have greatly outnumbered women, they tend to become meek and respectful, held in check by their own numbers.[23] When women vastly outnumber men, they tend to become more bold and aggressive, an interesting difference.

Garry is saved from the crowd of hungry souvenir hunters and presented to the Supreme Council. After the formalities, he is allowed to retire and clean up. While in the middle of a shower, Vana calls, causing him to discover one of the inconveniences of videophones.

Garry is given the honor of a review of the Air Forces of Terra. Later he also gets to review the Terran Navies. Colonel Zena, ("Just call me Zena, darling"[24]) takes him on an aerial trip to see one of Terra's modern cruise ships. However, when Garry does a little acrobatic flying in their air car over the seashore, Zena gets angry and leaves him stranded on the beach.

When Zena returns without Garry, she and Vana get into an argument. Soon they have picked up swords and, in the "tradition of our grandmothers" are dueling over Garry. Vana wounds Zena and then apologizes. Zena refuses to accept the apology and tells Vana that they are now enemies forever.

Garry, still on the beach, comes upon an old man who is fishing. The man turns out to be Professor Ohtho. He is over 150 years old and one of the few male scientists on Terra. Eventually, Garry returns to the city where he finds Vana and an angry Zena have just finished their swordfight

[23] A historically accurate portrayal of male behavior in an occasion where there was one woman for every 1,000 men is given in the movie, *Paint Your Wagon* (1969). The two partners' solution was similar to the Terran's; they both shared the opposite sex.

[24] During the space trip it was revealed that Terran women have not actually lost their romantic impulses. Vana and other Terran spies were conditioned to prevent fraternizing with Earthmen. Vana's conditioning was removed during the trip to Terra.

The Saucer Fleet

over him. Garry and Vana leave Zena stewing and go sight-seeing together.

Their sightseeing is cut short when they are seized by the Terran secret police and accused of espionage. Imprisoned, they stage a breakout with the help of Professor Ohtho. After a harrowing chase through the tunnels of Terra's capital city, they escape aboard Professor Ohtho's ship, the *Aquarius*.

In the Capitol city, Colonel Zena is called before the grand council. It has been discovered that she planted the evidence that led to Garry and Vana's capture and imprisonment. Zena confesses and is given a demotion as a reduced sentence.

Meanwhile, Vana and Garry are traveling aboard Dr. Ohtho's exploration submarine when saucers approach bringing news of Zena's disgrace and their pardon. However, they elect to stay with the Dr. Ohtho on the *Aquarius*.

Garry gets to see more marvels on the *Aquarius*. His room is completely convertible between being a bedroom, bathroom, or study and Vana has acquired some miniature pets. Then they have a sea battle with a giant squid that disables the ship. When they beach the ship for repairs, the horrible news arrives.

Terra has been attacked without warning. Thousands of huge globe machines have descended from space and are decimating her cities. (Figure 12)

The invaders turn out to be a vicious race of pygmies that are technically very advanced. Gary and Vana return to join in the fight for Terra's life. Garry and Vana "parachute" into largest city to spy on the pygmies. They find that the enemy is demolishing the city to install hives. They barely escape with the information.

Reporting their findings to the supreme Council, Garry wishes to send a message to warn Earth. The Council agrees and Garry sends his message. At the same time, the Terrans are retreating to their last remaining stronghold. Earth receives Garry's message but doesn't believe it. Garry tries to lead a counter offensive. Their attack against the pygmies" giant Globes takes out a few of them but the Terrans suffer a 40% casualty rate.

Unable to sustain this kind of losses the situation looks hopeless. Garry makes a radical suggestion. (Figure 13)

Thinking perhaps Earth will help Terra, the Supreme council of Terra agrees to Garry's plan and gives him authority to carry it out. For days, Garry and the Terrans work feverishly, making their saucer fleet ready for the long space voyage. Finally all is ready and, carrying most of the surviving population, the fleet of Terran saucers leave Terra bound for Earth. (Figure 14)

On Earth, giant telescopes catch sight of the fleet in space. Newspapers carry the headlines, "Spaceships or Asteroids?" Arriving at earth, Garry has the fleet remain in orbit while he takes the flagship down to convince the people of

Figure 12

Figure 13

Figure 14

Earth that flying saucers are real. (Figure 15)

Garry buzzes New York and then lands his ship in the parking lot of the United Nations. He makes his plea before the UN General Assembly. He asks for temporary refuge for the people from Terra in return for their advanced scientific knowledge. (Figure 16)

At first the UN delegates deny the urgency of the situation and refuse. They want to study the proposal in a committee. Garry takes them to the roof of the UN building to witness the size of the saucer fleet he has brought to Earth. At the sight of a thousand saucers hovering overhead, they call an emergency session.

While the UN debates, New York gives Garry and Vana a ticker tape parade, complete with a combined saucer and jet aircraft air show. In the middle of the air show, two jets collide, but an alert saucer captain catches the jet pilots in midair by adjusting gravity in their area. Unfortunately, the effects are also felt on the streets below and people and objects float upwards. Garry has the saucer reduce the effects gradually and everyone is lowered back to earth safely.

It looks like the UN may still be reluctant, so Garry shows them the movies of the destruction of Terra. He points out that they must prepare for the same thing and that the Terrans want to help. As a last demonstration, he brings out several captured pygmies. In the face of this evidence, they agree to let the Terran Saucer Fleet land. (Figure 17)

The fleet is divided into smaller units according to each nation's quota and saucers land everywhere on Earth.

A new economic day dawns for humanity on Earth. Terran technology allows huge flying cranes to lift ocean liners over mountain ranges. New solar electric power stations and transmitters are built. They generate and send solar power all over the world. People begin to farm the ocean for food. Gradually hunger is eliminated the world over. (Figure 18)

This has been but a brief synopsis of the first two years and dozen or so plotlines of the daily *TWIN EARTHS* strip. These marvelous, forward-looking visions were illustrated and presented in August of 1954. They were seen by people living in many countries. If there were ever a place to "end the movie" this would have been a good one. However, *TWIN EARTHS* was an adventure strip, so it continued onwards for years.

Plots over the next six or seven years would have many twists and turns. They would introduce new places and characters, most notably one Chris Cannon, a scientist, and Nan Daily, a reporter.

But, as they say in the action adventure strips, that is another story, "To Be Continued…"

TWIN EARTHS
Sundays

What finally appeared as *TWIN EARTHS* in Sunday newspapers across the nation differed significantly from what Lebeck and McWilliams had planned. They originally intended to use the same characters as in the dailies and follow another adult storyline, either paralleling or integrating it into the dailies story. The story line was created, the first month's Sunday panels were drawn and everything was submitted to the Syndicate for approval. It was rejected.

Figure 15

Figure 16

Figure 17

Figure 18

The Saucer Fleet

The business heads at United Features Syndicate insisted that the Sunday series "target" younger audiences, primarily pre-teenage boys. Their reasoning was simple. In *TWIN EARTHS,* they already had a highly popular adventure strip among adult readers. Now they wanted to expand their "market" and attract young readers (totally ignoring the fact that young readers already read and liked the dailies).

So they sent Lebeck and McWilliams "back to the drawing board" to develop an appropriate storyline with younger leading characters. The *TWIN EARTHS'* universe was going to stay the same, but little else. This is why Lebeck and McWilliams created "Punch" (no known last name), a young, preteen boy from Texas and the unlikely scenario that got him into the Sunday version of *TWIN EARTHS.*

Figure 19

On 1 March 1953 the first *TWIN EARTHS* Sunday adventure strip appeared. The story began with a flying saucer landing in a remote area of Texas. Two young boys, Tex and Punch, happen to be quietly watching from a ledge. Some people get out and go looking around. Who they are and why they are there, we never find out.

But, seeing the saucer momentarily unguarded, Punch decides that he's going to hitch a ride. He climbs down from his and Tex's hiding place and sneaks into one of the saucers" open hatches. (Figure 19) After Punch climbs onboard and hides behind some machinery, the saucer takes off. Why the saucer is so open and unguarded is never explained.

After a while, Punch decides it's time to let these alien people know he's on board. He leaves his hiding place and

climbs up into the control center only to discover that the saucer is being manned by "A bunch o' women."[25] The Terran ladies are as surprised to see Punch as he is to see them. Since Punch is quite young, he complains, "Shucks! I don't want to stay where there's nothing but girls!"

He is further dismayed to find that it is too late for him to get off. To prove it, they show Punch a view of the Earth through their "Fluoray projector", a device which makes solid walls transparent. Punch observes he is already 1,000 miles above the Earth. He becomes frightened when he accidentally steps into the "fluoray" and sees his legs disappear. They calm his fears by explaining that they're still there, just invisible.

As the saucer continues to climb, Punch sees they are approaching a space station. He wants to know if there are any men there. He is disappointed to learn that "men aren't permitted to 'take the risks of spaceflight.' " (Figure 20)

All artwork © Alden McWilliams Estate

Figure 20

When they arrive at the space station, the saucer slowly settles onto a docking tray on top the space station. He is told the space station has its own gravity and its own atmosphere. Punch has a lot of trouble adjusting to being upside down looking "up" at planet Earth.

Since the Terran women on board the space station don't have many men as guests – especially from Earth—they throw Punch a welcoming party. They give Punch a look at their home planet, Terra, on their largest TV screen. Punch is amazed at the 6-foot (2 meter), flat panel, Hi-definition, color TV screen and asks if their TV is the same as Earth's.[26] They reply, "It is in principle, just further developed." He remarks that it is just like looking through a window.

---

[25] This and all subsequent quotes directly from *TWIN EARTHS* Sunday adventure strips are copyright Alden McWilliams 1991. Used with permission.

[26] This would be an amazing prediction to any reader of the 1950's because, at this time, TVs were all black and white with 18 inch (46 cm) screens being about the biggest.

One of the ladies says "Now that you have had a look at our people on Terra, let them have a look at you." Punch's face is broadcast to Terra by a small handheld video camera.[27] His guide asks "Well, Punch you've made your first TV appearance on Terra. How do you feel?" Punch replies, "Mighty small, ma'am. I used to think we Americans were up to date and modern, but you people sure got ahead of us."

After a week of exploring the space station, Punch is bored. However, Terra's largest space cruiser is due to arrive soon. She's on her maiden voyage and is coming to the space station. Everyone is excited to see it. As the ship approaches, Punch remarks that she's bigger than the station and wonders how she's going to land on it. His guide tells him that the ship won't land. "They are going to lower their newly designed landing gear." But as they watch the gear being lowered, disaster strikes! (Figure 21)

"We're going to crash!" yells Punch!

In an attempt to dock with the space station, Terra's largest space liner collides with the upper ring of the space station. The landing gear tears a hole in the station's inner chambers. The gravity controls fail. Everyone, including Punch, starts drifting out into space. His Terran guide grabs Punch by the hand.

© Alden McWilliams Estate

Figure 21

She and Punch struggle toward an open hatch. Just as they reach the opening, she points outside to where some girls are hanging onto some broken beams. She's worried they won't be able to hang on much longer! Punch goes back inside the station and finds some rope. When he returns he is able to throw a lasso over the floating girls.

"Reckon this is the first time a cowboy has tossed a rope in these parts!" he exclaims.

---

27 Also an amazing prediction as TV cameras of the time weighed about 500 pounds (220 Kg), rolled on dollies and required two men to operate.

Meanwhile, the Station's upper ring slowly disintegrates under the impact. The larger lower ring of the space station is in trouble too.

The huge Terran space liner maneuvers back into position underneath the stricken station to lend support if necessary.

Back on the station, Punch is gasping for air, "I can't seem to get my breath back." He says. He slumps to the floor and passes out.

The Terran woman says, "They've restored gravity control but the air's getting thinner! The atmosphere plant must've stopped. Quick, we have to get to the crash shelter."

Punch and three of the station's crew have collapsed, unable to make the few feet to the temporary safety of the crash shelter. Moments later, an emergency squad spots them and they are dragged into the shelter. The crewmembers are revived but when they give Punch oxygen they find he is not responding. One of the emergency team reminds them of the basic tenant of Terran Law, "Allow not the males of Terra to diminish further in numbers. Sacrifice nine females if thou must if it saveth the life of one male."

Signaled by the emergency squad, an emergency craft descends to pick up Punch. They rush the boy to the "restorers of life."

Neither first aid nor artificial respiration has helped Punch. The rescue craft brings Punch's lifeless form from the badly damaged section of the space station. They hurry him to the medical center. They place him on a table beneath a vast, complex machine. As they work on him, his pulse begins to beat.

Slowly Punch fights his way back to life, however, he remains unconscious. While he is recovering, the doctors decide to attach a "thought communicator" to his brain.

The Saucer Fleet

"The boy will be sent to Terra. Our customs and language are strange to him. The mechanical brain will instruct him in learning our language while he's in his present trance," explains one of the doctors.

While they are working on Punch, a young man wearing a cape on walks in the door and asks to see the Earth boy. The doctors are upset at the interruption by "Prince Torro the Royal brat!" They tell him that he has to wait an hour longer until Punch has learned Terran language.

Finally, Punch regains consciousness. At first he thinks he is in the engine room. He is told that this is the Terran's medical machinery. Prince Torro looks in the door and asks if the Earth boy is awake yet. Punch responds in Terran language and the doctors smile. He has learned their language. The doctors release Punch but warn him not to let Prince Torro get him into mischief.

As Punch and Torro get acquainted, Punch is surprised to see the earth through a window—so far away that he thinks it's a full moon. Torro laughs at his ignorance. Punch makes a comment about Torro's dress and Torro takes offense.

"Well, on earth only girls wear skirts!" says Punch.

All artwork © Alden McWilliams Estate

Figure 23

Torro decides that he'll have to teach Punch some manners. He invites Punch to the gym to settle it. The two boys go to the gym where they have a lengthy wrestling match, much to the amusement of the women who are watching. Finally, the boys decide to call it a draw.

Torro is amazed that Punch can handle himself so well and asks him where he learned Judo tricks taught only on Terra. Punch says he doesn't know, he's never fought judo before. Torro realizes that they must have taught him more than just Terra's language with the brain machine.

"Could be, Torro. Since I hitchhiked on a flying saucer nothing surprises me anymore," says Punch.

Punch discovers that Torro is on board because he's a member of the royal family. Although kings have not ruled Terra for 300 years the descendents of royal blood have retained certain rights. Punch is glad to have another male on board. Torro tells Punch to stick with him because he'll get treated the same way as he is and they cannot treat Torro like any ordinary male.

Torro takes Punch on a tour of the ship. Down in the machinery room Torro shows him the gravity generation system and, as a prank, turns it off. Alarms ring and people float off the decks. (Figure 22)

By turning off the gravity generator, Prince Torro has shut down the entire space liner's internal gravity system. Inside the liner *S.L. Comet,* loose objects drift in every direction. Only the navigation officers, strapped in their seats, are not affected. An emergency repair crew equipped with magnetic boots hurries to fix the gravity system."

Torro tells punch that he has stolen a part from the gravity machine. He doesn't think they can fix it. Punch thinks Torro should put it back or they'll get into real trouble, but Torro just laughs.

The boys enjoy Torro's practical joke. They put on magnetic boots so they can walk wherever they want. They laugh as other people flounder around in midair. They go into the gym where people are hanging on to parts of the walls. Bragging, Torro starts to walk on water across the swimming pool. When he reaches mid-pool, gravity is restored and he falls in. The girls grab Torro and "give the Royal brat a royal dunking." Punch, as his partner in crime, gets the same treatment.[28]

As the giant space liner, *S.L. Comet,* approaches Earth's Moon, Punch and Torro are becoming good friends. Punch asks Torro why they're stopping at the moon and Torro tells him they have scientists living there. They have put an air bubble on it so people can live and the ship is bringing in supplies.

The huge ship comes in for a soft landing next to a large transparent dome (Figure 23). The ship's atmosphere plant puts out a layer of "heavy gas that weighs 5 tons per cubic foot."[29] Then air is pumped under this layer and the heavy gas stretches

---

[28] While Lebeck and McWilliams give a generally accurate view of weightlessness to readers of the '50s, they miss the fact that the water would not have stayed in the pool—perhaps on purpose so that Torro could fall in it.

[29] Easily the largest science error of the series, the substance is far too dense to be a gas. The drawings show it acting like a thick liquid, flowing out onto the moon's surface --an action impossible for a gas.

out like rubber. In that way they form a bubble on the surface of the Moon.[30]

Punch and Torro are among the first group to get on the elevator to go down to the surface. Punch is excited, but Torro isn't. He has seen many satellites.

Punch says "If only the folks back home could see me now. "Boy from Texas first to reach the moon." "Maybe we can send them a signal with a searchlight or something?"

Torro says he thinks that is a bad idea. He reminds Punch that Terrans" want to keep their existence a secret so that the people from Earth cannot make war on them. Punch protests that his people wouldn't do that. Torro points out that maybe Punch's wouldn't but from what he knows of Earth history, others would. Punch admits he may be right, but he's heard "talk of making the Moon a military base to keep people from making war!"

"See what I mean?" No, I'd rather have my people preserve peace up here and let yours do it down there!" replies Torro.

Punch and Torro look around the science station. Torro tells Punch that the reason his people came here is that they wanted to know if Earth's moon had once held life like what they had found once existed on Terra's three satellites.

In one of the labs, Torro shows Punch a "thought visualizer and recorder." Torro explains that eyes don't see pictures; it's the thought waves in the brain that forms the picture. Torro puts Punch in the machine and asks him to think of something. Punch looks up and sees himself riding a horse. Punch thinks it's a miracle but Torro says that it isn't any more than radio, radar or TV.

Torro suggests that they look at some of scientists mind recordings of "Moon samples" on the mechanical visualizer. The samples show that the moon developed an atmosphere early and with it, the first beginnings of life. However, the Moon's life cycle was short. It ended millions of years before life began to stir on Earth. Punch is curious about what kind of life was on the Moon. Torro tells him that for that they'd have to look at other recordings or dig up some evidence themselves.

So, Punch and Torro set out to dig into the Moon's surface for evidence of life. Torro cautions Punch to dig carefully as whatever they might find would have to be hundreds of millions of years old. For some time Punch

and Torro dig without results. Finally they find a piece of rock with insects in it and an unusually sharp stone. They show what they've found to one of the scientists. The Terran scientist says that they've found a prehistoric shark tooth and a fragment of a fossilized anthill. Punch is amazed that life was actually here on the moon. The scientist tells them that all planets and satellites are islands of life for a short period during their existence.[31]

They decide to do some more private excavating. When "Punch and Torro come upon a deep pit at the base

© Alden McWilliams Estate
Figure 24

of the crater rim," they are curious and decide to look into it. They don't bother telling anyone. (Figure 24)

© Alden McWilliams Estate
Figure 25

---

[30] Pushing air under something that dense would require enormous pressure. Obviously the material would have to be an engineered liquid with a high surface tension (like soap). It would have to be designed to resist boiling in a vacuum as well as easily expanding while preventing the air from escaping. If possible, this material would make an efficient way to make a quick environment for exploration of surfaces with no atmosphere.

[31] This strip was written, of course, more than a decade before the first moon landing, and this was a not-yet-discarded Victorian era theory of planetary evolution. The thinking was that all planetary bodies undergo similar evolutions, and would host life at some point. Those bodies smaller or further from the sun than Earth would cool faster; hence move through their "life stage" earlier. Wells used it in the opening chapter of *War of the Worlds* to explain why Martian civilization was dying. This is another example of adventure strips trying to be educational as well as entertaining.

Together they crawl into a narrow crevasse. After a distance the walls begin to slant, and as they get steeper, the boys lose their grip and start to slide. (Figure 25)

Punch tries to grab on but it's too late. Unable to stop their slide into the darkness, Torro and Punch slide until they land on a ledge some 50 feet down. They find themselves inside a large cavern and realize they could have fallen to their deaths. (Figure 26)

They notice they are in a limestone cavern which proves that the crater was not a volcano, but was made by a meteor crashing into the limestone. As they explore the cave with its columns of calcium they come to the realization they are stuck in the cave with no way out. Angrily, Punch kicks a stalactite.

Torro shouts "Watch out, Punch," and pulls Punch back from getting buried by the falling limestone.

They are trapped in the limestone cavern for some time. Punch and Torro start getting really hungry. Torro points out that it looks like it won't ever happen again. Punch doesn't think that's funny. He is worried that they are going to die. Torro says to cheer up. He thinks the scientists should be looking for them by now.

Meanwhile, above the surface, the Terran women are beginning to wonder where the boys are. They are not alarmed until someone mentions that they've already missed three meals (Figure 27, a B&W reprint of color page).

Sometime later a noise is heard in the roof of the cavern. The boys look up and see "a monster spider" coming down

at them. They are being rescued at last. Torro calls it a "tele-crab—a TV controlled electronic robot, used for work in inaccessible places." Above them the operators see them through the television screen and tell them to climb onto the robot's drill head.

As he climbs on the drill head, Punch says he is so hungry he could eat a horse. Up above the operators observe they are entering the drill shaft and know that they'll have them out shortly. One Terran laughs and remarks that this will make headlines on Terra. "Scientists excavate live boys on Earth's moon!"

Later the boys eat a hearty meal while the women comment that they don't even get thanked for rescuing them from being buried alive—all they do is eat!

"Well when they finished eating we could dump them back," retorts one.

"No but we could send them back to Terra!" says another. (Figure 28)

And so, Punch, from Texas, and Prince Torro, of Terra, board the huge space liner and leave the moon for Terra. Punch would like to stay longer, but Torro reminds him that it will be six months before the next Space liner stops at the science station again.

In preparation for takeoff the *S.L. Comet,* closes all its hatches and collapses its temporary air dome.

Punch remarks, "Well, were off."

"Yes," says Torro, "and when we get to Terra, I want to go fishing. After all that dry moon, I want to see lots of water."

Thus the boys leave the moon on the *S.L. Comet,* and head back toward Torro's home planet of Terra. (Figure 29)

Figure 26

Figure 27

Figure 28

Of course this isn't the end of their adventures together. This has been only the first half dozen. They have yet to meet Lahna, a girl just their age that will share their adventures.[32] They will meet her being chased by dozens of vicious pygmies on an abandoned pleasure cruise ship, rescue her and escape the ship on a "sport submarine." From that point on, they will be inseparable in the Sunday series. (Figure 30)

Figure 29

All artwork © Alden McWilliams Estate

Figure 30

Together the trio will battle pygmies, get lost in space, find themselves on a planet inhabited by giants (and become their pets), explore a planet, be attacked by "squimps" (squid-blimps), and travel back in time to name just a few of their adventures.

Altogether, the Sunday strip would last for five years until 28 December 1958, a pretty good run as adventure strips go.

The Vehicles

There are many types of flying saucers shown in *TWIN EARTHS*. Of course, there should be since Terra is another planet with a very advanced culture. It's natural that they would have a variety of spacecraft designs.

When one studies the saucers that are the major elements in the stories, a few significant types become evident. The first and most prevalent type seen are the saucers that the "Terran Secret Service" uses between Earth and the Terran Space Station orbiting the Earth. We've named this type the "Standard" class of spaceship, although it is not called that in the strip. It is seen on both Earth and Terra in the daily and Sunday series. It is able to carry up to about four dozen people and is able to perform the widest set of duties.

The second most important machine in both series is the Terran Space Station. Of course, it is not a complete spaceship since it has no propulsion. While not a saucer, it plays a central role in both the daily and Sunday series. It deserves study and understanding because it serves as an important link between the various types of Terran Saucers as well as between Terra and Earth.

The second class of saucer is the huge space liner type that is seen in both dailies and Sunday strips. In the daily strip it is represented by the *S.L. Galaxy,* and in the Sunday strip by the *S.L. Comet.* It is assumed that "S.L." stands for "Space Liner"

since it is called that each time the ship is introduced. The "Space Liner" class appears early in both series.

This type is much larger than the Standard class spaceship or even the Space Station itself. It is not known how many ships of this class the Terrans have, but they are present in several stories and are important to both story lines.

The third type of saucer, seen only in the Sunday series, is the "Baby Saucer." That is what the three teenagers call this type ship when they discover it. This is a much smaller spaceship than the Standard class and is designed to transport just three people. This makes it ideal for the three teenagers that are the heroes in the Sunday series. And the three, Punch, Torro, and Lahna are shown using this saucer almost exclusively as a vehicle to their various adventures.

As for others, close inspection shows that there are many other saucer types in the background in supporting roles. However, their role is primarily just that, support.

With that in mind, here is an analysis of the three major types of saucers and the one space station that represent the majority of Terran Spaceships in the series.[33]

Terran Saucer-*Standard Class*

Quickspec

Vehicle Morphology	Saucer
Year	1952
Medium	Adventure Strip
Designer	Alden McWilliams
Diameter	98.4 ft (30.0 m)
Height	28.7 ft (8.8 m)

The Standard class saucer can be seen in Figures 9 and 20. In Figure 9 it is one of the Terran Secret Service saucers making a delivery to the orbiting Space Station. In Figure 20 it is taking Punch to the Terran Space Station. The Standard class saucer saw service throughout the daily strip stories from the beginning to end. It was only seen in the Sundays in the very beginning where it picked up Punch.

We first see a saucer of this class when one is orbiting

---

[32] The one thing missing in this trio is romance. These three are otherwise healthy, good looking teens but are prevented from any such behavior by the mores of the times.

[33] Ignoring the alien Pygmy spaceships in the series, as they are globe, not saucer based.

The Saucer Fleet

the Space Station. There are several other times that it is seen in the sky (such as in Figure 9) or space around the station. These activities do not allow us to get any idea of its size, as there is nothing to compare it to. However, in the story where Soviet agents trap Vana, Garry, and Zena in the old mansion, an opportunity to scale the saucer presents itself.

Recall that in this story the trio is surrounded by Russian agents bent on killing them and stealing back the plans for some secret weapons that Zena has "retrieved" from them. During the gunfight, Colonel Zena has to call for help if they are to survive. She uses her two-way communicator[34] to call for back-up and just when it looks like curtains for our heroes, a flying saucer is seen hovering overhead, broadcasting orders to the Communists to drop their weapons.

Of course, one agent refuses and is turned into instant dust by a Terran disintegrator beam. The others surrender. Our opportunity to find out the size of a standard class saucer comes when the Terrans take the prisoners aboard the saucer. (Figure 31)

© Alden McWilliams Estate

Figure 31

There are also several illustrations of people standing beneath and next to the saucer during the loading that gives a good indication of the saucer's size. Further, the Terrans load several cars on board, giving us the ability to judge the saucer's size.

Additional detail information comes from the Sunday series beginning when Punch sneaks aboard a flying saucer. The saucer he happens to choose is a Standard Class saucer. The accompanying illustrations reveal many more details about this series saucer's characteristics. Happily a cut-a-way drawing is also included. All of this information was used in producing the data drawing and the following notes.

---

[34] Eat your heart out, Capt. Kirk! This 1952 flip-top communicator is identical to yours except it also has a video link.

By careful scaling the saucer against known objects like people and cars, I determined that the Terran Standard Class Saucer has a diameter of about 98 Feet (30 meters) and a height of 28.7 feet (8.75 meters) including the 1 meter tall fin at the top.

The cutaway drawing below gives a glimpse of the internal makeup of this saucer class. The internal arrangement has two decks of various sized compartments. In the center of the top deck is a central control room 23.5 feet (7.2 m) in diameter (assuming the room is circular, but there is no proof of this) and over 10 feet (3 m) high. Beneath the central control room is what appears to be the main "engine" room. It is also large at 30 feet (9 m) in diameter and 10 feet (3 m) high. There are several smaller rooms filled with machinery around the two central rooms. A room with an overhead "fluoroscope projector" is visible on the extreme right. Presumably this is where Garry Verth and later Punch would get their first look at Earth from orbit. (Figure 32)

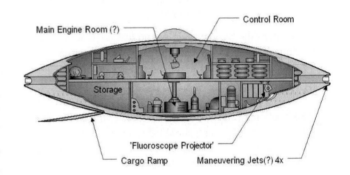

© Jon Rogers

Figure 32

There are no fuel tanks or any other type of energy storage. As Garry found out upon his visit to the Space Station, all energy to fly, maintain an internal gravity field and life support for all Terran space vehicles is obtained directly from the sun. Judging by the considerable amount of energy they use, this little technical secret alone is worth "a king's ransom", especially since there is no evidence of solar conversion cells anywhere on the ship. Likewise, there must be some way (never revealed) of storing the Sun's energy as the Saucers work just as well in darkness as they do in sunlight.

From the evidence available, the Terrans have enough fluoroscope projectors available for opening windows at 4 positions in both the top and lower skin of the saucer. Whether they have 8 projectors, 4 that pull double duty, or fewer that are movable to each window location is unknown. But the presence of the windows, both above and below, shows that they have the ability to open up several windows at a time.

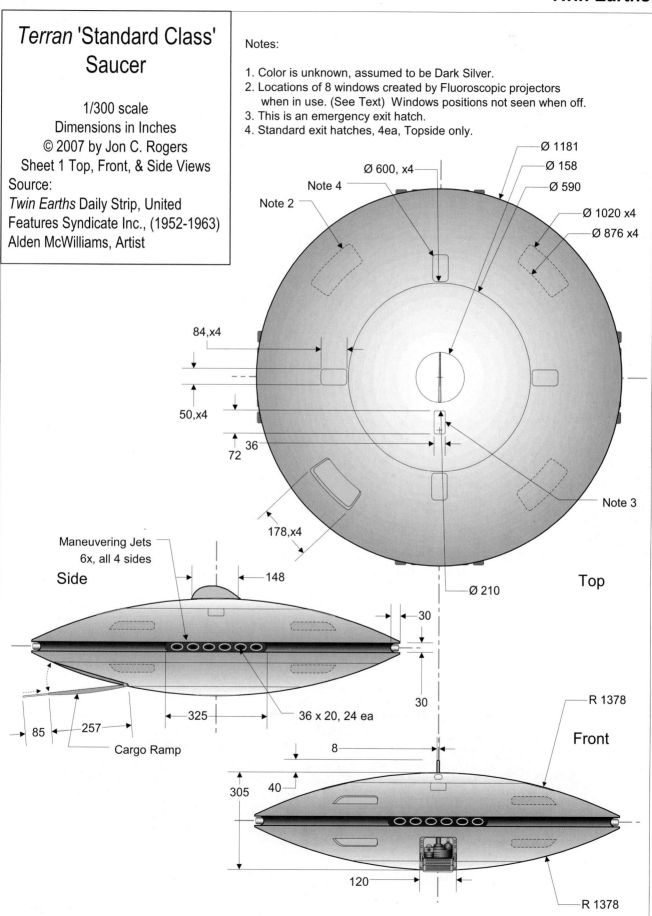

Terran 'Standard Class' Saucer

1/300 scale
Dimensions in Inches
© 2007 by Jon C. Rogers
Sheet 1 Top, Front, & Side Views
Source:
Twin Earths Daily Strip, United
Features Syndicate Inc., (1952-1963)
Alden McWilliams, Artist

Notes:

1. Color is unknown, assumed to be Dark Silver.
2. Locations of 8 windows created by Fluoroscopic projectors
   when in use. (See Text) Windows positions not seen when off.
3. This is an emergency exit hatch.
4. Standard exit hatches, 4ea, Topside only.

Ø 1181
Ø 158
Ø 590
Ø 1020 x4
Ø 876 x4
Ø 600, x4
Note 4
Note 2
84,x4
50,x4
36
72
178,x4
Note 3
Ø 210

Top

Maneuvering Jets
6x, all 4 sides
Side
148
325
36 x 20, 24 ea
85
257
Cargo Ramp

30
30
8
40
305
120

R 1378
Front
R 1378

The Saucer Fleet

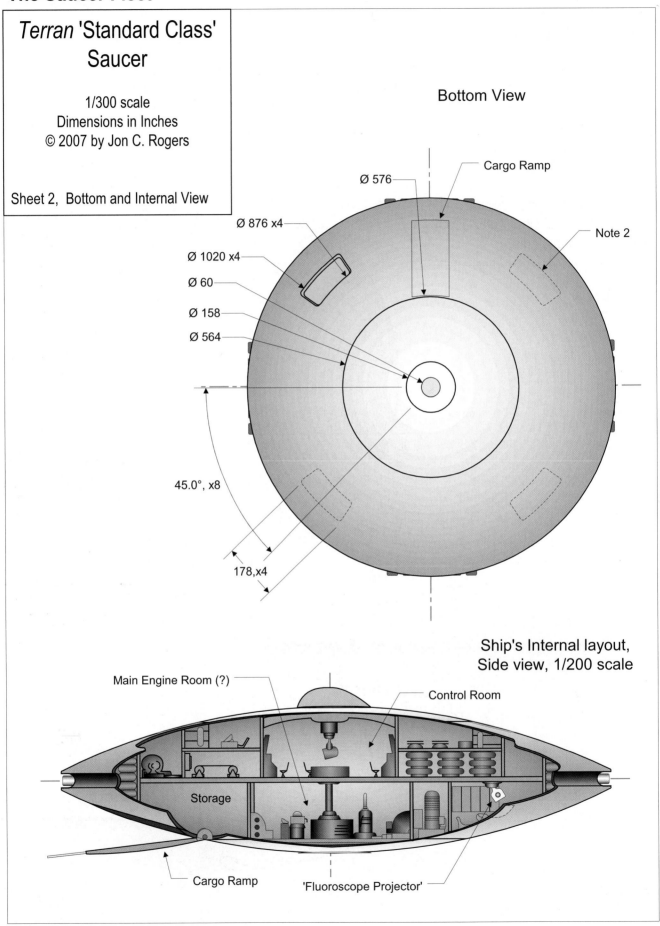

Terran 'Standard Class' Saucer

1/300 scale
Dimensions in Inches
© 2007 by Jon C. Rogers

Sheet 2, Bottom and Internal View

Bottom View

Cargo Ramp

Ø 576

Note 2

Ø 876 x4

Ø 1020 x4

Ø 60

Ø 158

Ø 564

45.0°, x8

178, x4

Ship's Internal layout,
Side view, 1/200 scale

Main Engine Room (?)

Control Room

Storage

Cargo Ramp

'Fluoroscope Projector'

The Standard Class Saucer also has one main cargo hatch (called a "landing ramp") on the lower front of the ship. It is used several times to load people and, at one point, cars. Whenever the hatch is opened and the ramp is extended, the saucer is vulnerable since the "magnetic barrier" must be turned off to offload cargo and people.

There are also four personnel hatches on the top of the saucer that are shown in use in the Sunday series. When he stows away, Punch enters the ship through one of these hatches.

Finally, there is also one small personnel hatch at the very top near the fin. It is shown being used once during an emergency in the attempt to take the Soviet agents back to the USSR. It is presumably for emergency escape only.

The Standard Class saucer also has maneuvering jets at four points of its circumference. It is not clear or explained when they are needed so they are assumed to be low power positioning jets. However, their presence is unmistakable.

This saucer class carries other equipment as well, like heat, gravity and disintegrator rays. Each of these is shown in use at one time or another during the series. Altogether these elements make this class of saucer a very versatile machine.

There is one final point regarding the color and insignia. While all Standard Class saucers have no color other than silver metallic and no insignia, there is one notable exception. In the story after the Terran saucer fleet comes to Earth, the governments (of Earth) obtain the rights to build their own Standard class saucers as part of the two planets combining strength to fight off the pygmy invasion. On these saucers—and no others—can be clearly seen USAF markings and Insignia. (Figure 33)

© Alden McWilliams Estate

Figure 33

Terran Space Station

Quickspec:

Vehicle Morphology	Wheel
Year	1952
Medium	Adventure Strip
Designer	Alden McWilliams
Dia. – Upper Section	168.5 ft (51.4 m)
Height – Upper Section	24 ft (7 m)
Dia. – Lower Section	296 ft (90 m)
Height – Lower Section	41 ft (13 m)

We see only a single Terran Space station in both series (another is mentioned but not shown). The interesting thing is that it is not circling above Terra. Since a space faring people develop space stations as an intermediate step in obtaining a permanent presence in space, logic dictates that there must also be a space station in orbit around Terra. However, there is never one shown or even mentioned in either series. Therefore, ironically, the only Space Station owned by the people of Terra exists above *a different* planet, Earth.

There is one point where they mention having another space station; however, it is only once when both Sunday and Daily plot lines cross each other. It was likely an unplanned occurrence and needed a quick explanation.

When the Sunday series began, Punch arrived at the space station in the 29 March 1953 strip (Figure 20). But that same week, in the dailies, Garry Verth and the Space Station were over Moscow being attacked by Russian MiG's. (Figure 10)

The *S.L. Comet* arrived to pick up Punch on Sunday 19 April and crashed into the Space Station. (Figure 21) It took until 24 May to get it repaired. Unfortunately, at that moment in the dailies the (supposedly only) Space Station was still hovering over Moscow rescuing Garry and Zena from the treacherous commies.

Having the Space Station ascending from Earth on 16 May (a Saturday), seeing a wrecked Space station in orbit on 17 May (Sunday), and then having Garry and Zena "back on the Space Station after a trying experience over Moscow" on Monday, 18 May, may have significantly confused readers! To add to the confusion, Zena announces on Wednesday, 20 May, that "One of our greatest Space Liners" is coming to pick up Garry Verth and company to take him to Terra (Just like in the Sunday Series).

On that Thursday, 21 May, they tried to clear up the confusion by showing the "healthy" Space station in orbit as the space liner approaches with this special caption:

> Great precautions are being taken... Only recently the latest of Terra's Space Ships collided with one of the interplanetary Stations.

"...one of the...'? It's curious because nowhere else in the strip do they mention [or show] another Space Station. Also, they always refer to it in the singular as "*The* Space Station."

The Saucer Fleet

Terran Space Station

1/700 scale
Dimensions in Inches
© 2007 by Jon C. Rogers
Sheet 1

Source:
Twin Earths Strip, United
Features Syndicate Inc.,(1952-1954)
Alden McWilliams, Artist

Notes:

1. Color is unknown, assumed to be White.
2. Main ring has four decks with 8 foot high ceilings.
   Upper ring has three decks.
3. Gravity generators maintain an atmosphere around station so it is
   unknown if entrance is an Air lock.
4. Shape of lower portrusions is a near parabolic curve as shown.
5. All interior details are speculative but best estimates.
6. One window per fin on 'leading' edge.

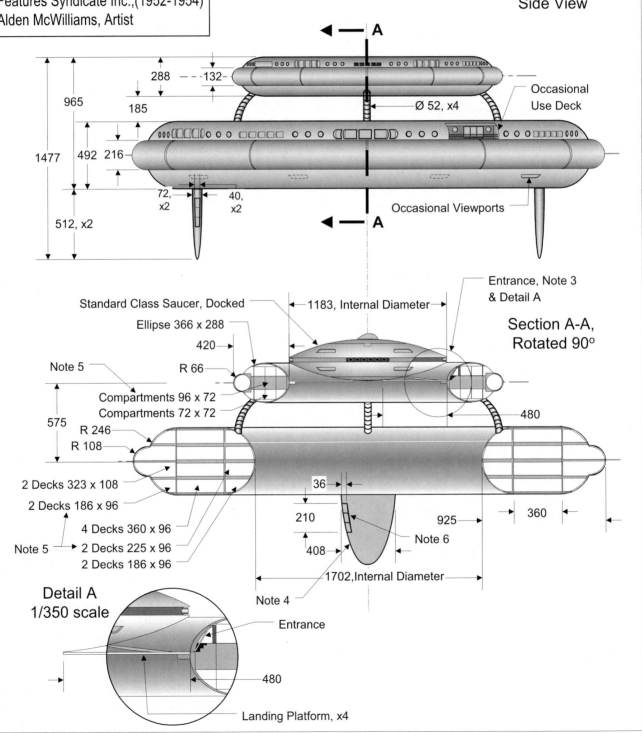

Side View

288
132
185
965
Ø 52, x4
Occasional Use Deck
1477
492
216
Occasional Viewports
72, x2
40, x2
512, x2

Standard Class Saucer, Docked
Ellipse 366 x 288
420
Note 5
R 66
Compartments 96 x 72
Compartments 72 x 72
575
R 246
R 108
2 Decks 323 x 108
2 Decks 186 x 96
4 Decks 360 x 96
Note 5
2 Decks 225 x 96
2 Decks 186 x 96

1183, Internal Diameter
Entrance, Note 3
& Detail A

Section A-A,
Rotated 90°

480
36
210
360
925
Note 6
408
Note 4
1702, Internal Diameter

Detail A
1/350 scale

Entrance
480
Landing Platform, x4

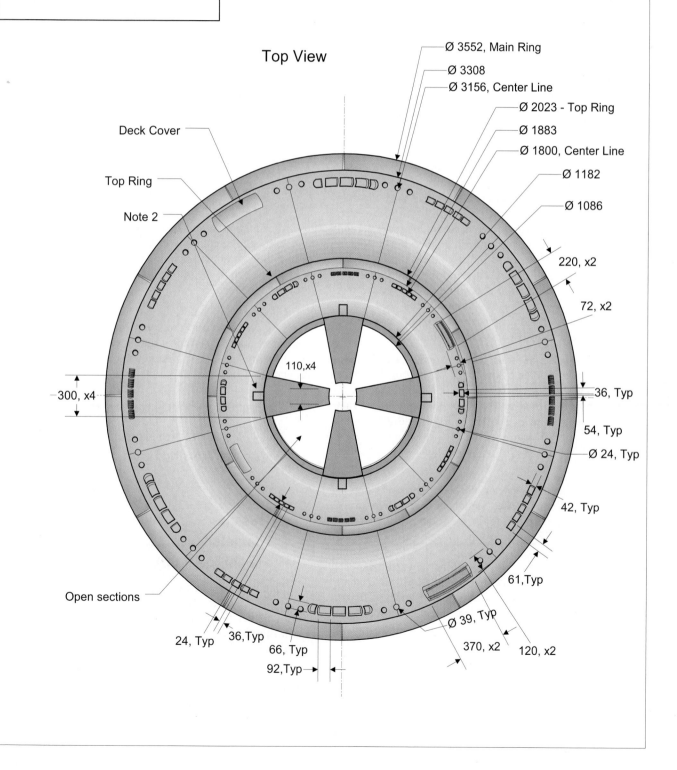

Terran Space Station

1/700 scale
Dimensions in Inches
© 2007 by Jon C. Rogers
Sheet 2
Top View

Notes:

1. The windows and port holes on the top side are dedicated,
   ie: real windows. Only on the bottom side do they use the Fluoroscopes.
2. The drawing shows four entrances. Only one is shown in use, so
   the existence of the other three is speculative.

Top View

Deck Cover

Top Ring

Note 2

Open sections

Ø 3552, Main Ring
Ø 3308
Ø 3156, Center Line
Ø 2023 - Top Ring
Ø 1883
Ø 1800, Center Line
Ø 1182
Ø 1086
220, x2
72, x2
110,x4
300, x4
36, Typ
54, Typ
Ø 24, Typ
42, Typ
61,Typ
Ø 39, Typ
370, x2
120, x2
24, Typ
36,Typ
66, Typ
92,Typ

The Saucer Fleet

The fact that there is a Space Station orbiting Earth was shown early in both series. The Space Station played a central role in the daily strip plot line, appearing again and again. However, it only appeared in the Sunday series when Punch stopped there, was involved in the accident, and consequently left for the moon. It is shown repaired in the 24 May strip and never seen in the Sunday series again.

On Monday, 25 May, the *S.L. Galaxy* arrived at the space station in the dailies. It lands perfectly and away goes Garry to Terra.

In the Daily series story line, it was obvious that the Terrans had established a station as a way point to sustain their "observation" of Earth. For the entire first year (and several adventures) the Space Station's purpose was espionage support. However, by November of 1954 after both planets formed an alliance against the invading pygmies, Terra turned the Space Station over to the United Nations.

There are several things that any science or Science Fiction fan may find familiar about Mac's design for a Space Station. For one, it was a wheel type, just like the Potocnik/Ley/von Braun Colliers designs that were the current cutting edge of technology at the time. However, unlike those scientifically derived designs, this space station did not need to spin to create artificial gravity. It had its own gravity generators that were so effective they could generate a full, Earth normal 1g field. This allowed an atmospheric envelope around the Station that was used to good effect many times. Of course, this also eliminated the need for the station to be circular and to spin, so we can only assume the reason Mac chose that shape was its familiarity.[35]

Something else SF fans may recognize are the two large "fins" that protrude below the lower ring of the Space Station. If you have seen the movies, *Close Encounters of the Third Kind* and *Independence Day* you may notice that the alien mother ships in both films have a design feature very reminiscent of Mac's space station design, i.e. the same pair of lower fins protruding beneath the main structure.

Whether intentional or not, these designs paid homage to Alden McWilliams' Space Station in their recreation of his 1952 design. These two lower protruding fins have also appeared in other SF art from time to time—usually representing an alien space vehicle of some kind. All my research indicates that Mac did it first. Therefore they stand as a silent reminder of a great artist who generated an icon of Alien Spacecraft that is still recognized today. If you thought it looked cool in the movies, now you know who to thank.

---

[35] Ignoring for the moment the problems that having an object with a 1g field (effectively another Earth) in low Earth orbit would cause (e.g. its effects on tides). Having the hero swim in outer space around the station is too delightful—a fantasy yes, but a delightful one.

There is still one mystery in regards to the fins. What are they for? We don't know. Their purpose is never mentioned in either series. Possibly they are for generating the 1g gravitational field that allows life to go on normally on the Space Station. It would be consistent with people walking upright inside and outside on the top of the station as they are shown to do. However, that is just an extrapolation from observed facts. We don't know positively what they are for.

Identifying the size of the space station, needs an object of known size near it. The best reference is the Standard class saucer described earlier. Saucers are seen around the station and also docking with the station, which allowed a good approximation of the size and shape of Terra's Space Station.

The first discovery was that it was not two circular tubes as first believed. Even though the upper section is shown as a circle in cross section when it is torn off during the collision with the *S.L. Comet* in the Sunday series, all the other renderings of it indicate it is actually oval. The lower ring is even wider in cross section than the top ring, making it a rectangle with semicircular ends.

The internal layout of the Space Station is a mystery. Our hero and heroines are shown throughout the station during the series but not in a way that allows one to get an idea of the plan of the station. Only the outside of the station is clear to the reader.

One external feature is a landing area for Standard Class saucers built into the upper ring. The station has permanent windows, portholes and several balconies. It also has "fluoray" generators for viewing through the bottom of the station.

Terran Saucer-Space Liner Class

Quickspec:

Vehicle Morphology.........Saucer
Year...............................1952
Medium.............Adventure Strip
Designer.........Alden McWilliams
Diameter..............394 ft (120 m)
Height (gear up)......110 ft (34 m)

There are two Terran Space Liners named in the strip, the *S.L. Galaxy* and the *S.L. Comet*. There seem to be others shown in the fleet during the great exodus from Terra to Earth during the pygmy invasion, but none are named or given as prominent a role.

When Garry is picked up and taken to Terra, he travels on board the *S.L. Galaxy*. Although other saucers are quite capable of making the trip, our hero is given the royal treatment. According to Zena, this class of space ship has super luxurious accommodations. Like a modern cruise ship, it

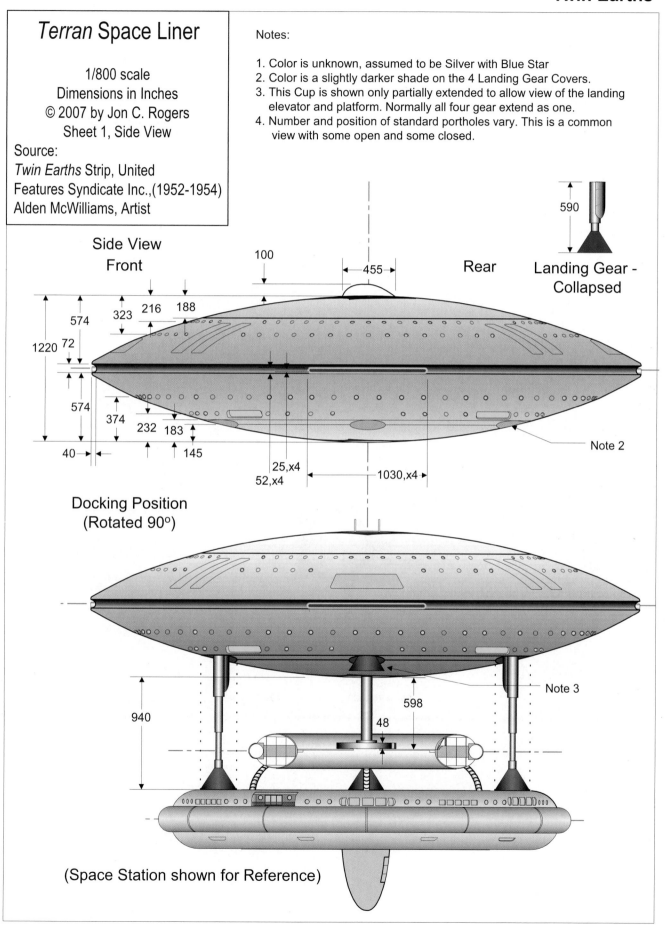

Terran Space Liner

1/800 scale
Dimensions in Inches
© 2007 by Jon C. Rogers
Sheet 1, Side View

Source:
Twin Earths Strip, United
Features Syndicate Inc.,(1952-1954)
Alden McWilliams, Artist

Notes:

1. Color is unknown, assumed to be Silver with Blue Star
2. Color is a slightly darker shade on the 4 Landing Gear Covers.
3. This Cup is shown only partially extended to allow view of the landing elevator and platform. Normally all four gear extend as one.
4. Number and position of standard portholes vary. This is a common view with some open and some closed.

Side View
Front

Rear

Landing Gear - Collapsed

590

100

455

574
323   216   188
1220   72
574
374   232   183
40   145
25,x4
52,x4
1030,x4

Note 2

Docking Position
(Rotated 90°)

Note 3

940   598   48

(Space Station shown for Reference)

The Saucer Fleet

Terran Space Liner

1/800 scale
Dimensions in Inches
© 2007 by Jon C. Rogers
Sheet 2

Top View

Notes:

1. Number and position of standard portholes vary. This view shows almost all possible locations with them open.
2. There are 8 hatches for the emergency vehicles. The locations shown vary. These are the most likely positons.
3. These 2 hangar openings are as shown. Their purpose is unknown.
4. There are 2 parallel vertical fins on the top instead of one.

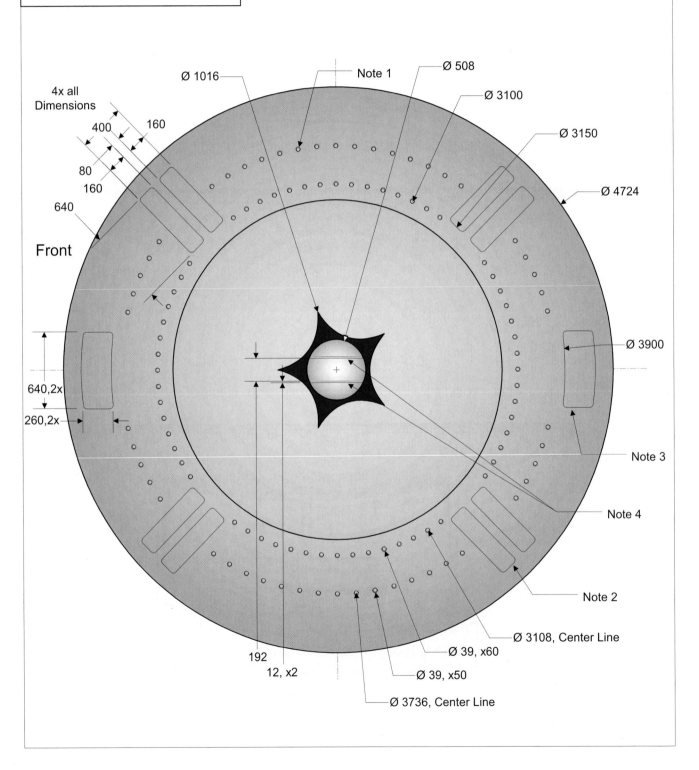

Terran Space Liner

1/800 scale
Dimensions in Inches
© 2007 by Jon C. Rogers
Sheet 3

Bottom View

Notes:

1. Number and position of standard portholes vary. This is the maximum number with some open and some closed.
2. Landing gear doors, 4ea. Slightly darker shade.
3. This feature and dimension is also for center pole of elevator.
4. This feature and dimension is for the elevator rim.

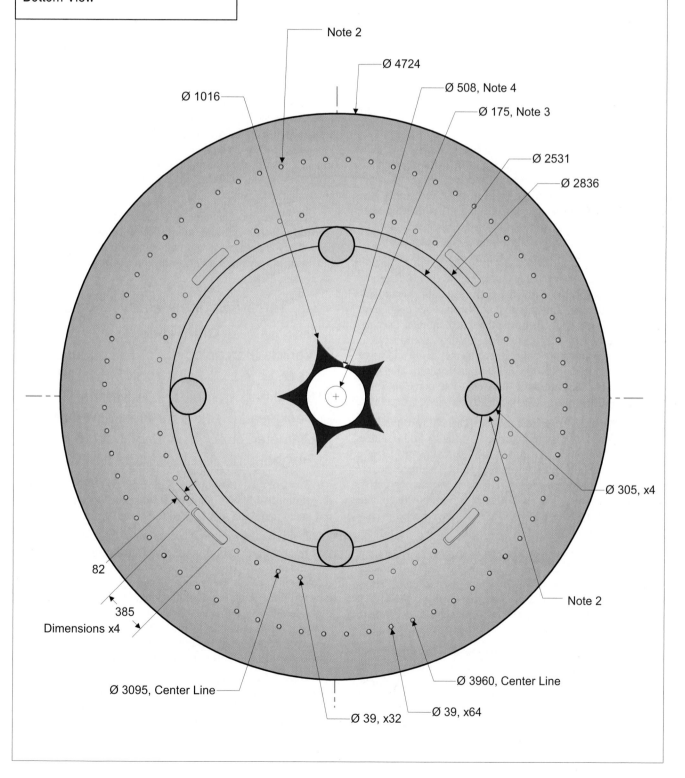

Note 2

Ø 4724

Ø 508, Note 4

Ø 175, Note 3

Ø 1016

Ø 2531

Ø 2836

Ø 305, x4

82

385

Dimensions x4

Note 2

Ø 3095, Center Line

Ø 3960, Center Line

Ø 39, x32

Ø 39, x64

boasts many conveniences like a gymnasium, theater, half a mile of promenade decks and a swimming pool.

Garry's trip is uneventful and we do not get to see him make use of these facilities. However, in the Sunday Series with Punch and Prince Torro aboard its equally luxurious sister ship, the *S.L. Comet*, Torro plays a "prank" by turning off the gravity generators and gets dunked in the swimming pool for his mischief.

Garry does get a tour of the Space Liners features. One of the first things they show him is the navigation chamber in the center of the ship. Since this series—like all Sci-Fi of the era—was written before the advent of transportable computers, the best-known solution to the very large problem of how to navigate in space was the Navigation Chamber. It was a large, spherical chamber and star chart. A super planetarium and "astrolabe" combined.

One interesting feature of these Space Liners was their landing gear and method of docking with the space station. This was a challenge as the Space Liner was bigger than the Space Station. Also, given the mass of the Space Liner, the "upper" ring was unsuitable for docking. This dictated four long legs to bypass the top ring and connect to the main ring of the station. What was surprising was the method of attaching to the Space Station. They used four enormous suction cups.

Wait a minute, suction cups in space? Well, yes. Remember that an atmosphere is provided by the station's gravity generators. Ok, but wait. You have to push suction cups down hard to evacuate the gas under them and the space station is in space with nothing to keep it from moving. So they would have to put a vacuum pump in the ship, drill a channel in the middle of the struts down to the center of the cups and suck the ship onto the station! This illustrates the distance we've come since the '50s and the difficulties you encounter in designing spacecraft.

Now that they've landed and the gear is down, the next design challenge was getting the people "down" to the station or ground. Remember, the ship is still approximately 80 feet (24 m) from the bottom of the landing gear. This time the answer is simple. Each space liner is equipped with an impressive extendible elevator. The last elevator ride down to the "ground", with nothing under them, must have been interesting.

But the Space Liner class of Terran spaceship was not out of surprises yet. Punch and Torro, aboard the *S.L. Comet*, landed on the airless moon. As shown in Figure 23, these ships carry their own special flexible material (heavy gas) so they can "blow up" their own pressure bubble on any airless surface with gravity.

Another feature that figures prominently in the Sunday series story line is the presence of eight hangars for repair craft. These agile little craft are used several times. The first time is when Punch and Torro, on board the *S.L. Comet,* encounter a mysterious derelict in space. The crew uses them to board and explore the vessel. After they discover its crew is dead and it has been abandoned, the small repair craft are used to dismantle the derelict for study.

Plot wise, perhaps the most important use is that, as the *S.L. Comet* finally approaches Terra, Prince Torro decides to steal one of the repair craft and go "adventuring." He convinces Punch to go with him and, after a quick stop in the capital city to call his mother; the two boys take off in the little craft for the "Forbidden Islands." This starts their main adventure stories that last throughout the entire Sunday series.

One thing that should be mentioned before leaving the super luxury liner class of spaceships is that several of their features are not identifiable. Normally, to do an outline diagram of the ship, several drawings or pictures of the ship in question are combined to show all of the features.

This doesn't work for the ships of Terra. The super liners, for example, are shown having from one to four visible thrusters depending on the drawing. Most of the time, there are two thrusters at 90° to each other, thus implying that they have four in all. However, their position and quantity are not consistent from drawing to drawing. Many other surface features are shown the same way.

Therefore, the data drawing is the best that can be made to recreate the character and style of these beautiful flying saucers. They are not consistent with all the drawings in the strip; but then, the strip's drawings are not totally consistent with each other, either.

Terran Saucer-*"Baby" Class*
Quickspec:

Vehicle Morphology……………..Saucer
Year…………………………………1955
Medium………………….Adventure Strip
Designer……………….Alden McWilliams
Diameter………………..35.5 ft (10.8 m)
Height…………………..13.9 ft (4.23 m)

Probably the most unique saucer in *TWIN EARTHS* was the diminutive "Baby" saucer. It was seen only in the Sunday series and wasn't introduced for several years.

Punch and Torro, now accompanied by Lahna, had survived a whirlpool in the sport submarine they had taken to escape the pygmies. They found themselves on one of the barrier reef islands just off the mainland. Seeing a pygmy spaceship pass over, they hide out until the danger had passed only to be hit by a surprise hurricane. After the storm passes, they are surprised to find one of Terra's ocean going ships washed ashore on their island. Examining the wreckage they discover the "Baby" Saucer in its hold, still undamaged after the storm.

Once they find the "Baby" saucer, a name given the ship by Punch when he first sees it, their adventures expand as they travel off Terra. The small ship is just the right size for the three of them. It carries them to several "out of this world" adventures over the next several years.

Terran 'Baby' Saucer

1/100 scale
Dimensions in Inches
© 2007 by Jon C. Rogers
Sheet 1, Top View, Side View
Source:
Twin Earths Sunday Strip, United
Features Syndicate Inc.,(1955-1957)
Alden McWilliams, Artist

Notes:

1. Color is unknown. Assumed to be white.
2. Engine Access Cover is for maintenance & repair.
   It is not an entrance to ship's interior.
3. The main Airlock is the only access into ship's interior.

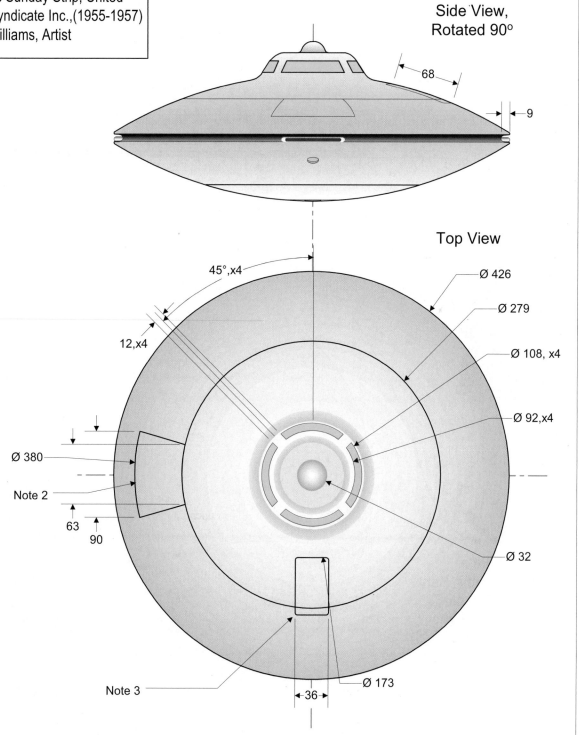

Side View,
Rotated 90°

68

9

Top View

Ø 426

Ø 279

Ø 108, x4

Ø 92,x4

45°,x4

12,x4

Ø 380

Note 2

63

90

Ø 32

Ø 173

Note 3

36

The Saucer Fleet

Notes:

1. Lower Porthole existence and size is definite. Location is a best approximation.
2. The ship carries three teenage occupants. The Earth boy, Punch, and two Terrans, Torro and Lahna. Lahna (4' 10") is shown for size reference.

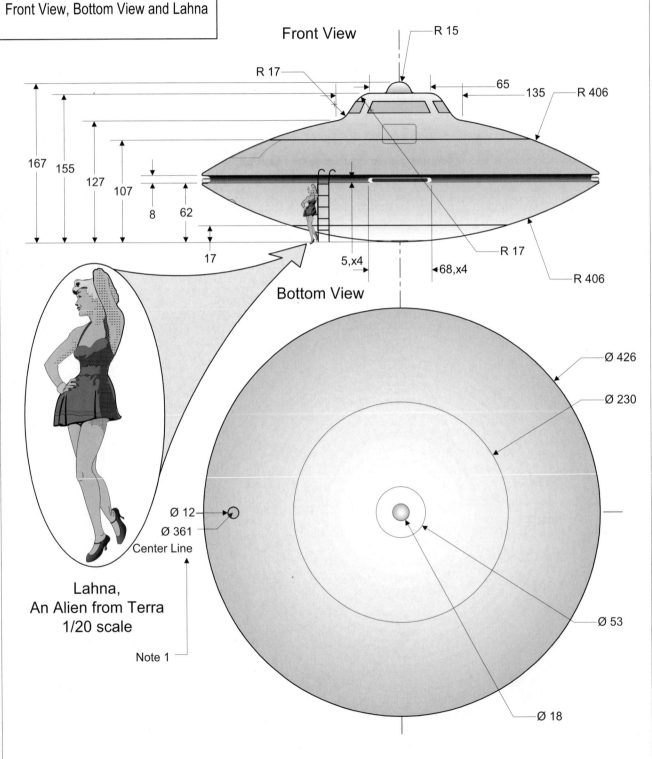

Front View

R 15

R 17

65
135
R 406

167 155 127 107

8 62

17

5,x4

68,x4

R 17

R 406

Bottom View

Ø 426

Ø 230

Ø 12
Ø 361
Center Line

Ø 53

Ø 18

Lahna,
An Alien from Terra
1/20 scale

Note 1

Of the Baby Saucer itself, not much is explained or detailed. It is the means to find adventure, not the source of adventure itself. This makes it difficult to document with regards to its exact size, let alone capabilities.

One of the difficulties stems from the fact that there were no full grown adults shown next to the saucer. The teenagers were of undetermined height, which makes judging the relative size of the saucer a problem.

The first challenge was to determine Punch, Lahna, or Torro's height. This too proved difficult, as they were also never shown next to any adults. They're seen with pygmies, giants, dinosaurs, "Squimps" and all kinds of strange animals, but no human adults. It was not until 1957 when they arrived back on Earth (but in the wrong century) and were captured by American Indians that the opportunity presented itself. In one cell, Punch and his Indian friend are shown next to a full sized adult Indian.

After working out the Baby saucer's size, the next discovery was that it did not follow the conventional Terran saucer proportions. It was much taller for its diameter than the usual saucers. This was probably due to the fact that the saucer was already a rather tight fit—even for teens. The taller height gives more room for equipment.

There were also still a few unusual characteristics to work out. One strange thing was that the saucer had no landing gear at all. This meant that each time it landed it must have been very unstable. It would have "teetered" back and forth whenever the three moved around in it or attempted to disembark.

Another feature was that it had only one hatch and that was on the topside of the saucer. The kids had to carry a portable ladder with them to climb down off the top of the saucer every time they landed. This must have been problematical, although they never complained about it.

One thing the kids never complained about was the long space voyages they took in the Baby saucer. According to the storyline, they not only went from Earth to Terra, a voyage that took many weeks, but they also visited a strange star system where the planets and people were giants. These trips took such an amount of time that they must have been very well balanced kids to withstand such tight quarters for such an extended period.

There is one thing about the Baby saucer that could not quite be documented completely. During the adventure with the "Squimps", Punch and Lahna are shown close up, using a porthole that is in the lower side of the saucer. However, no other drawing shows where that porthole might be located. We know its there, we just have to consider the position given on the data drawing to be arbitrary.

So in the final analysis what can we say of the Baby saucer? Well, it is cute, powerful, and somewhat enigmatic. Just what a good flying saucer should be.

Modelers' Note

There have never been any commercial models produced of any of the Saucers from *TWIN EARTHS*.

With this Data drawing, saucer modeling enthusiasts will be able to build their own accurate models of some of the most artistic and beautiful flying saucers ever designed, the saucers from Terra, our *Twin Earth*.

Epilog

It is customary in American culture to dismiss the comic strip, comic book and other forms of illustrated storytelling as unworthy of serious consideration either as art, or as storytelling. It is only after decades of neglect that this trend is beginning to be reversed. It is sad that it has taken us this long to "get it."

The fact is that a comic strip's ability to tell a story quickly and dramatically is a strength, not a weakness. The art they present can be powerful and dramatic in its own right. The stories they tell can be meaningful, influential and important. They aren't always so because of commercial needs to support the public's taste. However, when they do more than just follow popular trends, they contribute to our culture in many ways other than just being light entertainment.

In the case of *TWIN EARTHS* we have a series of beautiful artworks telling a story designed to make people think, to educate them, to open their minds. In many ways, *TWIN EARTHS* was ahead of its time. It spoke up for equality of the sexes, decades before that became a social theme. It educated people on science, both astronomical and natural. It predicted things to come.

To illustrate this last point, on 22 May 1957, almost six months before Sputnik was launched, a *TWIN EARTHS* main character, Commander Cee from Terra, was talking about "picking up one of the satellites that the United States and Russia are sending up" into orbit in order to track your [Earth people's] progress.

So it was not just the predictions of hand held videophones, flying cars or other far off inventions that made *TWIN EARTHS* prophetic, it was also its predictions of the immediate future.

But perhaps the most important influence it wielded was that for ten years, during the 1950s it kept flying saucers in the newspapers every day. Any day of the week people all over the world could pick up a major newspaper and see a clear picture of a flying saucer—just as they believed it might look—and read about Alien secret agents, Communist spies, space stations and trips to other worlds—all told and shown in clear, easy to understand pictures and text. Because of *TWIN EARTHS*, flying saucers never left people's consciousness for long.

For this we should be grateful to two forward looking, talented individuals, Oskar Lebeck and especially Alden McWilliams.

The War of the Worlds

[Author's Note: I started this chapter in December 2003 and primary writing continued through May 2004. That is the period when the twin rovers Spirit *and* Opportunity *landed on Mars and performed their initial explorations (and who would believe they'd still be running four years later!). It's hard to convey the bizarre state of mind created when you immerse yourself in "Mars lore" and the literary history of the planet while simultaneously watching the spectacular images and other data streaming back daily. You really do develop a sense of kinship with those authors of years past, and almost feel guilty that you get to see for real the place they could only create in their imaginations – JH]*

Every career has a highpoint. While George Pal went on to produce and/or direct many delightful films such as *The Wonderful World of the Brothers Grimm, The Time Machine* and *The 7 Faces of Dr. Lao,* there are many who consider *The War of the Worlds* to be his magnum opus. This is all the more impressive when you consider it was only his fourth feature film after *The Great Rupert* (1950), *Destination Moon* (1950) and *When Worlds Collide* (1951).[1]

Like any great film, *The War of the Worlds* had a torturous path to existence. It was started, or at least

©Andrew Probert. Used with permission

seriously considered, by some of the greatest producers and directors in history over a period of almost thirty years before Pal created the classic we all know and love. There's even a touch of mystery as to how Pal came to make it at all.

The movie is based, of course, on the story by H.G. Wells about an invading army of Martians using sophisticated weaponry and mechanical transports. It was originally published, in serial form, in *Pearson's Magazine,* April through December 1897; illustrated in a wonderfully dark and claustrophobic manner by Warwick Goble (image left). The book version was published the following year. Wells intended it to be a satire

on Victorian-era British imperialism,[2] but wound up creating an entire genre. The book was, incidentally, Robert Goddard's favorite, one that he treated himself to a re-read every year at Christmas.[3]

The choice of Mars as the origin of the invaders was an easy one. In addition to being the planet of the god of war, by the late 1890's, the world was "Mars crazy." Percival Lowell had set up his observatory outside Flagstaff, Arizona specifically to study the planet, and was writing what seemed to be almost weekly articles for newspapers and magazines speculating about the conditions on Mars, its possible inhabitants and their civilization. It was all based on sound scientific reasoning using the best data available at the time, but it really took off after 1894 when Lowell discovered periodic

---

[1] For a Profile of Pal and his career, plus detailed chapters on his three big "space movies" (*Destination Moon, When Worlds Collide* and *Conquest of Space*), see *Spaceship Handbook.*

[2] Wells was inspired by his revulsion at Britain's genocidal Tasmania campaign of the 1820's and '30s. There, an entire native race was almost totally eliminated by an unstoppable army that was technologically vastly superior. His intention was to put the British people on the receiving end of such a scenario. Indeed, despite the grandiose title, the entire story takes place in the towns and villages south of London and eastward to the sea. The title could easily have been "War of the Martians vs. Southern England" since the action encompasses an area barely 50 miles (80 Km) across, and at the end they receive relief aid from France (which was apparently uninvolved). The famous 1938 radio broadcast did the same thing with New York and New Jersey. It wasn't until George Pal's movie that it was truly a "War of the Worlds."

[3] Lehman, Milton, *Robert H. Goddard: Pioneer of Space Research,* De Capo Press, 1988.

bright flashes in the equatorial regions of the planet.[4] You would think that Lowell would take these flashes (which were verified throughout the 20th Century, and were proved to be sunlight reflecting off of properly aligned ice crystals in clouds or on the surface) as proof that there was an advanced civilization at work on the red planet, but he surmised in his journal something fairly close to what they actually are. Nevertheless, Wells used these flashes in the opening of the story, making them the muzzle flashes of a gigantic gun firing the invading cylinders towards Earth (à la Jules Verne).

Over in America, readers of Hearst publications were treated to two versions of the story. The original serial was published in *Cosmopolitan* magazine roughly concurrent with *Pearson's* in England. For those not well heeled enough to afford Cosmo, Hearst's Boston Post not only printed the story in daily installments, but did something unthinkable today. They rewrote it.

Members of the editorial staff took each installment and changed the locales, moving them from southern England to the areas in and around Boston (e.g. Horsell Common, where the first landing took place was changed to the common between Lexington and Concord). It appeared, starting in the 9 January 1898 issue, with its new title, *Fighters From Mars – or The Terrible War of the Worlds as it Was Waged in or Near Boston in the Year 1900*, and was a tremendous success. It was in this version that an adolescent Robert Goddard first encountered the story.

The series was so popular that before it was even finished, Hearst commissioned a sequel. Not from Wells, but from Garrett P. Serviss, a well known writer of science (mostly astronomy) articles and fiction books. When the original story concluded in the 3 February issue, the editors announced that they had "acquired" a sequel that would start the following Sunday. Sure enough, Post readers opened their papers on 6 February to find *Edison's Conquest of Mars*. Yes, that's the same Thomas Alva Edison of light bulb and phonograph fame, who in this story is not only an inventor and captain of industry, but also a pretty mean spaceship pilot and commanding general. It ran in installments every day for five weeks, finishing on 13 March.

The premise is based on a rumination that Wells put into the epilogue of the original story. He was wondering if the Earth was really saved, or if it was just a reprieve while the Martians planned a second attack. In this story, the good citizens of Earth didn't wait around to see, they decided to make a preemptive strike against the Martians. Edison is commissioned to invent an electric spaceship and disrupter gun (which he does in only a month or so), then oversees the construction of an invasion fleet of hundreds of ships

and thousands of disrupter guns (which takes another six months). Edison is, of course, chosen to command the fleet, which he does with great panache.

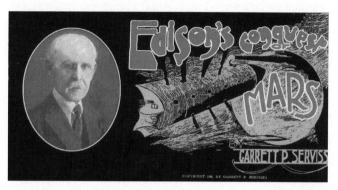

The story, which was recently reprinted by Apogee Books,[5] is an amazing example of early hardware/adventure Sci-Fi that would be the mainstay of the pulps of the 1920's and '30s. With spaceships, ray guns, alien abductions, outer space battles, etc., it was way ahead of its time. Still, for all of its prescience, it was a completely separate work, which today would probably qualify as "fan fiction" (as would, we suppose, the Post's rewriting of the original story).

The first producer to be interested in filming the story was the legendary Cecil B. DeMille at Paramount, who purchased the film rights in 1925. DeMille was famous for his huge, sweeping epics, and the story of an invading army from another world fit his style with the potential for "cast of thousands" battle scenes. Most film historians feel that had he made the film, he would have kept the late 19th Century setting, which, it must be noted, was only about 25 years in the past at the time.

Around that same time H.G. Wells' son, Frank, was attending Cambridge University where he met Ivor Montagu. Montagu was keen on getting into the movie making business, so in 1930, after both had dropped out of Cambridge, he formed a production company with the younger Wells. Montagu's uncle, Lionel, provided financial backing and Frank's father gave them the rights to three of his stories, including *The War of the Worlds*, either forgetting or ignoring the fact that he had already sold the rights to Paramount.[6]

Once in America, Montagu invited another friend of his, the brilliant Russian director, Sergei Eisenstein, to come to Hollywood. Among the properties they offered him as enticement, was *The War of the Worlds*. They soon discovered that the film rights to the novel belonged to Paramount, but fortunately, that studio was one of the few interested in Eisenstein. The Russian came to California and began to work on the film, but shortly after the script had been written, he abruptly withdrew from the project.[7]

---

[4] Thomas Dobbins and William Sheehan, *Sky and Telescope*, May 2001, pg. 115.

[5] ISBN 0-9738203-0-6

[6] *Cinefantastique*, V5, N4, spring 1977, pg. 6.

[7] Ibid.

The Saucer Fleet

Two years later, the story surfaced again. This time the producer was Robert Fellows, a protégé of DeMille. Since "CB" was no longer interested in doing the story himself, he encouraged Fellows to take it. Again a script was prepared and pre-production started, but Fellows accepted a job with Warner Brothers shortly after, and the project was once again shelved.

All of these false starts were unknown to H.G. Wells who was constantly being approached by directors in Europe to make film versions of his stories. In the 1930's Alfred Hitchcock, still early in his career, wanted to do a version of *The War of the Worlds*, but Wells felt the story too outdated by that time, and Hitchcock was unsuccessful in convincing him that it could be updated.

One producer who did get through to Wells was Alexander Korda, who, in 1934, offered to collaborate on a film project. Korda also wanted to do *The War of the Worlds*, but soon discovered that Wells no longer owned the rights. So instead, he bought the rights to Wells' latest novel, *The Shape of Things to Come*, and even hired him to write the screenplay for the film version, *Things to Come* (1936). Wells detested the future technocratic dystopia Fritz Lang had presented in *Metropolis* ten years earlier. He eagerly accepted the opportunity to provide his vision of a future technocratic utopia in order to "correct the errors of Fritz Lange (sic)."[8] Frank Wells also worked on the film as assistant art director. It was during production that the elder Wells finally met, for the only time, Cecil DeMille.

H.G. Wells (right) with Raymond Massey and Pearl Argyle on the set of *Things to Come*.

Near the end of the decade, another Welles, this one named Orson (no relation as you'll note by the spelling of the last name) gave the story it's most famous outing up to that point. In 1938, the book was 40 years old, nearly twice that of the young genius Welles, who was only 23. Earlier that year, Welles had brought his "Mercury Theater" drama company to the airwaves with *Mercury Theater on the Air*, broadcast on CBS. To do a Halloween "spook show," Welles bought the radio rights to the story and instructed staff writer Howard Koch to create a radio play version.

Koch's treatment of the story did what H.G. Wells had told Alfred Hitchcock could not be done; he brought it up to date. He incorporated all of the high points of the story, from the "bright flashes seen on Mars" in the opening, to the long dialog with the crazed artilleryman near the end. Koch moved the action from rural Victorian England to rural present-day New Jersey. To pick the exact location of the initial landing site, he used an old writers' trick:

> I spread out the map [of New Jersey], closed my eyes and put down the pencil point. It happened to fall on Grovers Mill. I liked the sound, it had an authentic ring. Also, it was near Princeton where I could logically bring in the observatory and the astronomer, Professor Pierson, who became a leading character in the drama.[9]

To thoroughly modernize the production, he did the first half (where the Martians are first detected, land and start their attack) as a series of radio bulletins and remote broadcasts. After that, it transitioned into direct monitoring of military radio communications with artillery crews and bomber pilots. Finally, in the last act, it dropped all pretense of being a radio broadcast and became a first-person story told by Professor Pierson,[10] played by Welles, in his struggles for survival. The production was clearly announced at the beginning, before and after the central commercial break, and at the end as a dramatization by Mercury Theater.

The now-legendary broadcast took place on 30 October 1938. Many listeners, especially those who tuned in late and missed the opening disclaimer, thought they were listening to an actual broadcast of Armageddon. So authentic sounding was Koch's dialog, and so skillfully staged was the performance, that thousands panicked. The switchboards at radio stations, newspapers and police departments across the country lit up with concerned citizens calling in with questions and wanting to know what to do next. Many packed a few belongings into the car and took to the road in terror, completely choking the highways around New Jersey and New York. By the time the broadcast ended, the studio control room was full of police waiting to take Welles and the other Mercury Theater principals into custody.[11] It was

---

[8] *The Science Fiction Film Reader*, Limelight Editions, 2004, pg. xvi.

[9] Howard Koch, *The Panic Broadcast,* Avon Books, 1971, pg. 13.

[10] "Pierson" could be a reference to the magazine (*Pearson's*) that first published the original story, but Koch makes no other reference to this in his book, so it is probably a coincidence.

[11] *The Panic Broadcast*, pg. 86.

for their own protection, not detention (they were not arrested or ever charged with anything).

In the days following the broadcast, fear turned to anger:

> The brunt of the public outcry fell on Orson. For days he was pursued by reporters and threatened by outraged citizens. Law suits for injuries and damages were filed against CBS and Mercury Theatre, running into millions of dollars, but none of these came to trial since there was no applicable precedent for such actions under those circumstances.[12]

But finally:

> As the shock was absorbed and the excitement died down, the tide of public opinion reversed itself. Mail continued to pour into radio stations, but the vast majority [was] favorable, even congratulatory.[13]

Welles was an overnight sensation, and there was a brief effort at Paramount to revive the *War of the Worlds* film project to capitalize on the publicity. This added one more script to the four previously written.[14] But, the notoriety quickly faded, and the studio, like all others in the country, was soon completely absorbed in the war effort as World War II broke out.

H.G. Wells died in 1946, and it would seem that all hope of producing a film version of the book died with him. However, a young Ray Harryhausen had been thinking about doing a film version of the story as far back as 1942 while he was in the Army. He thought the giant three-legged fighting machines and tentacled Martians would be the ideal subjects for his developing stop-motion expertise. By the time he was discharged from the service in 1944, he had a detailed outline written. He had been profoundly influenced by the Orson Welles radio version, so he also set his treatment in present-day New Jersey and New York, if only so he could destroy the Brooklyn Bridge, the Holland Tunnel and the Statue of Liberty in the process![15]

He put the project aside for a couple of years to work on his first big feature film, *Mighty Joe Young* with his mentor, Willis O'Brien (which won the Academy Award for best special effects for 1949). In 1950 he created some detailed concept drawings to go with the previous storyboards and even made about a minute of test footage.[16] The scene he chose to animate was the Martians unscrewing the hatch and emerging from the landing cylinder for the first time.[17] His efforts to get the film produced were frustrating:

> With my outline, drawings and 16mm test reel, I set off to see Jesse Lasky, Sr.[18] He seemed impressed and indicated that he could see possibilities in the project. He took all of my materials and, over the next six months or so, attempted to generate interest in the project. But none of the executives or the studios, including Paramount, seemed in any way interested. I even wrote to Orson Welles, whose name I hoped would stimulate an interest in producing the project, but I never received an answer.
>
> In October 1950, I visited Frank Capra in his office.[19] We talked about George Pal's film *Destination Moon*, which had just been released. I realized that he would be the best man to approach about selling my *War of the Worlds* to Paramount.[20] In mid-October, I visited him at the studio. We discussed the idea of Wells' story for several hours. Eventually he said that even though he had heard that Fox and RKO were working on the idea, I should leave everything with him and that he would present it to the front office.
>
> A few days later, I received a letter from Willis O'Brien saying that Pal was already negotiating to make the film with Paramount. It was apparently a fight between him and another producer, Robert Fellows, who had been planning the film for

[12] Ibid.

[13] Ibid.

[14] *Cinefantastique*, spring 1977, pg. 9.

[15] *Ray Harryhausen, An Animated Life*, Billboard Books 2004, pg. 45.

[16] Ibid., pg. 299.

[17] Three of the concept drawings, and about 20 seconds of the footage, can be seen in the Richard Schickel documentary *The Harryhausen Chronicles*, only there the animation is erroneously described as being the scene at the end where they succumb to bacteria.

[18] Lasky was one of the founders of Paramount, but also worked with the other studios.

[19] Harryhausen had served in Capra's film production unit during the war.

[20] Before the war Harryhausen had worked with George Pal at Paramount on many of Pal's *Puppetoon* shorts.

The Saucer Fleet

Both images ©Ray Harryhausen, used with permission

Preproduction art (above) and frame from test reel (below) of the Martian emerging from the cylinder.

two years. I was not happy that my material had been used by Pal, but I decided that if it got the project made, I would be happy.[21]

It's difficult at this late date to sort out the individual timelines of just what happened and when. What is clear is that when Harryhausen visited him, Pal was well into production of *When Worlds Collide*, (another property Paramount had owned since the '30s, originally licensed by C.B. DeMille) and was probably starting to look for his next project. That Fox and RKO were considering it could have been due to Lasky's efforts shopping the project around the studios earlier that year. O'Brien's letter saying that Pal was already fighting with Fellows over the production is confusing. Fellow's departure from Paramount (described on page 78) was in 1934, and research has not uncovered any mention of his returning and getting involved in the project again. Whether he was at Paramount or not is academic, since he left (wherever he was working) in 1951 to partner with John Wayne forming their "Batjac" production company.

Did Harryhausen's proposal influence Pal to make this film? We'll probably never know. It's possible that Pal was one of the producers that Lasky approached early in 1950, but if so, why didn't Pal mention it to Harryhausen when they met personally some months later?

The referenced *Cinefantastique* article doesn't mention Harryhausen at all. According to Steve Rubin, Pal just co-incidently came across all five scripts for *The War of the Worlds* in the Paramount story department while searching for a follow-on project to *When Worlds Collide*. Whatever the initial motivation, Pal took the scripts and hired noted screenwriter Barré Lyndon to review them.

Lyndon was the perfect choice for the job. An expatriate Englishman, he had been living and working in the US since 1938. He was a great admirer of Wells, and collecting Wells-related memorabilia was one of his hobbies. When Pal hired him, Lyndon had just completed working with C.B. DeMille on *The Greatest Show on Earth* (making yet another "CB" connection to this story).

Pal next hired Byron Haskin to direct and the three of them set out to modernize the story yet again. As Pal recalled in 1953:

> The *War of the Worlds* was no longer as ancient as Wells had once believed. With all the talk about flying saucers, it had become especially timely. And that is one of the reasons we updated the story to the present and placed it in California. The three of us decided that we should do as much as we could to make the audience feel that they're actually witnessing an attack and that the Martians are really here. The success of the film on this level was [because] Byron and I decided that we would never show the point-of-view of the Martians, despite the pleas of the front office, which kept demanding that we shoot something of how they see us.[22]

Writer Bob Stephens concurs with a remembrance:

> The movie changes the setting from Victorian England to Los Angeles in the middle of the 20th Century, so the cinematic battle was more unnerving to Southern Californians like myself because it was fought in our own neighborhood. I grew up in Oildale, a little place outside of Bakersfield that was just a three-hour drive from Los Angeles. The new E.C. comics *Weird Science* and *Weird Fantasy* came out with stories on the flying saucers' reconnaissance of our civilization, perilous dog-fights with the Air Force, and dramatic intrusions into the lives of ordinary citizens, especially in rural settings. By the time *The War of the Worlds* made it to the River Theater in my home town, I was ready.[23]

Pal and his director also made a storytelling decision that was transparent to almost everyone when watching the film, but was probably absorbed at a subliminal level:

> To add realism, ease the logistics and simplify the effects, we had Los Angeles always in the west and the Martians always in the east. All of the movements between the Army and the invaders were east to west. This made a complicated story easier to understand visually.[24]

---

[22] *Cinefantastique*, spring 1977, pg. 9, quoting an earlier article by Pal during the movie's original release.

[23] *Outré* #22, pp. 49-51.

[24] *Cinefantastique*, spring 1977, pg. 9

---

[21] *Ray Harryhausen, An Animated Life*, pg. 47.

Since the movie would open in the semi-desert region east of Los Angeles (which is where most of the flying saucer "activity" of the day was concentrated as well), Barré Lyndon drove out to scout locations. About an hour out of the city, he came to the farming town of Linda Rosa. He felt that this small rural town (population 2,000) in the Chino Hills would make the perfect California equivalent of Grovers Mill, New Jersey or Horsell Common, England.[25]

Barré Lyndon

After finding the setting for the story, he wrote his first draft of the script and turned it in to Pal at the beginning of June 1951. While recognizable to fans of the film, it is considerably different from what was eventually seen in theaters. The opening is still the same with its rapid-fire review of mechanized warfare in the first half of the 20th Century, followed by a narrative over the montage of Chesley Bonestell paintings explaining the Martian's rationale for attacking Earth. The narration was eventually performed by Sir Cedric Hardwicke, whom Lyndon had worked with previously as a narrator on the British film *Sundown* (1941), but the screenplay simply describes it as being "delivered in a Wells-like accent."

After that, though, the original screenplay diverges sharply from the eventual film. The opening action sequence introduces us to quite a different main character. It is Air Patrol pilot, Greg Bradley, who spots the first cylinder fall while he is on duty at 15,000 ft. Bradley is a veteran WWII fighter pilot, which allowed Lyndon to sprinkle the script with heroic aerial combat sequences when confronting the Martians. Likewise, his original female lead was named "Sylvia Ashton," an aristocratic socialite from a wealthy east coast family. He describes their first meeting thus:

Greg eyes her. She isn't his kind of dame. She's an orchid out of a hot house. Spoiled. Snooty. She looks at him. She's seen his type around her father's docks. Crude. Rude.[26]

Pal didn't care for this characterization at all. He wanted the science and strategic thinking of the humans to be the focus of the film, not the battle scenes. Originally, he wanted to use Wells' unnamed protagonist, who was a happily married man that is separated from his wife early in the invasion, and worries about her the entire rest of story. Unfortunately, there were complications from the front office:

I was fresh at Paramount and they forced on me this ridiculous story of a scientist going on a fishing trip, meeting a girl and then, after only one evening, he begins this tremendous search for her, from church to church.[27]

The pressure was coming from Don Hartman, vice president of production. Hartman was the creator of the Hope/Crosby "road pictures" and a firm believer in the trite boy-meets-girl formula as a box office draw. Therefore, not only could the male lead not be married, but they had to create a believable romantic chemistry between him and the female lead during a planetary invasion.

Taking these orders directly from the top, and to address Pal's objections about the characterizations, Lyndon reworked the male lead into the character of Dr. Clayton Forrester, a nuclear physicist from Pacific Tech.[28] The spoiled socialite, Sylvia Ashton, was changed to Sylvia Van Buren, a small town girl who is not quite as innocent as she first appears. She has attended college, earned a masters degree in Library Science, and now teaches it at the University of Southern California (it must be summer break if she's back in Linda Rosa). The fishing trip was left in, but to make Forrester's agonizing search for Sylvia at the end more believable, Lyndon has the two of them endure several harrowing days together on the run, one step in front of the Martians as they work their way towards Los Angeles. This echoes the book's cross-country trek of the main character and the Curate.

With his hero now a scientist rather than a military man (which also brings it more in line with the radio play where the main character, Professor Pierson, is an astronomer from Princeton), Lyndon began refocusing the story on the scientific aspects of the fight rather than sweeping battle scenes. In any case, even if Pal had wanted to make a full-blown confrontation in the DeMille style, the budget just wasn't there. The only character retained from the book, and then only in one brief scene, is the astronomer, Professor Ogilvy, who is shown giving a radio interview speculating on the nature of the Martians at the beginning of the invasion. Before the scene was filmed, though, the character was

[25] In the *Cinefantastique* article, Steve Rubin relates, in considerable detail, the story of Lyndon's scouting trip and his "discovery" of Linda Rosa. Unfortunately, searches of California maps, both modern and period, web searches, help from reference librarians and a professional California historian all have failed to find any town of that name in the Chino Hills, or anywhere else in California, for that matter. It was most likely a pseudonym for Yorba Linda, a farming town in the same area, which has been around since the Spanish era 200 years ago. Lyndon probably changed the name in the same way that Jack Finney changed Santa Maria, California to the fictional "Santa Mira" for his story *Sleep No More* (which was later made into the film *Invasion of the Body Snatchers*) and Rubin never picked up on the ruse.

[26] *Cinefantastique*, spring 1977, pg. 10.

[27] Ibid.

[28] A pseudonym for Cal Tech (California Institute of Technology in Pasadena). This use of a screen name for a well-known institution reinforces the probability that Linda Rosa is also not real.

changed to a Canadian meteorologist and the name was changed to Professor McPherson in an apparent homage to the Orson Welles character.

With the script now on the right track, Pal turned his attention to creating the "look" of the film, i.e. how all of these fantastic concepts will appear on the screen. This is the job of the art director, and for this critical task Pal picked Albert (Al) Nozaki with whom he had just finished working on *When Worlds Collide*. To supervise the special effects, Pal turned once again to effects wizard Gordon Jennings, who had also worked on *When Worlds*

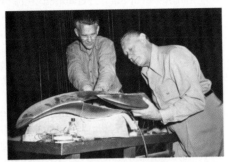

Gordon Jennings (right) and unknown technician work on one of the (dusty) war machines

Collide, creating the Oscar-winning visuals. Sadly, as it turned out, *The War of the Worlds* was Jennings' last film as he died of a heart attack just after its release. His work on the film earned his third Academy Award for special effects, his second in a row, which was awarded posthumously.

Lyndon finished his story update by the end of June. While feeling it still needed a lot of detail work, Pal was generally pleased with the way it turned out and sent a copy to Don Hartman for final approval. He went to Hartman's office a few days later, expecting him to just green-light the project. Instead, he was blindsided by a totally unexpected reception. "I almost lost the project entirely," he said in 1977. "Hartman was a former writer, but he had no appreciation whatsoever for science fiction." Steve Rubin relates what happened at the meeting:

> "George," Hartman began, "this script is a piece of crap and this is where it belongs." Hartman then threw the script into a nearby wastebasket.
>
> Pal stormed around Hartman's desk and proceeded to call him everything he could think of.[29] While Hartman suffered through a tirade of expletives, Frank Freeman, president of Paramount, walked into the room. He first looked around for a movie crew thinking he may have accidentally stumbled onto some sound stage, but there being none, he demanded an explanation. Both Pal and Hartman told their sides of the story and

when they were finished, Freeman turned to Pal and said, "Go ahead and make your film. Do anything you want."[30]

Unbeknownst to Pal, Freeman had earlier spoken with C.B. DeMille, who had put in a good word for the new (to Paramount) producer, and had complete confidence in Pal's ability to make a film from what was originally his (DeMille's) property. Freeman gave Pal an initial budget of $1 million,[31] a very large amount for a science fiction film in the early '50s, which normally had budgets well under half a million dollars. Unlike *Forbidden Planet*, which started off as a "B" picture, but moved incrementally into "A" movie territory through budget overruns (see page 157), *War of the Words* started out as a "big" picture and was given first class treatment from the start.[32] The budget eventually inflated even further to $2 million during production, of which $1.3 million went to the special effects.[33] This made it, in fact, the most expensive Sci-Fi film produced to date.[34]

With the harrowing close call with Hartman behind him, Pal went to La Paz, Mexico, for a one-week working vacation where he could create the production plan without interruption. It was then that he began to realize the enormous task he'd taken on in visualizing Wells' Martian machines. His fears were confirmed when he returned and met with Nozaki, Jennings and director Haskin.

They originally wanted to use Wells' concept of a three-legged stilted walker crashing over the countryside.[35] From Nozaki's early sketches, it was clear that this would be impossible to film convincingly. As Nozaki explains:

> When you draw illustrations in a book you can make the Martians do anything. And in our early production drawings,

---

[29] This wasn't the first time Pal had crossed swords with Hartman. Earlier in the year, before he'd chosen *War of the Worlds* as his next project, Pal was trying to get approval to do *20,000 Leagues Under the Sea*, but Hartman wanted the story to be updated to the current day (like *War of the Worlds* eventually was). While they were bickering, Harper Goff over at Disney was starting his own treatment of the Verne novel for an animated feature, which eventually was made as the famous live action feature.

[30] *Cinefantastique*, spring 1977, pg. 16. This quote has been edited down quite a bit from what appeared in the magazine. In the original, Rubin has Pal grabbing Hartman by the lapels and cursing in Hungarian. This has been largely discredited by later film historians, and even the heated discussion depicted here is considered apocryphal. What is certain is that Hartman did not like Pal's original script at all and it took intervention by studio head Freeman to resolve the issue.

[31] Ibid., pg. 34.

[32] *Starlog*, #195, pg. 68.

[33] *Cinebooks Motion Picture Guide*, News America Publishing, Inc.; as presented in Microsoft *Cinemania '95*.

[34] It only held that distinction for about a year until Disney's *20,000 Leagues Under the Sea* was released. With cost overruns (mostly due to the giant squid sequence) that landmark film cost $4.2 million, making it the most expensive film ever made up to then in any genre.

[35] In the book, Wells explains that the Martians had never developed the wheel, so that their mechanical transports mimicked their own physiology with multi-jointed tentacles and three awkward legs for walking. Note that George Lucas freely adapted this concept for his "ATAT" (four legged) and "ATST" (two legged) walking war machines in *The Empire Strikes Back* and *Return of the Jedi*.

you saw the machines walking over rough terrain or crashing through buildings. However, as Gordon [Jennings] pointed out, if you tried to do things realistically, in miniature as we planned, you were going to run into definite problems. We wanted the machines to be highly mobile, but with stilted marching machines it was impossible to be smooth while crashing across gullies and wrecked streets.[36]

Byron Haskin concurs:

> Although we were afraid to desert the entire Wells concept, we eventually decided that anything he may have written about water tanks and towers walking slowly across meadows in rural England was now ridiculous in a film sense.[37]

Even though everyone agreed that a new direction was needed, no one could come up with any better ideas to guide Nozaki in designing an updated Martian war machine, so he went home and began "sketching and sketching to keep the ideas flowing."[38] Some of his intermediate concepts retained the three-legged stance, but did away with the articulated legs by making them supporting beams or jets which caused destruction in their own right while transporting the machine. The body, too, was updated to bring it in line with the flying saucer craze currently going on. It was flat and circular, but still not terribly different from the ovoid bodies of the original in the book, only with glowing sections added around the rim and underneath. There was a large dome-

shaped window on one side that gave it a sense of direction, and suggested the armored "hood" in Wells' design where the Martians operated the machine (although the hood was located on the top in the book).

It should also be noted that in the book, the machines don't emerge fully functional out of the landing cylinders, but rather take a day or more to assemble using "handling machines." This represented the most advanced building techniques of the late Victorian era.[39] This idea was abandoned as well at this stage, although they did leave in the fact that it took several hours to extract the machines from the cylinders.[40] We weren't shown what was going on, other than weird sounds and a green glow surrounding the "nest" as they dig themselves out.

Nozaki with original war machine model. Note that the Heat Ray arm emerges from the tail, like the stinger on a Manta Ray.

After weeks of fruitless sketching, it was early in July 1951 when Nozaki had the inspiration for the version of the war machine we ultimately saw. Said Nozaki in a 1977 interview:

> It came as an inspirational flash on a Sunday afternoon. It came from something like the Manta Ray and originally that cobra like control arm was coming out of the rear of the machine, like the tail of the Manta Ray. It was one of those ideas that instantly you know is right.[41]

Nozaki took his rough sketch of the machine into the studio the next day and had the special effects department

Nozaki's 3-legged saucer concept

[36] *Cinefantastique*, spring 1977, pg. 16.

[37] Ibid.

[38] Ibid.

[39] Using machines to build machines was quite novel. Steam shovels and large powered cranes were just starting to become the dominant construction tools for building the first skyscrapers and large earthworks, like dams and canals.

[40] The cylinders in the film aren't cylindrical at all, but shaped more like a loaf of sourdough bread. They are elliptical in both vertical cross sections, and basically round with a tapered tail when seen from the top (see page 107). The best view of it in the movie is the establishing shot for the scene where the three guards approach it for "one last check."

[41] *Cinefantastique*, spring 1977, pg. 16.

create a small model to show Pal, Haskin and Jennings. It was an instant hit with all of them. Haskin recognized that it would "read" well on camera and Jennings knew that it would be far simpler to manipulate in the effects sequences than something with articulated legs. Pal, especially, loved it's multiple, subliminal messages. It combined the familiar, yet threatening organic shapes of the Manta Ray and the King Cobra, while simultaneously looking convincingly alien. He also hoped that the shape would convey the "flying saucer" image.[42]

©CBS Paramount Network Television
Our first view of the war machine shows the copper skin and green nose dome. Barely visible are the arcing "legs" which are very difficult to capture in a single frame since the FX animators did a skilful job of strobing them between frames to look more random. Andy Probert's painting on page 76 gives a better impression of the way the rays are seen on screen.

Full scale clay sculpture from which the props were made.

Jennings took Nozaki's model and brought it to Ivyl Burks in the prop department. Burks had one of his sculptors make a 42-inch (107 cm) wide clay version from which they pulled a mold to build three effects miniatures. The final miniatures were made of copper (its reddish hue suggesting Mars), and were all equipped slightly differently. While all three had cobra-headed Heat Ray projectors that could swivel left and right, only the "master" unit could actually dip up and down (such as when it incinerates the three guards and Pastor Collins). Likewise, only one of the others had the hole in the bottom where the TV camera-like scanner emerged on its tether. The only significant change made during the whole fabrication phase was to move the Heat Ray projector from the back of the unit to just forward of center. This was done partially for aesthetics, and partially to take advantage of the extra room in the central body for the wires, motors and other mechanisms that operated the arm.

The large green forward nose section is the mutated "control dome" from Nozaki's earlier saucer-like concept, thus double removed from the "hood" of the original. There are matching green "Skeleton Beam" projectors on the tips of each

"wing," and the three-legged stance is now only suggested by the three magnetic support beam generators on the bottom (also green[43]). They experimented for a time with using high voltage arc discharges of one million volts to create visible energy beams for the legs. Jennings tested prototypes that worked fine, but the concept was abandoned for safety reasons. Not only could it have electrocuted someone, they were concerned that it could set fire to the fine dust and other combustibles on the set. All that remains of the concept is some ghostly, animated energy discharges seen in the scene where the machines first rise up out of the nest and advance on Pastor Collins.

With the all-important props safely under way, Pal next turned his attention to casting. Generally, he preferred to use unknown actors[44] since they made the movies seem more

©CBS Paramount Network Television
Ann Robinson and Gene Barry

---

[42] Ibid., pg. 34.

[43] The green color for all of those parts seems a bit too cheery for such a menacing machine, but it was Wells' own choice for the color associated with their technology. The meteors have a green glow and trail a green mist as they descend, and the cylinders emit green smoke as the Martians build the machines. Then there's the description of "scintillating green flashes" emanating from the Heat Ray projector in operation. Ironically, the Heat Ray was the only part of the updated war machine to not have any green associated with its operation.

[44] *Starlog*, #195, October 1993, pg. 68.

real (just like Julian Blaustein didn't want to use Spencer Tracy in the role of Klaatu for *The Day the Earth Stood Still*). For the role of nuclear physicist, Clayton Forrester, Pal picked New York stage actor Gene Barry whom Paramount had recently signed to his first movie contract. He had only acted in one film previously, ironically playing a nuclear physicist in *The Atomic City*.[45] To play Sylvia Van Buren, Pal had Barry read with several young actresses from Paramount's "Golden Circle" of young starlets.[46] Both Pal and director Haskin picked Ann Robinson, a 24-year-old actress who had already performed in *A Place in the Sun* and the western *Cimarron Kid*.

Les Tremayne

A major plus for the film was casting Les Tremayne in the pivotal role of General Mann. After two decades on radio, and half a decade on TV, this was Tremayne's first film and he did a phenomenal job. As Pal recalls:

Les Tremayne gave a speech about the Martian strategy which must have read very boring, that scene had to be acted out to convey interest. It was difficult to convince executives like Don Hartman that this material would make an involving film. But once you had Les Tremayne on film, in uniform, sweeping a piece of colored chalk across a blackboard, we convinced everyone.[47]

Tremayne based his performance of General Mann on Dwight Eisenhower[48] (Eisenhower as the WWII allied commander, not president. He had not yet been elected when this movie was filmed).

Paul Frees

To give the film an authoritarian voice, if only in a couple brief scenes, Pal cast Paul Frees. Already a well-known voice actor, Frees did the rapid narration over the pre-credit newsreel footage of the Earthmen's war machines, and later, in one of his very rare on-screen appearances, he plays the part of the radio reporter speaking into a portable tape recorder just before the A-Bomb is dropped.

"I'm George's good luck charm," he said in 1977, "and he has had me in every one of his films."[49]

As mentioned earlier, Sir Cedric Hardwicke was hired to do the film's narration at three points: the opening planetary tour using modified Chesley Bonestell paintings,[50] the "rout of humanity" sequence in the middle, and the film's final dénouement explaining why the Martians were dying.

Probably the best remembered character in the film, certainly the one with the greatest impact, was only on screen for a few seconds. That was, of course, the Martian that comes into the wrecked farmhouse and puts his sucker-tipped fingers onto Sylvia's shoulder. To create this horrific creature, Nozaki turned to Charlie Gemora in the prop department. Gemora was an expert sculptor and painter, but also designed special makeup appliances. He took Nozaki's sketches and built a latex rubber suit that incorporated many special features such as air bladder "veins" that could pulsate and wire-operated fingers that could open and close.

Charlie Gemora

As designed by Nozaki, the Martian was a spindly creature over six feet (two meters) tall, so that's what Gemora built. But when Pal and Haskin stopped by the day before the shoot, they had a completely different idea. In the planned camera setups, the Martian was only supposed to about four feet (120 cm) tall, so Gemora and his daughter, Diana, (who was only 12) worked through the night building a new, shorter body and head from scratch. There was no time to re-build the elaborate arms, though, which is why this Martian has arms nearly twice as long as its body.[51]

The latex was still wet when it was time for the cameras to roll, but now they had a different problem. The new Martian costume was too small for the stagehand who was supposed to operate it. Since Gemora was quite short, Pal asked him if he'd like to work it, an offer Charlie eagerly accepted. Of course, Gemora wasn't **that** short, so he knelt on a dolly while stagehands pushed him in and out of the shots.[52] For the close-up,

[45] *Starlog*, #315, October 2003, pg. 48.

[46] *Starlog*, #195, pg. 66.

[47] *Cinefantastique*, spring 1977, pg. 34.

[48] Bob Dorian, *American Movie Classics*, 1999.

[49] *Cinefantastique*, spring 1977, pg. 34.

[50] Incidentally, all of the paintings used in this sequence were originally published in Bonestell's 1949 book *Conquest of Space* (see *Spaceship Handbook*). He executed new versions of the paintings for use by the Paramount Special Effects department to add animation (e.g. blowing snow on Mars, the moving clouds of Saturn and Jupiter's lava falls). Most of Cedric Hardwick's narration came from the captions of those paintings in the book.

[51] Documentary, *The Sky is Falling: Making the War of the Worlds*, 2005.

[52] *Starlog*, #195, pg. 70.

his daughter was lying beneath camera range operating the "veins" with squeeze bulbs.[53]

With so much money being spent on the special effects, there was no budget for expensive foreign location shoots. To do the globe sweeping "rout of humanity" sequence, Pal and Haskin scoured Paramount's newsreel department for stock footage they could use, a technique used to great effect by Ray Harryhausen two years later in *Earth vs. the Flying Saucers* (see page 199). This had the added bonus of turning up an extended piece on the Northrop B-49 "Flying Wing" bomber (from Paramount's "World of Tomorrow" series), footage from which was also used in the film.

There were a few location shots needed, though, which were all done before the principal photography started. The first one turned out to be mandatory. To show the military surrounding the initial Martian nest, Haskin contacted the California National Guard (CNG). The Guard often supplied men and materiel to movie studios as "training exercises" for the troops. They had, in fact, just been used earlier that year to surround Klaatu's ship in *The Day the Earth Stood Still* over at Fox (see page 16) standing in for the regular Army, who didn't want to be part of what they perceived as an anti-war film. This time, though, it was the CNG who didn't want to be depicted as being completely helpless and ineffectual against the Martians, so Haskin went across the border to Arizona where their National Guard had no problem with depicting the USMC in their hopeless struggle with an otherworldly foe. So in early December 1951 Pal, Haskin and a small crew spent 10 days outside of Florence (southeast of Phoenix) shooting military operations, hoping that the Sonora Desert looked enough like the Mojave Desert that viewers wouldn't notice the difference.

After returning from Arizona, they spent a couple of days up in the Simi Valley hills photographing the exodus of people out of Los Angeles, and the crowd watching the A-bomb blast.

The final location shot was done, in comparison, right near the studio on Sunday, 30 December. To get the images of Dr. Forrester running though the deserted streets of Los Angeles looking for Sylvia, Haskin worked with the city's film board to have the police cordon off several blocks of Hill Street downtown between midnight and 7 AM. They photographed Gene Barry driving the Pacific Tech truck and running though the streets, in shots ranging from close-ups to extreme long distance taken from the tops of the buildings.[54]

As 1952 opened, the crew started preparing Stage 18 on the Paramount lot for principal photography. Despite the comparatively large budget, the movie only had one dedicated stage and part of another. Al Nozaki had designed an exterior set for Stage 18 that could be redressed for several different scenes. It was the gully where the first "meteor" landed, the Marine command post where Forrester and Colonel Heffner watch the Martians advance for the first time, and the field in which Forrester crashes his plane. It was later, after considerable rework, used for several of the large-scale miniature scenes (no, that's not an oxymoron) such as the war machines hovering over the bean field (just barely glimpsed as Forrester's plane crashes), the farmhouse exterior, and the set of downtown LA that the Martians lay waste to.

Gordon Jennings (right) supervises his crew in setting up the "bean field" shot which wound up on the screen for only three seconds .

The interior of the farmhouse was built on a different stage since it had to be destroyed in the course of shooting. To save money, they built it out of regular lath and plaster rather than making a sophisticated "collapsing" set. This meant that they only had one chance to get the shots of the house falling down all around Gene Barry and Ann Robinson as yet another Martian cylinder plows into it. The interior of the dance hall was built on the same stage while its exterior was part of the standing "New York street" backlot. One end of the street was redressed to make the "A-bomb bunker" from which the military and scientists watch the blast. The "Pacific Tech" exterior was simply the Paramount studio gate off of Van Ness Avenue with a prop sign over the entrance.

Principal photography started on 14 January 1952. It was not an auspicious start as Gene Barry's house in Benedict Canyon had been involved in a huge mudslide over the weekend. Since the roads were washed out, the studio sent one of the tracked military vehicles on hand up to get him.[55] Things didn't improve any after that. Only two days later they had yet another setback. As Steve Rubin relates:

> Haskin was in his third day of shooting when the head of the legal department came running onto the stage. "You have to

---

[53] Documentary, *The Sky is Falling.*

[54] *Cinefantastique*, spring 1977, pg. 36.

[55] *Cinefantastique*, spring 1977, pg. 38.

stop shooting—right now!" Pal, who was nearby, came rushing up. What's the matter?" he asked. "There's been some sort of oversight, Mr. Pal. Paramount doesn't own the **talkie** rights to *The War of the Worlds*."

"What?!" shouted Pal. "That's right," the nervous lawyer repeated, "DeMille only purchased the **silent** rights in the '20s. We have no legal right to do a [sound] picture based on the Wells novel."[56]

Pal told Haskin to keep shooting while they contacted the Wells estate. Frank Wells was very cordial, and enthusiastic that someone was finally making a film that he himself was once involved with. He sold the rights to Paramount for a very reasonable $7,000. Pal got more than just the movie rights, though. This introduction to Frank Wells eventually lead to Pal making *The Time Machine* some nine years later.

Principal photography wrapped the third week of February. As soon as the live actors were out of the way, the special effects crews took over. They rebuilt Stage 18 to create the miniature sets described above. They did run into one major problem with photographing these sets. The film was shot using the 3-strip Technicolor® process and the cameras were massive. Even sitting on the floor, the lens height was nearly three feet (1 meter) off the ground. This meant that every shot on the miniature set would look like it was from the perspective of a giant. To get the lens height down, effects cameraman Wallace Kelley devised an inverted periscope of two first-surface mirrors that got the point of view right down on the ground so that we would be looking up at the war machines, making them properly menacing. Also, all of the effects scenes were shot at 96 frames per second, four times the normal rate. This "overcranking" slows down the action and smoothes out the somewhat jerky real-time motions of the models.

One thing the film had in abundance was explosions. In the initial confrontation with the Marines, over 200 explosive charges were used. These were supervised by veteran "powder

man" Walter Hoffman, who was 81 at the time (!). To simulate the electromagnetic force fields that the Martians use to resist the attack, the prop department made up three large plastic domes to fit over the war machines. Unfortunately, it was impossible to operate the wire-suspended miniatures through the plastic, so the domes were photographed empty against the explosions and the machines optically matted in later.

Preparing to blow up city hall the first time. Note the bright sky backdrop to the left. The effect wasn't dramatic enough, so they rebuilt the miniature and blew it up again against a darker, smoke-stained sky.

There were hundreds of more explosions used as the machines make their way across Los Angeles, most famously the destruction of the LA City Hall, a clip used endlessly in other productions (including *Earth vs. the Flying Saucers*). The biggest explosion of all, of course, was the attempt to destroy the Martians with an atomic bomb, "the latest thing in nuclear fission."[57] For this Jennings consulted

Filming the "downtown destruction" scene. Note the large camera and lighting equipment in the foreground.

---

[56] Ibid.

[57] The atomic bomb is analogous to the book's battle between two war machines and the cruiser HMS *Thunder Child*. The *Thunder Child* was an ironclad ram, and represented the most advanced weapon on the planet at the time. It was all-metal, steam powered with turret-mounted guns and a torpedo launcher in the bow. It proved more effective than the A-bomb did against the movie Martians, taking out the two of them by the simple expedient of ramming them in a most heroic fashion, even after being heavily damaged by the Heat Ray. It was finally destroyed by the Heat Ray of a third war machine that joined the fray. See the *Modelers' Note* later for photos of a wonderful diorama of this scene.

several pyrotechnic experts who came up with a custom mixture of colored flash and smoke powders that would create a massive smoke cloud. The explosives were set off inside Stage 7 against a neutral background, and they boiled themselves up to the 75-foot (23 meter) high ceiling in a perfect mushroom cloud. This footage was later matted into the scene of the Martian machines at the base of the San Gabriel Mountains.

©CBS Paramount Network Television

Comparison of the flash powder "A-bomb" on Stage 7 (right) and as matted into the scene.

Once the live effects photography was finished in August, the film moved into its final phase, postproduction, where the optical and sound effects are added, and the music is composed to match the scenes.

The optical effects team was headed by Paul Lerpae, and they had many new optical techniques at their disposal. Since the film was shot in 3-strip Technicolor® it was easy to use the newly-developed "blue screen" process to perfectly matte the miniature war machines into a live action shot. They had new optical printers with micrometer-controlled film gates that could position images within .0001 of an inch (2.5 microns) in the frame (this was especially handy when matting the war machines into their magnetic force-field shields in the opening attack sequence). Finally, they used good old hand-painted animation to add many lighting effects. They painted between three and four thousand frames[58] to create the red flashes surrounding the Heat Ray (the sparkly Heat Ray itself was created by burning welding wire in a high-speed air jet to blow the sparks vigorously in a narrow stream), and the green Skeleton Beam energy bolts. The most famous of these is when Col. Heffner is hit by a bolt, and you see his skeleton (hence the name) briefly as he disintegrates (see photo sequence on page 105). It took 144 separate mattes to complete that effect.

The sound effects were equally creative. Department head Tommy Middleton had his lead technicians Gene Garvin and Howard Beal create some groundbreaking audio effects that set the standard for decades to come, some of

which are still in use today. Primarily they made excellent use of audio oscillators in designing the alien sounds. While there was nothing new in the technology (the use of oscillators in film was pioneered by Disney Studios for *Fantasia* some 14 years earlier), the way they combined them was almost overwhelmingly effective. Many of the effects were created by wiring the oscillators through custom circuits, in much the same way that Louis and Bebe Barron would use three years later for the sound effects and "music" in *Forbidden Planet*.

The first oscillator effect we hear is the soft purring of the Heat Ray projector as it emerges from the cylinder and begins scanning for its first victim. It's soon joined by the bright, sparkly tambourine-like sound of the projector pulsating in standby mode. There's the almost majestic, whirring of the war machines as they rise up out of the "nest" and advance on the Marine's field HQ. Once the firefight begins, the dominant effect is the soft, rapid-fire staccato bursts of the Skeleton Beam emitters. The attack and decay[59] on all of these sounds were tuned to give them a forceful, organic feel, perfectly in keeping with the organic shapes of the war machines. Other oscillator effects include the purposeful warbling sound of the electronic eye, the teeth-on-edge "skritching" of the machines as they march through Los Angeles and the electric death-whine as they start to fall. Of all of them, though, it's the Skeleton Beam sound that proved the most popular with film producers over the years, being used endlessly in movies and TV, right up to the present day (it was even used in the "Spongebob Squarepants" movie in 2004).

Perhaps the most effective sound effect was not generated with an electronic oscillator. The completely indescribable alien sound of the Heat Ray in operation was created on, of all things, an electric guitar. Many tracks of overlaid sounds created by scraping and strumming the strings (some played backwards) were mixed with an old soundman's trick of scraping a contact microphone over a block of dry ice. The high-pitched squeal of the dry ice sublimating under the mic added the final, unearthly touch. One other guitar-produced effect was fairly straightforward. The "thum-thum-thum" of the Heat Ray building up to a discharge was simply the E-string on an electric bass guitar being plucked. The groundbreaking homage film *Sky Captain and the World of Tomorrow* (2004) used the Heat Ray sound effect as, well, the heat rays of the robots attacking New York City.

By the time Middleton and his crew were done, electronic oscillators said "science fiction sound" as solidly as the Theremin said "science fiction music." However, for the music in this film, Pal wisely eschewed that cliché, and turned to his long time collaborator, Leith Stevens.

Stevens was a classically trained pianist who moved into film score composing after the war. After spending

---

[58] *Cinefantastique*, spring 1977, pg. 42.

[59] Attack and decay are sound engineers' terms for how fast a sound rises and then falls off.

some time at RKO and Universal, he moved to Paramount where he met Pal.[60] Before *The War of the Worlds*, he had composed the score to every Pal feature, including *The Great Rupert*, and would go on to do several more. For *Destination Moon*, Pal's first Sci-Fi film, Stevens showed his creativity in producing otherworldly effects from an orchestra, most notably the "outer space sounds" (something of an oxymoron) as the crew opens the hatch for the first space walk.

In *The War of the Worlds*, he creates a driving opening theme under the titles that is militaristic, but not overtly so. It is intended to depict an army on the march, but a desperate one embarking on a mission to save its existence, at the expense of ours. The title theme builds to a crescendo and then fades into the opening scene of the planetary montage.

Leith Stevens

Here we are introduced to the "Martian theme," a quiet, soulful melody underscoring their desperation to find a new home before their race becomes extinct.[61] Stevens was equally effective in depicting the plight of humans in the "Flight from the Cities" played as the people evacuate Los Angeles, and in producing tension as the first cylinder "unscrews" itself and the Heat Ray projector appears. In the final scene we hear a reprise of the Martian theme as Forrester cautiously approaches the crashed war machine. A failing heartbeat on tympani is superimposed over the theme as the spindly Martian arm inches its way towards the opening and dies. It is a bittersweet moment as we are told, musically, that even though the Earth has been spared, the entire Martian race, and all of its knowledge, is now extinct.

When a rough cut of the film was ready, Pal took it to a children's matinee at the Paradise Theater in Westchester, California to gauge audience reaction. Apparently the visual and sound effects people had done their job a little too well. About half of the parents found the advancing Martians so terrifying that they left the theater with their children before the film was over![62] It would seem that 15 years after Orson Welles created a panic on radio, the story, properly told, still had the power to overwhelm.

[60] *Outré* #22, pg. 51.

[61] Ibid, pg. 80.

[62] *Cinefantastique*, spring 1977, pg. 46.

Pal premiered the film in February 1953, but Paramount didn't put it into general release until the autumn. The film was a critical and commercial success, making back several times its huge budget. It was nominated for three Academy Awards (Film Editing, Sound Editing and Special Visual Effects) and won for Special Effects. As mentioned earlier, the award had to be presented posthumously as Gordon Jennings died of a heart attack shortly after the film's premier.

Byron Haskin worked on dozens of films in his 45-year career as cinematographer and director, but only a handful were science fiction. After *The War of the Worlds*, he directed *Conquest of Space* (1955) for Pal and later the ill-conceived *From the Earth to the Moon* (1958). The most notable thing about that film, probably the most awkward retelling of the Verne story, is its re-use of the sound effects from *Forbidden Planet* in the sound track. He returned to familiar ground in the much-better *Robinson Crusoe on Mars* (1964), which used a variation of Nozaki's war machines, this time showing them both on their "home turf" and much more energetic. He worked with George Pal again on his very last film (which was Pal's penultimate film), *The Power* (1968). Genre fans might also recognize him as the director of several episodes of *The Outer Limits*.

Al Nozaki moved directly from *The War of the Worlds* to work with C.B. DeMille on *The Ten Commandments*, a project that took over two years. Leith Stevens had a very eclectic career writing music for film and television. In addition to Sci-Fi, he wrote the music for *The Wild One* with Marlon Brando, and *The James Dean Story*. On the TV side, genre fans are usually surprised to find he wrote the incidental music to *Lost in Space*. Tragically, he died of a sudden heart attack in 1970 after learning that his wife had been killed in a traffic accident.

Gene Barry made a few more films, but discovered that his true medium was television where he found greater success in series like *Bat Masterson*, *Burke's Law* and *The Name of the Game*. Les Tremayne became a "B" movie staple in science fiction, even encountering Mars again in *Angry Red Planet*. He did the opening narration to *Forbidden Planet* (uncredited), and outside the genre he worked for Alfred Hitchcock in *North by Northwest*. He also appeared in several episodes of the *Alfred Hitchcock Presents* TV series.[63]

Ann Robinson had a career about evenly split between movies and TV. Her TV roles were largely westerns with guest appearances on *Cheyenne*, *Rawhide*, and a starring role on the series *Fury* (as the school teacher, Helen). In the Sci-Fi world, her notoriety from *The War of the Worlds* led to a recurring role on *Rocky Jones, Space Ranger* as the alien queen Juliandra (and also Juliandra's evil twin sister, Noviandra). She even reprised her role as Sylvia Van Buren in three episodes of the *War of the Worlds* TV series

[63] *Starlog*, #132, July 1988, pg. 24.

(described in the Epilog). Robinson was also instrumental in organizing the 25th, 40th and 50th anniversary reunions of the cast and crew. Of the film she says, "I've gotten more mileage out of *The War of the Worlds* than Vivien Leigh did with *Gone With the Wind*!"[64]

George Pal, despite producing three Oscar-winning SF movies in a row, could not convince Paramount to do any more past *Conquest of Space*. He moved to MGM where he produced and directed *The Time Machine*, *Tom Thumb*, *The Wonderful World of the Brothers Grimm* (in which he only directed the Puppetoon sequences) and the unfortunate *Atlantis*. In 1970, after the success of *Star Trek* on TV, he proposed a TV series based on *The War of the Worlds* (that would have again re-used Al Nozaki's war machines). This time, though, the invaders were from Alpha Centauri. Nothing ever came of it, for which we should probably be thankful.

We'll be talking much more about the social impact and eternal longevity of this story later in the Epilog.

The Story

After a brief and rousing fanfare over the Paramount logo, our ears are bombarded by a frantic military tattoo being beat on a snare drum. A call to war!

A title card appears proclaiming "**World War I**" and a breathless narrator informs us:

> In the First World War and for the first time in the history of man, nations combined to fight against nations using the crude weapons of those days.

While he's speaking we see images of early fighter airplanes, tanks and horse-mounted cavalry.

A second title card, "**World War II**" appears and the narrator continues:

> The Second World War involved every continent on the globe and men turned to science for new devices of warfare, which reached an unparalleled peak in their capacity for destruction!"

Now we see images of landing craft approaching the beaches on D-Day, Seafire fighters launching from an aircraft carrier, and shots of the carrier task force from the air.[65] This cuts to a shot of a V2 taking off as the narrator concludes:

> And now, fought with the terrible weapons of super-science, menacing all mankind and every creature on Earth, comes…The War of the Worlds!

Titles in bold primary colors, sometimes flashing like lightning (or exploding bombs), splash across a black

©CBS Paramount Network Television
Jupiter?

screen. We segue directly into a montage of Chesley Bonestell paintings depicting the various planets as the oddly soothing voice of Cedric Hardwicke describes them. The introductory portion of the narration uses snips from the first pages of the book, but the planetary descriptions, containing some outrageous errors,[66] are from the Bonestell/Ley book *Conquest of Space*. The concluding section is back to lines from the novel, including the poetic description of the Earth: "…green with vegetation, bright with water,[67] and possessed a cloudy atmosphere eloquent of fertility." He explains, "Mankind had no idea the swift fate that was hanging over us…until the time of our nearest approach to the orbit of Mars during a pleasant summer season…"

We dissolve to an evening shot of the mountainous southern California desert. A large fireball passes over. It is spotted by two forest rangers in a fire watchtower on Pine Summit, and by some folk leaving a movie theater in a nearby town. They speculate on what it might be. "Maybe it's a comet," says Alonzo, a local businessman who will later be the Martian's first victim. So it goes.

It lands with a muffled "thud" just over the hill. "C'mon, let's go find it!" says Zippy, a local gadabout,

©CBS Paramount Network Television

---

[64] *Starlog*, #195, pg. 71.

[65] Interestingly, even though the narration says that science provided weapons at an "unparalleled peak in their capacity for destruction," there's no shot of an atomic bomb, the ultimate WWII weapon. Perhaps Pal and Haskin didn't want to tip their A-bomb hand this early.

[66] Even in 1952 when the script was written, they knew that Saturn and Jupiter had no solid surfaces and that Earth is the closest neighbor to Mars, not Jupiter. And why'd they leave out Venus?

[67] The book says, "grey with water," obviously written before anyone had seen the Earth from above.

who, with a bunch of his friends, rushes off. Meanwhile, the rangers in the watchtower call in the location of the meteor since it started a fire where it landed. Fire crews respond and, with the help of the locals who came out to see the meteor (now technically a meteorite), quickly have the fire under control. The ranger in charge calls back to the District Office to say the fire is under control, but adds that "somebody ought to check on it," referring to the meteor. The DO responds with, "There are some fellas fishing at Pine Summit. I think they're scientists."

The second ranger from the watchtower (the goofy one who cheats at cards) drives up to the campsite. As he arrives, three men are sitting around a camp table, eating dinner. The ranger greets them with, "You're the guys from Pacific Tech, ain't you?" After confirming that they are, he goes on to describe the "meteor" that came down. They invite him to have some of the fish that they'd caught, and he goes on to help himself to most of the rest of their meal as well; even pulling a cigarette right out of one of the scientists' pocket!

©CBS Paramount Network Television

He doesn't let a little thing like eating slow down his delivery of the message. Mouth brimming with bread and fish (possible foreshadowing of the religious themes to come), he continues, "Came down about 10 or 12 miles from here, over by Linda Rosa." "Are they sure it's a meteor? It didn't come down like one," says the dark haired scientist. "Well, you fellows have to figure it out, you're scientists!" says the ranger, expressing the unassailable faith in science common in this period.

We dissolve to the next morning at the gully where the meteor landed. It's still smoking with little flickers of flame rising from its surface. The townsfolk are all discussing it and come to realize it could be an economic boon to the town as a tourist attraction. Buck, a mechanic at the local garage, slides down into the crater and starts poking at it with a shovel.

As all of the men folk rush off to watch him, we see an attractive auburn-haired woman in her early '20s. She had

been in the background listening to everyone, but rather than follow, she gets her purse out of her car and pulls out a cigarette. As she's digging through her purse for a lighter, a yellow convertible pulls up, driven by the dark-haired scientist from last night. The two strike up a conversation,[68] first about the meteor (and Buck's efforts to excavate it), but it quickly turns to the woman explaining about the scientist coming in to study it. She waxes eloquent that it's "Dr.

Clayton Forrester, the man behind the new atomic engines," obviously a hero of hers. The scientist looks at her, somewhat askance. "He isn't that good," he growls. "Well now, how can you say that when you

©CBS Paramount Network Television

don't even know him!" the woman replies defensively. "Well, I do know him, slightly," says the scientist, who then reveals that he is, in fact, the great Dr. Forrester about whom she is so knowledgeable (but didn't recognize in person).

©CBS Paramount Network Television

After an awkward recovery that lasts only a second, she introduces herself as Sylvia Van Buren, a teacher of Library Science at USC (the University of Southern California). She admits she didn't recognize him in the beard and glasses. "They're really for long distance," he says. "When I want to look at something close I take them off," which he does and looks straight into her eyes. She gives a coy smile and shyly turns to walk towards the meteor. He follows doggedly, giving us the first hint of an attraction between them.

Meanwhile, Buck is still poking away with the shovel. He gives it a good whack and a piece of the crust falls off to a hiss of escaping gasses and a blast of heat. Retreating back to the crater rim he says, "Boy, you could fry eggs on that thang!" revealing his colloquial roots by using a cliché that was tired even then.

Sylvia introduces Forrester to her uncle, Matthew Collins, pastor of the local church. They'd barely exchanged pleasantries when Sheriff Bogany interrupts them. "Watch'a got in here, fella? It's ticking like a bomb!" He's pointing to a small metal box in the back seat of Forrester's car with a meter and a single lamp that flashes every time the box ticks. "This is a Geiger counter for detecting radioactivity," Forrester ex-

---

[68] Asking for a light was one of the few socially acceptable ways at the time for a young woman to start a conversation with a strange man. Lyndon used this plot device to get Sylvia and Forrester talking, but there was no further indication for the rest of the movie that Sylvia actually smokes.

plains, as he uncoils the detector. He aims the detector at the meteor and the counts go up enormously.[69] "It's radioactive?" asks Pastor Collins. "Maybe we ought to keep people away from it," says the ranger, the master of understatement. Forrester's scientific curiosity is piqued by radioactivity, and the fact that the meteor seems far too light for its size. He decides to stay "until it cools off," and Pastor Collins invites Forrester to stay at his house. He then mentions that there's a square dance at the Social Hall in the evening for entertainment.

We move straight to the dance where Forrester is cutting it up with the locals. The caller of the dance is our fiend, the funny, pudgy ranger. After only a brief glimpse of the festivities, we move back to the gully. Three men have been left behind to make sure the meteor won't start any more fires, Alonzo, who we saw earlier; Salvatore, a Mexican farm worker; and Wash. They've decided that it's now safe and start packing up to leave. Salvatore, though, hears something strange and shouts, "It's moving!" Sure enough, a circular section of the top of the meteor is turning around. The three men back away slowly, speculating wildly. "It's a bomb," says Alonzo. "It's an enemy sneak attack,"[70] counters Wash. "It don't go off last night, maybe it's gonna go off now!" says a very worried Salvatore. "Wait a minute," says Alonzo, "bombs don't unscrew," as the circular section continues to twist around, rising on a fine-pitch thread.

Another fast cut back to the dance where Forrester is bringing some bottles of soda to the table. There is much merri-

©CBS Paramount Network Television
The Martians first victims: Wash, Salvatore and Alonzo

ment and he admits he's having a good time. He makes a lame joke about gathering all the energy expended in the square dance to send the meteor back to where it came from.

As everyone laughs, we cut back to the gully where the top of the meteor finishes unscrewing. The lid slides off to the ground with a metallic clang, revealing a green flash followed by an ominous red glow from the interior. A metallic snakehead emerges from the hole and begins to slowly move in a circle, scanning its surroundings accompanied by an eerie, rhythmic pulsing.[71] The men, now hiding

©CBS Paramount Network Television

behind their cars, speculate further and come to the conclusion that the inhabitants of the meteor are from Mars since it's close to the Earth right now (that's quite a leap of logic since all they've seen is the snake head!). They further decide that they'll be famous if they are the first ones to make contact with the "men from Mars." Alonzo fashions a white flag while Salvatore asks, quite reasonably "What are we going to say to them?" Wash turns this over in his mind, and finally, with great effort, decides "Welcome to California!" is the perfect greeting. They approach the meteor while waiving the white flag and admonishing the occupants to "open up" and "we welcome you!" The snakehead spots them, stops scanning and draws a bead on them. The searching electronic beat changes to a menacing "thum-thum-thum" and the head bursts forth with an

©CBS Paramount Network Television

---

[69] He aims the detector the wrong way at the meteor. Geiger tubes sense radioactive particles coming in the side of the cylinder, not the end. If you watch closely, Gene Barry correctly slides the shield out of the way exposing the Geiger tube, but he aims it like a gun barrel at the meteor rather than holding it sideways like it's supposed to be. This was probably more understandable to audiences than the correct way of using it.

[70] Actually, he's right!

[71] Steven Spielberg, a huge fan of this movie, incorporated many references to *War of the Worlds* into his early films. This particular scene is acknowledged in the opening scenes of *E.T. The Extraterrestrial*. While the E.T.'s are out collecting samples, the camera takes us inside their ship where we see the large collection of specimens already picked up from other worlds. One of them is a fiddle-head plant that has a cobra head shaped flower with a red and yellow glowing center.

enormous fusillade of heat and light. The men stagger back as we fast-cut back to the dance where the lights suddenly go out.[72]

"No smoochin' in the dark, folks," says the half-witted ranger who was calling the dance. He looks out the window and notes that the power is out over the whole town. As people light candles, Pastor Collins asks Zippy to call the power company, and he discovers the phones are out as well. Forrester notes that the phones are powered separately from the lights, so they shouldn't have gone out. An elderly man says that his hearing aid (battery powered) has also stopped working. Pastor Collins comments that it's about time to end the dance, but when he pulls out his watch, he notices it's stopped running as well. One-by-one they all discover that their watches have stopped, and "all at the same time," notes Sylvia.

Forrester instantly seems to understand, and borrows a pin to demonstrate that the watches are all magnetized. "How could it happen to everybody's watch all together?" asks Sheriff Bogany. Forrester takes the Sheriff's pocket compass and they see that it's not pointing north, but rather at the gully where the meteor is.

They're interrupted by the wail of a siren as a police car pulls up. The Sheriff and Forrester hop in and they drive to the gully where a huge fire is burning. There they find a horrific sight: the three men left behind have been reduced to three vaguely human-shaped piles of ash in the middle of a blackened triangular swath. Before they can do anything, the cobra headed Heat Ray rises up again, building to a discharge. The policeman panics and drives off, abandoning Forrester and Bogany. They dive for cover as the Heat Ray goes after the car, incinerating it. "What is that gizmo?" asks Bogany. "That 'gizmo,'" says Forrester, "is a machine from another planet. You'd better get word to the military!"

©CBS Paramount Network Television

©CBS Paramount Network Television

Professor McPherson gives an interview sounding very much like Professor Pierson (Orson Welles) in the radio broadcast

We next see a huge mobilization of Marines from El Toro base moving men and materiel into position. Mortars, machine guns, field artillery, tanks, and even rocket launchers are set up. A crowd of civilians has formed outside the gully. A reporter from KGEB radio is describing the scene. He starts interviewing Professor McPherson who describes the places around the world where the meteors are coming down, but McPherson defers to Forrester when it comes to speculation on the nature of the Martians. Reprising a technique used so well by Robert Wise in *The Day the Earth Stood Still*, we are treated to a montage of folks from all walks of life listening to their radios while Forrester speaks; even two bums on a sidewalk outside an appliance store. (The two bums are, incidentally, George Pal and studio chief Frank Freeman, and there is some evidence that the scene was originally shot for *When Worlds Collide*, and inserted here as an in-joke.[73])

©CBS Paramount Network Television

The "two bums," Paramount studio head Frank Freeman (left) and George Pal in their big cameo appearance (redshirt in the back is unknown)

Back at the gully, the interviewer turns to Colonel Heffner, the officer in charge of the Marine operations. He describes how an Air Force plane is going to drop a flare so that they can photograph the Martians. Right on cue, the plane shows up and a brilliant white light envelopes the gully. The Heat Ray rises up and shoots at the plane. It then lowers and sweeps across the ground (and directly at the viewer!) scattering all of the people, including the radio reporter who's distressed that his broadcast truck has been destroyed.

The Marines continue to pour soldiers and weapons into the area. At the field HQ, Colonel Heffner orders his unit commanders into position. Forrester is there, as is Sylvia (in a Red Cross uniform) serving coffee and doughnuts.

---

[72] This is actually a fairly faithful representation of the scene as described in the book, only in that case the three men with the white flags are the scientists who initially studied the meteor, and the crater ("pit" in the book) is surrounded by more than 100 people; many of whom are taken out in the first Heat Ray blast.

[73] *Starlog*, #195, pg. 68.

The Saucer Fleet

Her uncle walks in with Sheriff Bogany just ahead of General Mann, head of Army Intelligence. The military men snap to attention, but he bypasses them, goes straight to Forrester and greets him warmly. "Clayton Forrester! I haven't seen you since Oak Ridge!" he exclaims (referring to the Manhattan Project laboratory in Tennessee that produced enriched uranium for the first atomic bombs).

©CBS Paramount Network Television
"I haven't seen you since Oak Ridge!"

Mann reads aloud the intelligence reports from around the world. They are coming down all over, but the military can't determine any sort of plan because "once they [the Martians] begin to move, no more news comes out of that area." He goes on to mention destruction and massacre in Bordeaux, France by a "ray of undetermined nature" where "nothing remains" afterwards. The photo taken earlier has told them that this is the original pilot cylinder. Showing confidence in the military's ability, Mann says, "This is the only place where we've had time to surround them with sufficient force to contain them. What happens here will be a guide to all other operations."

Dawn breaks and a forward observer calls in to report, "There's something moving in the gully!" With an ominous whirring, we see a Heat Ray projector rise up over the ridge. As it continues, we finally get to see what it's attached to: A graceful, copper colored manta-ray shaped machine with green tips on the "wings" and a matching green dome in the front (see photo, page 84).

"Look at it, will you!" implores Mann as he hands the binoculars to Forrester. The scientist instantly understands that the machine isn't flying, but rather supported from the ground by magnetic rays. He lets his scientific detachment slip for a second when, in the middle of his analysis, he bursts out, "This is amazing," almost giggling like a schoolgirl.

©CBS Paramount Network Television
"This is amazing"

Col. Heffner tells all command posts to stand by to fire. "But Colonel," says Pastor Collins, "shooting's no good. We should try to communicate with them first and shoot later if you have to." Heffner lowers his binoculars

©CBS Paramount Network Television
Pastor Collins pleads with
Col. Heffner for restraint

just long enough to give a sidelong glace to the pastor as if to say, "What kind of nut are you?"[74]

With the courage of his convictions, Collins quietly leaves the HQ tent while the military targets the enemy. Despite a panicked effort by Sylvia to stop him, he slowly walks towards the Martian machine, which is soon joined by two others. He removes his hat in a gesture of respect and holds up his pocket Bible to show them he is a man of God all the while reciting the 23rd Psalm. With perfect dramatic timing, the lead machine bears down and, just as he finishes, incinerates him with the Heat Ray. That's all Heffner was waiting for. Turning to his second officer he orders, "Let 'em have it!" in a somewhat un-military manner. The captain hollers "Fire!" into the field phones and a stupendous barrage of artillery, mortars, rockets, tanks and Howitzers is let loose.

©CBS Paramount Network Television
"I will dwell in the house of the Lord…"

---

[74] Imagine how different this scene would play today with our "understand the enemy" pacifist attitudes. Today he'd be the hero, but in this scene, Collins is shown to be a well meaning, but completely delusional optimist. The military was just preparing to defend us against this proven threat, and Collins paid for his naiveté with his life.

©CBS Paramount Network Television
Shields Up! The Martian's protective magnetic "blisters"

The Martians are completely unaffected. Illuminated by the flashes of exploding ordinance, we see that each of the machines is protected by a barely visible transparent dome. Forrester tells General Mann that it's a protective electromagnetic "blister." They watch with mounting disbelief as the full might of modern weaponry proves completely impotent.

©CBS Paramount Network Television
Nothing's getting through

After about 30 seconds, the Martians let fly with a retaliatory attack. In addition to the Heat Ray, they also emit green energy bolts from the wingtips. This "Skeleton Beam," as General Mann calls it, neutralizes mesons, effectively disintegrating its target into nothingness (we'll have a lot more to say about the Martian weapons in "The Vehicle" section later). Forrester immediately grasps the significance of this technology as he grabs Mann by the lapels and shouts, "This kind of defense is useless against that kind of power! You'd better let Washington know, fast!"

©CBS Paramount Network Television
This defense is useless!

Admonishing Heffner to "hold 'em as long as you can!" Mann and his aide make a swift exit. The first thing Heffner does, strangely, is order a full retreat. With the command bunker on fire from a Heat Ray blast, he starts driving everyone out to turn the fight over to the Air Force. Forrester and Sylvia are among the last to leave, but like a captain of a sinking ship, Heffner stays to the last. Just as he turns to leave, a bolt from a Skeleton Beam hits him and he fades to nothingness from the outside in (see the sidebar on page 105 for a detailed examination of this scene and the skeleton beam).

A wing of F-86's soars overhead as Forrester and Sylvia dodge the retreating tanks. They find an abandoned observation plane (it was established back at the fishing camp that Forrester is a pilot) and fly, literally, at treetop level to avoid the jets. Sylvia is terrified and with good reason. Forrester cuts it too close and clips a tree. As the plane banks to a crash landing, we catch a glimpse of a pair of war machines on the march.[75] The plane skids to a stop and the two tumble out, unhurt. They waste no time diving into a ravine as the machines open fire on the countryside.

We cut to downtown LA. An elderly woman is hawking newspapers with headlines screaming, "Attack from Mars!" "Killer Rays!" and simply "INVASION" while she sits there, Madam Lafarge-like, calmly knitting. Her chant of, "Read aaaaalll about it!" is interrupted by sirens as a police escort brings a large limousine and a plain Army jeep to City Hall.

A brown uniformed figure hops out of the jeep and starts up the steps. From inside, we see it's General Mann. He's surrounded by an entourage of military and civilian authorities, but they can't stop him from being mobbed by reporters as he fights his way to the briefing room. Once inside, he addresses the civilian agency heads: police, fire and Red Cross. He warns them that "the enemy is about 25 miles outside Los Angeles," and that they should be ready to evacuate the city. He is interrupted by a phone call from his superiors in Washington. The assembled officers listen with mounting concern as Mann describes (into the phone) the total failure of the military efforts to stop the Martian machines.

---

[75] Any viewer who cares to single-frame through this sequence will see the care with which this elaborate scene was created (see also production photo, page 86) for only a few seconds on the screen. As seen in this frame, on the ground under the war machines are three bright spots where the support rays are grounding out into the Earth.

©CBS Paramount Network Television

The Saucer Fleet

After hanging up, he talks briefly with reporters before leaving. The first question is about what Dr. Forrester "thinks about all this." They're asking Mann because Forrester never showed up at Pacific Tech. The last we saw, he was cowering in a ditch with Sylvia, protecting her from the advancing war machines.

After a brief montage of wild animals fleeing from the Martians, we see that they're still in the ditch. It's been some time because Sylvia is asleep in his arms, covered by his jacket, while he keeps

©CBS Paramount Network Television

vigil. He looks down to wake her, but stops for a second giving her a paternal half-smile. She slowly wakens, looking as beatific as a Botticelli painting until the horror of the situation comes flooding back. She sits up in a panic. "That machine!…"

Forrester calms her down and mentions there's a farmhouse nearby and maybe they could find something to eat. We dissolve directly to bacon and eggs cooking. Sylvia has made a traditional hearty breakfast from what she's found in the abandoned house. They sit down and discuss their pasts briefly before getting back to the problem at hand. "If they're mortal, they must have mortal weaknesses," says Forrester confidently. "They'll be stopped, somehow." Sylvia relates how she had once wandered off as a child, and had gone into a church because she felt safe there. Remembering that it was her uncle Matthew that found her brings back the horror of the past few hours.[76]

Their pleasant conversation is interrupted[77] by an explosion from outside. Forrester runs to the window. "Get down!" he shouts and they dive under the table. Another cylinder has come down and plows to a stop, half collapsing the house with the wall of dirt it pushes up. Back inside the house, Sylvia treats an unconscious Forrester with a towel wet by a broken water pipe. He startles awake. "How long was I out?" "Hours!" is her singular reply. (This tells us that the Martians have had time to extract the machines from the cylinder.)

From outside we see a machine lower down next to the house, its support beams crackling and burning the earth. A tether is lowered from a small round port in the bottom. At its end is an odd, almost Art Deco looking scanner head (patterned after a Roman helmet attributed to their god Mars). It, too, is copper and has three lenses in the three primary colors red, blue and green.

©CBS Paramount Network Television

They successfully hide from it for a minute until it closes its shutters over the lenses and withdraws. Sylvia catches a brief glimpse of something in the dark. "It was… one of them!" she gasps. "We're right in the nest of them. I've got to get a good look at them!" says Forrester, his scientific passions aroused (and also overpowering any shred of common sense for self preservation). He tries to get out through the collapsed doorway, but notes, "They've pushed earth or something all outside."

They move to the stairway up to the attic and begin removing debris. Sylvia hears a soft oscillating sound behind her, and turns around only to come face-to-face with the scanner! Forrester picks up an axe and hacks it from its tether/umbilical. The umbilical is rapidly withdrawn, bundles of raw wires hanging out. He places the severed head aside, carefully, and they go back to clearing a pathway to the attic. Unseen, an alien shadow plays over the wall. As Sylvia is diligently tossing debris aside, a large, but

©CBS Paramount Network Television

spindly hand with three sucker-tipped fingers comes down lightly on her left shoulder.[78] Without looking, she recog-

---

[76] Pal and Haskin used this scene as a screen test for Ann Robinson, since it required some subtle acting. That she was a good screamer was a bonus!

[77] …and so is their meal. The sharp-eyed viewer will notice that neither of them has eaten a single bite.

[78] Spielberg incorporated elements of this scene into two of his biggest early films. In *Close Encounters of the Third Kind*, he used the alien lights from outside, shining through the blinds as the ship lands before abducting Barry. In *E.T.*, he duplicated the hand-on-shoulder scene when ET puts its fingers gently on Elliot's shoulder, but in that case, it's a gesture of affection. ET's entire appearance, in fact, is based on Pal's Martian with the same height, color, skin texture, extra-wide head and spindly arms. The arms, though, attach to a more conventional location below the neck rather than to the sides of the head.

©CBS Paramount Network Television

nizes it for what it is and, absolutely frozen with horror, can't even scream. Forrester grabs her away and pulls a flashlight from his pocket. The dim (to us) light blinds the creature that makes a futile attempt to block it from its three-lens eye with its hands.

Forrester tosses the axe at it, and it runs off screaming. They remove the last obstruction and run up the stairs carrying the mechanical eye. In the attic, Sylvia drops her scarf. Forrester picks it up and hands it back to her as he wraps the scanner eye with his jacket. While he's doing so, she discovers that her scarf is wet with Martian blood. She loses it completely for a few seconds until Forrester "shakes some sense into her." They squeeze out an attic window (which is on the ground level since the house is collapsed) and get away just ahead of the Heat Ray blast that incinerates the house in a tremendous fireball.[79]

The story pulls back to show that what was happening in California was also happening, indeed had already happened, all over the world. The narrator details the Martian's war plan attacking all of the major cities and population centers. He describes the bravery and resolve of the military forces of all the nations when fighting the Martians, along with their total failure to have the slightest effect on slowing the Martian advance. The result was the "rout of civilization," as whole populations fled in panic.[80] The montage is an effective mixture of stock newsreel footage (mostly from WWII), travelogue footage, and new detail shots of frantic government officials and fleeing crowds. The sequence ends with the revelation that Washington D.C. was the only unassailed capital.

We dissolve to a large strategy map of the world where aids are pushing around clusters of black triangles, representing groups of war machines. The background chatter is terse military callouts of coordinates and status reports coming though in many languages. At one end of the room, General Mann is giving a briefing on the Martian's strategy.

He illustrates with colored chalk on a board: the Martians form a series of interlocking triangles, and after establishing a perimeter, they wipe out everything within it. All branches of the military are present, including some foreign services. An imposing man in civilian clothes listens intently and asks a few questions. It's the Secretary of State and he finally says, "Alright, I've seen enough. There's only one thing that will stop the

©CBS Paramount Network Television

Martians. The White House will confirm an order to use the Atom Bomb!" The first drop will be on the initial landing site outside Los Angeles (the Army Chief of Staff pronounces it "Los Angle-es") so that the Pacific Tech scientists can monitor the effect.

We cut to the entry gate of the "Pacific Institute of Science and Technology." Forrester and Sylvia enter a room full of scientists busy packing up. In typical emotionless scientist fashion, they don't seem especially happy or relieved to see their lead guy, who's been missing for days, back safe. Forrester does some quick introductions then shows them the Martian lens head and the scarf with the Martian blood. The lens looks more than just clean, it is freshly polished. The Martian blood is given a quick analysis by Dr. Duprey (apparently the

©CBS Paramount Network Television

only woman on the staff) who declares it to be the most anemic she has ever seen. "Physically, they must be very primitive," she concludes.

Forrester relates the Martian's reaction to the flashlight, to which Dr. Bilderbeck readily expounds on the nature of sunlight on Mars (thus illustrating the depth of knowledge available from the Pacific Tech scientists).

Dr. James hooks the Martian eye up to the epidiascope[81] and cranks it up. It starts humming and warbling, just as we heard in the farmhouse as it projects an image on the screen. "There's how the Martian's see us!" he declares (thus satisfying, a little, the Paramount management's insistence on showing things from the Martian point of view). The image is unreal and (literally) otherworldly. Not only is the image physically distorted, as if by a fisheye lens, but

---

[79] This entire scene is another that is fairly faithful to the book. Wells has his protagonist and the Curate eating a meal from what they found in an abandoned house when a cylinder lands and plows into it, collapsing the house around them. The protagonist is knocked unconscious and wakes to the Curate wetting his face. The Martians later explore the house with their mechanical tentacles. This is also where we get our first detailed look at the Martians. The only difference is the time scale. Wells' hero is trapped in the house for two weeks.

[80] The general feel of the flood of humanity matches, metaphorically, the book's description of the population of London fleeing in panic. The last line ("The rout of civilization, the massacre of humanity") is lifted directly, although Wells uses the formerly generic phrase "mankind."

---

[81] The epidiascope, despite the fancy name, is simply a projector that can project either opaque or transparent objects onto a screen. The name comes from the Greek words describing how the object is viewed, either from above (epi) or through (dia). The real question is how such a simple passive device can power up and control this alien technology.

©CBS Paramount Network Television

the colors are distorted as well. Despite using the three primary colors in their lens, the image shows people with purple clothes and green skin (perhaps a winking reference to "little green men," only seen from the Martian side). Other color shifts include Sylvia's bright red lipstick, which shows up a ghastly green-cyan. [Note: Since the film was shot in 3-strip Technicolor® this effect was actually quite easy to do. The lab just left out the magenta negative when printing and this is the result.]

After some "nostril shots" of those mugging the lens, they prepare to leave. Forrester tells Duprey to make a quick analysis of the Martian blood, which Dr. James dismisses with a sarcastic, "You can get all you want after the plane drops the Bomb!" followed by, "The 'Flying Wing' is going to carry it."

Ah, yes, the Northrop B-49 "Flying Wing," by far the most advanced and exotic warplane of the postwar era. It taxies down the runway and roars to a takeoff, right over us. As it heads to the target, we cut to the forward observation post where the scientific and military personnel are getting ready to monitor the drop.

Courtesy Northrop-Grumman

A radio reporter, played by Paul Frees in a very rare on-camera performance, gives us the background. They're in the Puente Hills, east of LA, where more of the cylinders came down the night before, apparently massing for the attack on the city. We also learn that this bomb is "the latest thing in nuclear fission," ten times more powerful than anything exploded before.[82] As if we need to be reminded of the importance of this blast, he tells us that "the fate of all humanity depends on what happens here."

A military controller announces that there are only a few minutes left before the blast, but that's enough time to speculate how long mankind has, should the A-bomb fail. Everyone gets into position as we hear the Flying Wing soaring overhead. Just as the bomb is released, the Martians (dozens of them in a very tight cluster) activate their "protective blisters." The con-

troller counts down the last 15 seconds, and the bomb detonates in a classic flash followed by a gigantic blast wave. As the mushroom cloud boils heavenward, forward observers down on the valley floor peer out though slits in their bunker while blast debris hits them square on. "What can you see?" pleads

©CBS Paramount Network Television

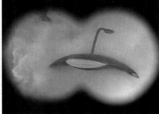

©CBS Paramount Network Television

General Mann by field phone. The observer makes out a shape moving right in the middle of the nuclear inferno. A war machine! It glides peacefully out of the turbulent cloud, which is all that remains of humanity's biggest explosion. "They haven't even been touched!" he reports.

Unbelieving, General Mann strips off his blast goggles and checks for himself over the sandbags. "It didn't stop them," he says dejectedly. "Guns! Tanks! Bombs! They're like toys against them!"

"It will end only one way," intones Bilderbeck, "We're beaten."

©CBS Paramount Network Television

"No, not yet," says Mann with gritty determination. He then relates how his orders (if the bomb failed) are to evacuate the city and then "establish a line and fight them all the way back to the mountains!"[83] And finally, "Our best hope lies in what you people (the scientists) can develop to help us."

Forrester, his head and face outside of the goggle area white with residue from the blast, suggests analyzing the Martian blood further to take a biological approach. "We know now that we can't beat their machines, we've got to beat them!" (which both repeats his observation from the farmhouse and neatly foreshadows the ending).

As they leave to pack up their instruments to head for the Rocky Mountains, we do a wiping dissolve to Los Angeles. Speaker trucks roll through neighborhoods telling people to evacuate. We see massive numbers of people poring onto highways (including a 4-level interchange on the still-under-construction Hollywood Freeway). In an homage to Sergi

---

[82] Interestingly, this film was released only a few months before the first H-bomb was detonated (see photo page 105), producing an explosion nearly 100 times bigger than any previous.

[83] With what? Nothing, not even mankind's biggest stick, has had the slightest effect, while the Martian's weapons have been 100% effective at killing or destroying everything anyone in the world has thrown at them.

Eisenstein, who started to make the film in the '30s, we see a child's ball bouncing down the steps of a hospital, à la the

©CBS Paramount Network Television

baby carriage in *Battleship Potemkin*. The crowd eventually becomes an endless stream of people pushing their way on foot up into the mountains.

Back at Pacific Tech, the scientists have packed up the last of their gear and head off. Forrester is alone in the last truck as he had to run back into the lab to grab the "biotics" from the refrigerator. He drives frantically down the deserted streets of Los Angeles.

Rounding a corner, he runs straight into a mob of vagrants and thieves who stayed behind after the initial evacuation to loot, but now have realized their foolishness and are commandeering anything on wheels. Forrester is dragged from the cab and they start throwing out the instru-

ments to crowd themselves on. He pleads with some Civil Defense guards (who are engaged in open fistfights with the mob) to help him, but they sum up their helplessness with, "The law's no good now!" He makes one more attempt to retrieve the truck and is soundly beaten.

©CBS Paramount Network Television
One of the mob coldcocks
Forrester and steals the truck

As the truck drives off (and the rest of the mob leaves on foot) Forrester learns from a looter, still gathering up money and valuables into a case, that the mob got the earlier Pacific Tech trucks as well. Forrester begins a frantic search on foot, running up and down the deserted streets. Just as he finds an overturned truck with "Pacific Tech" on the door, the sky erupts in explosions. The Martians have arrived!

©CBS Paramount Network Television

Next to the truck is the placard from the bus that Sylvia was driving. Forrester presumes that the same thing that happened to him, happened to them, so if they're alive, they must still be in the area. As he runs off, we see the Martians attacking downtown. With the

city in flames in the background, they are heading towards City Hall. Moving down a street they are indiscriminately using the Heat Ray and Skeleton Beam on everything. The Ray sweeps across, hitting a water tank on a tall frame (a nod to Wells' original design) that explodes as the water in it boils. A large gas storage tank is next, which goes up in a tremendous fireball.

©CBS Paramount Network Television

A lone military jeep with MP's races down the deserted street and comes upon Forrester standing behind a bus stop bench. They plead with him to leave and offer him a ride. He rambles on about looking for the Pacific Tech professors, the mob and the broken instruments. They take him for a nut case and give him one last chance to leave with them. Just then, a war machine moves on City Hall, famously reducing the central tower to rubble with a Heat Ray blast. Forrester waives them on with a cryptic "I think I know where she'll be…"

©CBS Paramount Network Television

He runs down a street, dodging flaming debris falling from nearby buildings as incessant Heat Ray blasts, Skeleton Beam discharges and explosions fill the air. He comes across a church where, inside, the minister is holding a vigil, praying for divine intervention. The people attending are the ones who couldn't leave with the evacuation, and had nowhere else to turn.

Not finding Sylvia or the Pacific Tech people, he leaves and heads to another church as the Martians wreak incredible devastation. At one point, he runs right across the path of a pair of machines coming up the street, Heat Rays blasting.

The Saucer Fleet

He makes it to a Catholic church where a priest is conducting a rosary in Spanish. There he comes across Bilderbeck and Duprey. Bilderbeck is lying in a pew, badly injured with his head bandaged. He relates how Duprey pulled him from under the

truck when the mob attacked it (there is some sort of unspoken connection between these two as they have some affectionate gesture in almost every scene). They didn't see what happened to the others.

We see more Martian destruction, larger, more violent. Through the smoke, Forrester sees another church, right in the path of the machines. A plain-clothed minister, looking remarkably like Billy Graham, is beseeching God for intervention. Those inside know the end is near as the Martian weapons sound louder. Forrester continues to call, "Sylvia!" which is finally answered by a frantic, "HERE!" They fight their way across the packed parishioners and fall into each other's arms as the first Heat Ray blast hits the church. A stained glass window shatters while lamps and debris fall from the ceiling. Outside, we see the war machine about to deal the fatal blow, but something's wrong! The formerly smooth sounding machine now has an unbalanced-sounding electric whine. It starts listing to its right, and the Heat Ray gives only a few furtive spurts and stops. The machine continues to slide to the right, eventually crashing into the building across the street from the church. Its left wing hits the ground with an enormous "thud" and an eerie silence falls.

Inside the remains of the church, Forrester and Sylvia are still holding each other as the dust begins to settle. Other parishioners slowly realize that they are not about to die. They get up and ever so tentatively

Amid the flash from an explosion, a war machine crashes into a building, drops to the ground and stays motionless.

This shot is only on the screen for five seconds, yet it tells us quite a bit about the battle both directly and symbolically, so it needs more than a simple caption. The view is from underneath the "wing" of a war machine (almost certainly a matte painting). The hatch in the process of opening is to the right. There is a magenta line running from it off to the left suggesting another opening. The machine is no longer pristine, but shows the stains and blast effects from the destruction it has been wreaking, demonstrating the intensity of the battle, even if it has been one-sided. The shape of the opening formed by the wing and ground suggest an eye, implying the Martian point-of-view (another bone for Paramount management) and seems to say that even in death they are "still watching us."

move towards the open door thorough which the fallen machine can be seen across the street. Eventually, Forrester notices everyone moving toward the doors and he fights his way to the front of the crowd (after finally finding Sylvia, he sure leaves her quickly enough when something interesting comes up!).

Sitting on a pile of rubble, a hatch on the underside of the war machine slowly opens with an electric whine and a loud hiss. A magenta light floods out as a long, spindly Martian arm starts to creep down the hatch, dragged by its three suction tipped fingers. Forrester moves cautiously forward, followed closely by the crowd. Before they can reach it, a loud crash to their right shows a second machine crashing to the ground. "Something's happening to them!" declares Forrester, master of the obvious. The arm stops moving and suddenly vitrifies. Touching the wrist (is he checking for a pulse?) he concludes, "It's dead."

©CBS Paramount Network Television
What does he see?

He peers up into the hatch,[84] then turns around. All he can say is, "We were all praying for a miracle," and lifts his eyes heavenward. As the other join him in his upward gaze, the narrator informs us, with jubilant church bells ringing, "The Martians had no resistance to the bacteria in our atmosphere to which we have long since become immune. Once they had breathed our air, germs, which no longer affect us, began to kill them. All over the world, their machines began to stop and fall." We see the machines grinding to a stop amid the ruins of famous structures all around the world: The Eiffel Tower, the harbor at Rio, the Taj Mahal (we'll be commenting about this odd mix of science and religion in this film later in the Epilog).

As the multitudes of refugees in the San Gabriel Mountains look down on the city (now, strangely, no longer on fire), the choir swells with the end of the

©CBS Paramount Network Television

hymn we heard them singing at the first church, "...in this

world and the next," followed by a great "Amen!" over a shot of a cathedral silhouetted against a magnificent sunset reminiscent of the American flag. We fade to the Paramount logo, leaving us exactly where we started with the Earth safe and in our possession.

The Vehicle

As mentioned in the production section above, George Pal and director Byron Haskin made a deliberate decision, in direct opposition to the wishes of Paramount management, to never show the point of view of the Martians. Their reasoning, which proved correct, was that the Martians would be much more threatening if they stayed an unknown menace. So, without first-hand observations, where does that leave us regarding knowing how the war machines work? As it turns out, we're not as bad off as one might think.

Pal and writer Lyndon used a plot device that worked quite well in *Destination Moon*, and to a lesser degree in *When Worlds Collide*, namely the "explaining character." In both of those other movies, there was a character (Dr. Cargraves in *Moon* and Dr. Hendron in *World*) that explains all of the technical stuff to other characters in the film (and, thus, to us in the audience). In *The War of the Worlds*, the explainer is, of course, Dr. Forrester. A crack nuclear physicist and Manhattan Project alumnus, Forrester is constantly analyzing the Martian devices and reporting his observations to General Mann or to other scientists. His dialog provides all that we are going to learn about Martian technology, so we have to presume his guesses were right. Before we get to his observations, though, let's review the original Martian technology from the book for a basis of comparison to give us some appreciation for how it was updated.

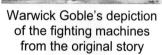

Warwick Goble's depiction of the fighting machines from the original story

Wells' original Martian machines[85] had ovoid bodies with multiple tentacle-like arms and three extremely tall legs for walking. They came approximately five to a landing cylinder, and the

---

[84] What does he see up there? This tantalizing tease fires the imagination, much like Spielberg's homage ending of the original *Close Encounters of the Third Kind* when Roy Neary looks up into the Mother Ship (you'll note that the Mother Ship has a remarkably similar hatch, differing only in scale). Spielberg ruined it in the CE3K "Special Edition" when he showed us what Neary saw. No matter how magnificent, it's now his (Spielberg's) vision, not ours.

[85] Wells calls them "fighting machines" in the book. Actually, he usually just calls them "machines" and only uses the adjective if there's more than one kind of them around (fighting machines and handling machines). The term "war machine" is only used once at the very end. It is also used only once in the movie, during Cedric Hardwick's narration in the "rout of civilization" sequence. The rest of the time they're also just "machines."

The Saucer Fleet

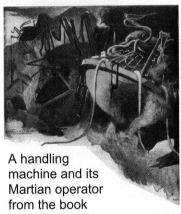

A handling machine and its Martian operator from the book

Martians spent a day or two assembling them with secondary machines called handling machines. The fighting machines are described as "vast spider-like machines nearly a hundred feet [30 meters] high, capable of the speed of an express train."[86] The machine's main weapon is an invisible beam of intense heat. Wells' protagonist, who is the "explainer" in the book, conjectures as to the ray's operation:

> … in some way they are able to generate an intense heat in a chamber of practically absolute non-conductivity. This intense heat they project in a parallel beam against any object they choose, by means of a polished parabolic mirror of unknown composition, much as the parabolic mirror of a lighthouse projects a beam of light."[87]

This is a pretty good concept of how a Heat Ray might work given the Victorian-era technology that Wells had to work with. However, it could have been a CO_2 laser, which also operates in the infrared, and can easily set things on fire across long distances. It could also have been some device as unimaginably beyond our technology as lasers would have been to Wells.[88]

The Heat-Ray, as Wells punctuates it, is projected from a lantern like device mounted on a gimbal at the end of an arm attached to the front of the machine, leaving it free to aim in any direction. It is completely silent except for the sizzling of whatever it hits, and an occasionally heard low humming sound. The beam, being infrared, is also com-

Martian Weapons hanging off the left side of the machine: The Heat-Ray (top) and Black Smoke launcher (bottom). The "Control Hood" is the triangular shaped structure on the top.

pletely invisible, although "scintillating green flashes" can be seen coming from inside the projector.

They had a second weapon as well, a heavy, black poison gas:

> The Martians are able to discharge enormous clouds of a black and poisonous vapour by means of rockets. They…are advancing slowly towards London, destroying everything on the way. It is impossible to stop them. There is no safety from the Black Smoke but in instant flight.[89]

The Martian operating these weapons, along with the rest of the machine, sits at the top; in a swiveling, armored "hood" (what today we might call a command turret).

From this fertile starting point, Pal, Nozaki and Jennings created a completely updated and frighteningly effective new machine that paralleled the original in all important aspects. There is one thing that makes the war machines difficult to document, though. The clay master created to make the mold for the prop units (see photo page 84) was made directly from Nozaki's small model, using it as the only reference. No fabrication drawings of the war machines were ever produced and after filming, the three props were donated to a Boy Scout scrap metal drive.

They kept the Heat Ray, of course. It just wouldn't be a Martian War Machine without one! However, as terrifying as an invisible, silent beam would be in reality, (how would you know which way to run?), it would be a cheat on a movie audience. When comfortably ensconced in our theater seats, seeing a character raise a prop weapon and, without audio or visual clues, simply having something across the screen catch fire is not fear inducing; it just looks cheap. Lasers were still seven years in the future, and even if they had been around, they didn't want to just copy Gort's narrow destructive beam of a couple years earlier. Instead, they went in a completely different direction producing a Heat Ray that was vigorously dynamic, spraying the countryside with vast showers of heat particles. And far from being silent, it produced the most perfectly alien of sounds.

---

[86] *The War of the Worlds*, Berkeley Highland edition, 1964, pg. 74.

[87] Ibid, pg. 27.

[88] In another case of life imitating art, in early 2007 the US Army introduced a prototype of a non-lethal "heat ray" weapon whose description sounds remarkably like Wells'. The system uses millimeter-length electromagnetic waves, which only heat the top layer of skin, just enough to cause discomfort. The device can be used from more than 1,500 feet (500 meters) away, and anyone hit by the beam feels a sudden blast of heat over their body. While the 130°F (55°C) heat is not painful, it does make the target think their clothes are about to ignite. It could theoretically be upgraded to a lethal weapon that matches Wells' description by changing from milliwaves to microwaves (such as used in common kitchen appliances) and increasing the power significantly.

Courtesy DARPA

---

[89] *The War of the Worlds*, pg. 79.

Its operation must have been completely intuitive since it's the one Martian technology that Forrester never comments on. We have no clue as to whether the Heat Ray projector was emitting actual physical burning material, like an incendiary bomb (seems unlikely since there isn't much room inside the machine to store it), or if it was purely energetic, creating little "plasma packets" out of the surrounding air. We do know that, however it works, it creates an enormous electromagnetic pulse (EMP) since the first time it was used, it magnetized everyone's watches, stopping them all at the same time. Curiously, this effect is never directly mentioned again in the film, even though an EMP that strong would render many devices, such as the portable tape recorder used by Paul Frees' character, completely useless. On the other hand, as the machines are advancing through LA, we see all the power lines next to a machine start smoking and sag as it lets go with a Heat Ray burst. This is exactly what would happen if a massive current were suddenly induced in a wire by a strong magnetic pulse.

©CBS Paramount Network Television

Using poison gas was a bit more problematical. While Wells' prediction of the horror of this indiscriminant weapon was born out some 18 years later in World War I, there had been another World War since, and anything to do with the first one, no matter how horrific, seemed old fashioned to an early '50s audience. That viewpoint is clearly hammered home in the prologue where we hear that WW-I combatants used "the crude weapons of those times." The Skeleton Beam was the answer for a replacement threat. At least we know the principle behind this one. Recall Forrester's impassioned analysis during the heat of the first battle:

> It neutralizes mesons somehow. They're the atomic "glue" holding matter together. Cut across their lines of magnetic force and any object will simply cease to exist!

That's a pretty accurate description of subatomic physics, at least as it was understood in mid-century. There are two small problems with it, but both were necessary deviations since this film was popular entertainment, not an educational film for a college physics class. To understand the problems, though, requires a little side trip into the fundamental nature of the universe. Don't worry; this shouldn't take long.

[Warning, educational material follows. Those of you who don't want to risk learning something should skip the next few paragraphs.]

In our universe there are, according to physicists' "Standard Model," four fundamental forces: Gravity, Electromagnetism, and the so-called Strong and Weak nuclear forces. Gravity is the weakest, but reaches across the entire universe. It is the mutual attractive force that all matter has and causes it to clump together into stars and planets. It also causes them to orbit one another forming solar systems and galaxies.

Electromagnetism is the only one of the four that we can sense directly[90] since it is the force of heat and light. Its reach is as far as gravity since we can see starlight from billions of light-years away, but most of the electromagnetic spectrum is imperceptible to us, ranging from radio waves up to gamma rays. It is also the force behind

chemistry, since chemical reactions take place between the electron shells of atoms.

The appropriately named Strong nuclear force is the strongest of the four. Its job is to bind quarks together to form protons and neutrons in

---

[90] Please don't write to say we can sense Gravity. We can sense the effects of Gravity, such as the pressure on the soles of our feet when we stand, or the muscle strain we exert when lifting something, but we can't sense Gravity itself. Astronauts in orbit are still well within Earth's gravitational field, but they can't sense it because they are removed from its effects while "weightless."

the nucleus of atoms. It is also the force that overcomes the mutual repulsion of protons (since they're all positively charged, they repel each other) to pull them, and the neutrons, together to form the nucleus itself. When you set off a nuclear bomb, this is the force that's released. The Weak force is subtler and less obvious. It works to cause a certain type of nuclear decay called "Beta decay" where a neutron spits out an

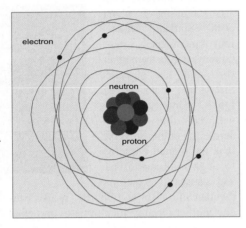

electron and becomes a proton. It is between Gravity and Electromagnetism in strength. Both of the nuclear forces only operate over very short distances. Once you get outside the nucleus of an atom, they have no effect at all.

These forces work by means of energy packets called "field particles." The field particle for Electromagnetism is called the photon, which most people have at least heard of at some point. The field particle for Gravity is called a graviton, something that has been theorized for most of a century now, but not yet detected. In the 1950's the field particle for the Weak nuclear force was the unimaginatively named "W-Particle," and the particle for the Strong force was called (ta-da!) the meson.

That's what things were called in mid-century, anyway, around the time the movie was made. As you can imagine, ever-evolving subatomic theories have changed enormously since then,[91] and almost everything has been redefined and/or renamed. The field particles are now collectively called "bosons" and as it turns out, the meson did not carry the Strong force holding the nucleus of atoms together as was thought fifty years ago. The photon is still the field particle (boson) for electromagnetism. The graviton is still the boson for gravity (and still theoretical), but the other two have been renamed The Weak force now has two bosons associated with it (the "W-boson" and the "Z-boson") and the new boson for the Strong nuclear force is the "Gluon." And yes, that's exactly why it's called that. Even though they had the wrong particle, it seems that Forrester's observation about being "atomic glue" was right on the nose!

Now that you're all caught up on how the universe works, you can see that the first problem with Forrester's

statement is that it's not "lines of magnetic force" but rather "lines of Strong force" that the Skeleton Beam is neutralizing. That certainly wouldn't have made much sense to anyone in the audience, in fact, it sounds phony. Saying simply "lines of force," while perfectly correct, sounds bland.

The second problem is that they left two words off of the statement, "…any object will simply cease to exist," namely, "as matter." Remember that matter and energy cannot be destroyed, just changed from one form to another. If the Skeleton Beam is going to be disrupting things at the quark level, the objects they hit could flash into pure energy according to Einstein's now-cliché formula $E=mc^2$. If the mass in the average human body, say Col. Heffner at about 220 lbs (100 Kg) with his helmet and other army gear, were all released as energy it would produce 9×10^{18} joules, or roughly 2,000 megatons! That's just about the yield of all of the nuclear weapons ever detonated combined, and it's the result of "neutralizing" only one body. They were also destroying trucks and tanks weighing multiple tons, which would result in blasts thousands of times bigger (for comparison, the bomb that leveled most of Hiroshima converted approximately *one gram* of matter, about the mass of a paper clip, into energy). Rather than subduing this planet, they'd be blowing huge chunks of it out into space with this weapon. For those with some scientific background that want to delve a little deeper into the physics behind this concept, please see the sidebar on the next page.

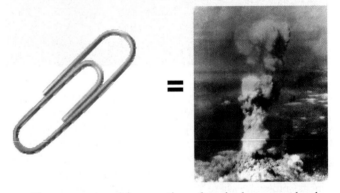

For support and locomotion, they had to completely abandon the stilted legs for reasons described previously in the production section. No more 100-foot (30 meter) tall mechanical walking machines that could move with "the speed of an express train." Instead, they devised something equally as advanced and technologically impressive to the mid-20th Century audience as Wells' mechanical walkers were to his readers.

When we first see the machines, they have no visible means of support. General Mann expresses what the audience is thinking when he asks, "Is that some kind of a flying machine?" to which Forrester replies, "No, it's supported from the ground by rays, probably some form of magnetic flux like invisible legs. They must keep the opposing poles in balance and lift the machine." Lifting

---

[91] And continue to change. The 2004 Nobel Prize in Physics was awarded to three US scientists for their work describing how the Strong nuclear force works.

(Continued on page 106)

Mesons and Gluons and Quarks (Oh, my!)

Since particle physics is hardly my forte, I thought I'd better have a real expert review my ideas on how the Skeleton Beam worked. Dr. Joyce Guzik, a physicist at Los Alamos National Laboratory, was kind enough to review my analysis and had this to say:

If you neutralized gluons, you would certainly cause a huge explosion as the excess of positively charged quarks packed closely together in the nucleus would violently repel each other, causing a huge energy release. But, there would not be a total energy conversion as you describe. When the beam is turned off, the quarks (which retain their mass) would quickly recombine back into protons and neutrons.

Extrapolating a little further, the free neutrons would decay in about 10 minutes by the Weak force into a proton, an electron and an anti-neutrino. The free protons would capture an electron and turn into hydrogen gas in the explosion cloud. Would this be likely to react with oxygen and cause a chemical explosion, creating water and releasing more energy? Probably, but it would be a lot less than the explosion generated by the charge particle repulsions. The velocity of the neutrons or protons in the explosion might be high enough that when they hit some of the surrounding atoms of nitrogen, oxygen, concrete, whatever, they could be captured by nuclei creating some radioactive isotopes, or cause a few fissions as they were slowed down. The free neutrons would certainly be a radiological hazard.

Okay, so the explosions wouldn't be the monstrous ones I described in the text, but they would still be gigantic, and not just a simple "dissolve" of the object as shown in the film.

Dr. Guzik continues:

You can roughly calculate the amount of energy released in the following way: Take the electromagnetic potential energy between two particles, $(k \times p_1 \times p_2) \times 1/r$, where k is Boltzmann's Constant, p_1 is the charge of a proton, p_2 is the charge of the rest of the nucleus, and r is the distance between them. Calculate for an initial distance between them of one femtometer (10^{-15} meters, the typical size of a nucleus), and then calculate again for a distance of 1 meter (essentially infinity as far as these particles are concerned). The difference between the two values will give you the energy released from one proton escaping from the nucleus. You could multiply for the rest of the protons in one nucleus, and then again for the number of nuclei (or atoms) in the object, and see how much energy it will actually generate.

Gee, Dr. Guzik, I didn't know you were going to make me work! Okay, leaving the actual grinding out of this equation as an exercise for the reader, we'll use poor old Col. Heffner again as our test subject. Using this technique, we get an energy release of 1.8×10^{16} joules, or 4.2 megatons. That's only ¼ of 1% of the total energy conversion I thought would happen, but it's still an awesome amount of energy.

Finishing up with one more clarification, Dr. Guzik concludes:

Mesons were once thought to mediate the Strong nuclear force between protons and neutrons, but this was found not to be true. The meson is made up of two quarks, while protons and neutrons are made up of three quarks each. A gluon, on the other hand, is not made of quarks at all, but is a massless particle. Free quarks cannot exist in nature (quarks are "confined" particles), and the Strong force gets stronger as you try to pull two quarks apart.

=

Courtesy DoE
Ivy-Mike Test, the first thermonuclear bomb.
Yield ~ 10 megatons

x 200
(Complete Conversion)

Or

÷ 2
(Gluon Neutralization)

things with magnetism is nothing new, but for us to do it requires an external mechanism. Science museums have long had "levitation machines" consisting of two electromagnets of opposite poles kept in balance (just like Forrester said) that will make a steel ball "float" in the air.[92] However, to work, the magnets have to be outside the device being lifted. To raise something magnetically using only an internal mechanism is difficult indeed. Of course, if they can use magnetism to counteract gravity, it means that the Martians apparently have discovered the graviton and how it interacts with photons (now aren't you glad you read the educational section above?).

They kept the three-legged stance by having three support-ray projectors on the bottom, but this isn't very obvious past the scene where we see the machines for the first time. There, we see the ghost-like discharges of their support beams in operation. While we don't see the actual beams after that, we do see their effects as flashes on the ground as this massive energy flux discharges into the Earth. In addition, far from moving with express-train speed, these war machines inch forward at barely a walking pace. Of course, that only adds to their menace. Even though they are just creeping along, they are as unstoppable as an advancing tide.

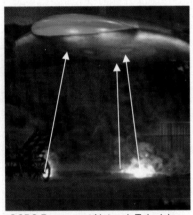

Support Ray generators (arrows) and the beams intersecting the ground.

It's not recorded just who came up with the idea of supporting the movie machines on invisible rays, but the idea had been around for at least two decades by that point as shown in this Buck Rogers daily panel from 1929.

Which leads us to the last piece of Martian technology, their protective force fields. As the initial barrage of Earth weapons are brought to bear against the Martians, we see both Heffner and Mann's faces expressing a mixture of confusion and disbelief that nothing is happening to the enemy. As usual, Forrester comes through with an explanation, "Those shells can't get through to them. They've put up some sort of

In the book, the Earthmen get a lucky hit and destroy a fighting machine (left), but in the movie all a direct hit does is light up the Martian shield.

electromagnetic covering, a protective blister."[93] This is a new facet of the Martian's capabilities. In the book, the machines were heavily armored, but they weren't invulnerable. In fact, one of the artillery companies firing at them scored a lucky hit right into the control hood: "The hood bulged, flashed, was whirled off in a dozen tattered fragments of red flesh and glittering metal."[94] The machine staggered off, driverless, for some distance smashing though the village of Shepperton before tripping and crashing into the river in an enormous steam explosion as the ultra-hot source of the Heat-Ray hit the water.[95]

The Earthmen were not so lucky with the movie Martians. Absolutely nothing, not even the biggest (by ten times) atomic bomb ever detonated had the slightest effect. They were, by any definition, completely indestructible. While such absolute perfection in protective technology was unprecedented, the basic concept was hardly new. Science fiction had long proposed the "force field" defensive device. Dating back to the "gadget" fiction of the '20s pulps (crystallized by Philip Nowlan's *Arma-*

---

[92] We can see one of these in *Earth vs. the Flying Saucers* in the scene where Dr. Marvin tests his second-generation weapon (see page 212).

[93] Blister, in this case, is an obsolete term for any sort of rounded, protective covering. They were usually found on airplanes to form a streamlined covering over pieces of equipment that stuck out beyond an airplane's skin.

[94] *The War of the Worlds*, pg. 61.

[95] This suggests that the machines had some sort of automated mechanism controlling the walking motions so that the operator only had to tend to higher-level guiding functions. This is an extraordinarily advanced concept for the period. Automated machine controls, such as the autopilot of an airplane, were still many decades in the future. Your typical Victorian era machine had no automated controls whatever, outside of maybe a speed governor, and required an operator to be constantly attending all of the levers and valves to keep it operating correctly.

geddon-2419 A.D. in 1928, the original Buck Rogers novel), it survives to this day in the ubiquitous "shields" seen in *Star Trek* that always seem to be failing in percentages just as they're needed most.

So, how are all these marvelous technologies powered? Wells never says explicitly, but he drops strong hints that his Martian machines are steam powered. They occasionally emit green smoke from their joints as they moved, and they wash down the "Black Smoke" poison gas with blasts of steam like a locomotive doing a boiler blow-down. Some of Warwick Goble's illustrations even show smokestacks on the top or sides, and the ma-

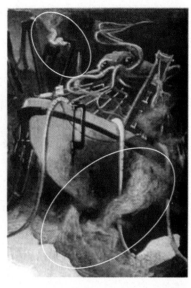

chines were always surrounded by "smoke and vapour." By the 1890's they were more than a century into the steam age and the machines powered by it were highly developed. It seemed reasonable that it was the ultimate power source that would continue indefinitely into the future, even on other worlds. What could possibly replace it?

By the mid-20th Century we had a new ultimate power source: atomic energy. All advanced designs in the post-war period were atomic powered. At the time this film was made, the Navy was well into development of the nuclear submarine, the Air Force had a proof-of-concept "A-Plane" flying, and even the Army had, in addition to nuclear bombs and artillery, a small portable nuclear reactor under development for powering advanced field headquarters. The drawback to all of these, especially for the Air Force, was the massive shielding needed to protect nearby personnel. They address this in a clever bit of dialog when we get to hear another Forrester analysis, but not directly. When General Mann is briefing the civil authorities in Los Angeles, he is interrupted by a phone call from his superiors in Washington. He relates Forrester's observation that "they generate atomic force without the heavy screening we use" (he also refers to their force field somewhat quaintly as an "electronic umbrella"). Beyond that, though, we don't know anything. It could be fission or fusion, but whatever it is, it's mighty compact. It was probably their power plant idling that Forrester's Geiger counter picked up in the scene where we first see the cylinder.

Finally, we even know a little bit about the landing cylinders. In the radio interview montage near the beginning, Forrester and Professor McPherson are speculating on the Martians. After hypothesizing on their physiology (e.g. the ability to "smell colors"), Forrester discusses the cylinders and opines, "They're probably controlled by jets after they enter our atmos-

phere and navigated by some form of gyroscopic mechanism." This is important because it means they aren't just coming in ballistic, but can actually target a specific landing area.

Later, in the War Room scene, just after General Mann finishes his "chalk talk" of the Martian offensive strategy, someone off camera (not Mann) says, "We know that there are three cylinders to each group and three machines to each cylinder." While this abandons Wells' capacity of five machines per cylinder, it reinforces a subliminal aspect of the Martians worked into the story by Pal and Lyndon.

Just in case we hadn't picked up on it, in the very next scene Dr. Bilderbeck notes, after seeing the Martian scanner eye for the first time, "It's curious how everything about them seems to come in threes." This is probably an oblique religious reference to the Trinity, which we'll be discussing in the next section, but there **are** a lot of threes associated with them. In addition to three cylinders a group and three machines per cylinder (which, incidentally, is directly contradicted by Al Nozaki's production drawing of the cylinder that clearly shows room for only one machine inside), their offensive strategy is interlocking triangles, and the Martians themselves have a three-pupil eye and three fingered hands. It's not clear how Bilderbeck came to his conclusion since at the time he had seen none of this except for the three-eyed camera. The only "three" associated with the machine that we can see is the three force beams that support it off of the ground, which came directly from the book machine's definitive tripodal stance. Wells had no particular dominant number associated with his Martians. They had one large shapeless body/head with two eyes (single pupil each), one mouth (used only for breathing) and one ear, or at least a tympanic membrane, in the back of the head. They had 16 tentacles in two groups of eight sprouting from either end of the head.

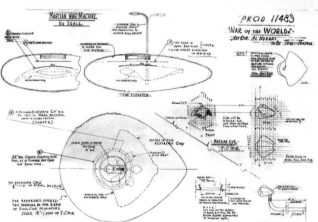

Three machines per cylinder?

As effective as Wells' Martian war machines were in the book, Pal and his production team came up with the definitive alien invasion craft that set the standard for a generation.

After Jon's Archeological report on the design (starting on the next page), we'll continue by following the influence that this film has had over the years.

Archeological Report: The Martian War Machine

By Jon Rogers

In any scientific attempt to recreate something that no longer exists, a historian must accurately determine the original object's correct size and document it. As it turned out, the design of the War Machine seen in *The War of the Worlds* would be relatively easy to scale. The first hand information I had to work with was the film itself and one surviving studio production drawing.[1] However, upon closer inspection of the two sources, several major questions arose.

The first problem was interpreting the studio drawing. Having worked with drawings that were barely readable, I was happy to discover that this one was very clear. However, unlike many blueprints I had worked with, this drawing had some ambiguous scale references.

The first scale reference was that the drawing had no scale at all. As you can see in Figure 1, the title of the drawing indicated "No Scale". Also in Figure 1, observe the note that said the "Exterior of meteor to match full size meteor." This note was redundant and confusing unless it meant that the exterior of the meteor *model* must match the full size meteor. Still, this is an assumption and must be suspect until proven by other evidence.

As you can see in Figure 2, there was another scale reference on the drawing. It indicated that the meteor was drawn 1/8" to 1' in scale. And, right below it is another scale indicating that this drawing is also "1 ½"= 1 foot of the F. S. min [Full Scale miniature]." They clearly wanted this drawing to scale to both the real meteor and the full-scale model of the meteor.

Unfortunately, because we do not have the original drawing but just a photographic copy of it, we cannot just measure the drawing to know what dimension the meteor -and its miniature were. We know the scale but not the actual measurement--yet.

However, in the center of the drawing of the meteor is the outline of the War Machine and on this drawing there were three circles labeled 22", 20", and 14" in diameter respectively to define the details of the opening from which the Martian heat ray first appeared. Here is where the drawing is very helpful. If we scale the drawing of the meteor and War Machine to these measurements, we have the following conclusions:

1. The actual meteor is about 70.5 feet long, 58.4 feet wide, by 20 feet high. The full scale model would have been 8' 9.75" long, 7' 3.6" wide, and 2' 6" high as indicated by the drawing.

2. The War Machine is about 35.2 feet long by 41.5 feet wide by 8.5 feet high. Its model would have been 4' 4.8" long, 5' 2.25" wide, and 1' 0.75" high as indicated by the drawing. .

Figure 1

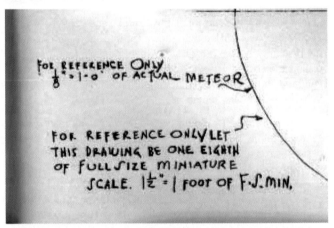

Figure 2

Figure 3

---

[1] The full studio sketch is shown at reduced size on the previous page with the caption, "Three machines per cylinder?"

These dimensions seem very clear and straightforward. However, in the film, does the meteor or War Machine actually confirm these dimensions? This is what I had to ask myself in order to be sure.

For the first example, in the film, we see a meteor crashing into a farmhouse (Figure 3). The farmhouse we see is a scale model of a typical, two-story farmhouse. A farmhouse this type is nominally about 25 feet high. Using this scene as a reference, and carefully scaling the meteor to the farmhouse, it appears to be about 70 feet long and 20 feet high. So far so good.

So our problem is that, while the drawing says that it is "No[t drawn to] Scale," it seems that it does accurately represent the size of the meteor seen in the film. Then what about the War Machine itself?

I began a careful analysis of the War Machine as seen in the film. The first problem was to find something in the film that could be used as a reference in judging its size. This turned out to be fairly difficult. In the entire first half of the movie, we see the War Machine*s* ravaging the countryside, blowing up trucks, Army tanks, buildings, and people. But, we do not see them get close to any of these objects. There are many scenes where there are close-ups of them, or pictures of them overlaid on things being destroyed, but none of these scenes give any opportunity for an accurate reference of their size.

Then, in the final scenes in the city streets of Los Angeles, we see them coming down the street, blowing up everything in their path. The opportunity to finally establish the size of the War Machine comes when one of the machines passes directly by a four-story building with a telephone pole in front of it (Figure 4).

There are two objects in this frame that can be used to accurately scale the intended size of the War Machine. The building has the standard size windows in each of its stories. Also, the telephone pole standing in front of the building is a good reference.

Using the fact that telephone poles are a standard 40 feet (480 inches) tall, we can check the pole in the scene against the building to insure it was modeled correctly. Then by choosing the moment that the War Machine is directly alongside the pole we should be able to scale it accurately.

By comparing the telephone pole to the building's window, we see that the window measures 8 feet (96 inches) in height. This is correct for a 10-12 foot high story, typically found in an older building of the style shown. This corre-

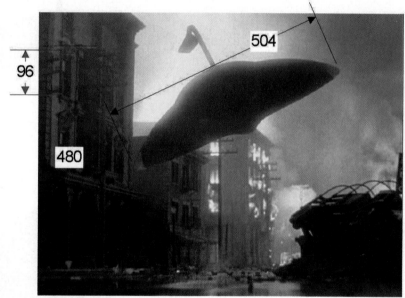

©CBS Paramount Network Television

Figure 4

lates with the telephone pole's height of 40 feet. Since the War Machine is directly opposite it, the one accurate measurement will be the width, which we can now show to be 504 inches or 42 feet. So now we know that it is, in reality, about the 42 feet width indicated by the drawing. From this one dimension, we can now create the rest of the War Machine with accuracy.

Thus, this is one of those rare cases where we have additional data that supports what is shown in the film. As always, the film is our final reference.

After establishing the correct size of the War Machine we can continue analyzing its features. One of the first notable features was its color. As seen in the film, the War Machine*s* were of one color-copper. However, as the film story proceeded, they acquired darkened streaks of black on their underside. Presumably this was from the heat of their heat-ray weapon and the considerable amount of fire they went through during their campaign to wipe out the human race.

Their graceful shape belied their destructive nature. Their smooth skin had almost no features to interrupt their stream-

---

Quickspec: Martian War Machine	
Vehicle Morphology	Other
Year	1953
Medium	Theatrical Film
Designer	Albert Nozaki
Length	36 ft (11 m)
Span	42 ft (13 m)
Height	8.5 ft (2.6 m)

The Saucer Fleet

Martian War Machine

1/150 scale
Dimensions in inches
© 2005 by Jon C. Rogers
Sheet 1

Sources:
War of the Worlds, Paramount Pictures, (1952)
Production Drawing # 11489 by
Al Nozaki and Teal Senter

Notes:

1. Body color is Copper. The large front dome, two wing tips, and three support beam projectors all glow green as shown.
2. Three models were made (see text). This drawing is a composite.
3. Size and scale is based on external references (see text).
4. Dotted line refers to the approximate outside dimensions of the landing 'cylinder'.
5. Although no hatches are visible on the models, one (port) hatch is shown in the final close-up scene. Since it is offset from the center of the machine, a second, matching (starboard) hatch is speculated.

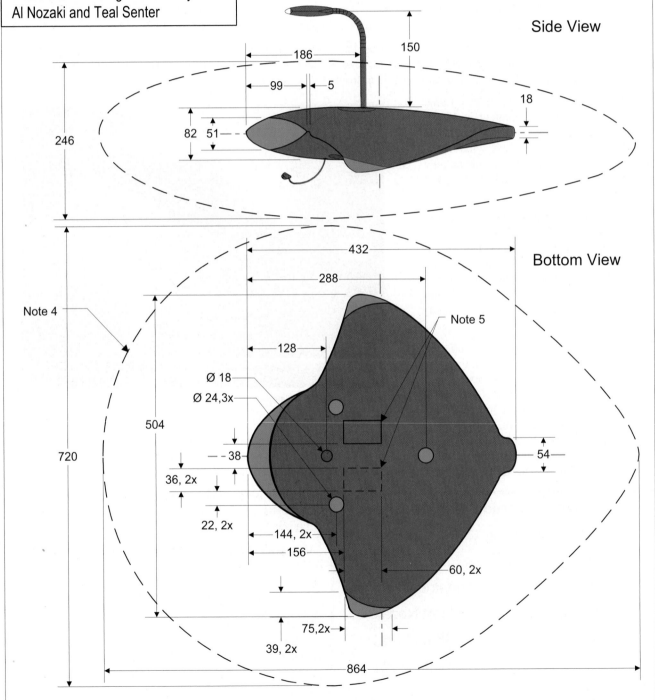

Side View

Bottom View

Martian War Machine

1/150 scale
Dimensions in inches
© 2005 by Jon C. Rogers
Sheet 2
Front view, Top view, and Details

Notes:

1. Electronic eye Colors are: Body is Copper. Lens glows Red, Blue, and Green. Individual shields cover each lens when off.
2. The body is not entirely a single structure. It has an oval fuselage and the wings attach at this point and then blend into the fuselage.
3. Upper body curve is not single arc but has 3 different radii as it goes from center to edge.
4. Umbilical length in film would have to be at least 40 feet to exit machine and enter farmhouse through side window. This drawing does not show the full length.

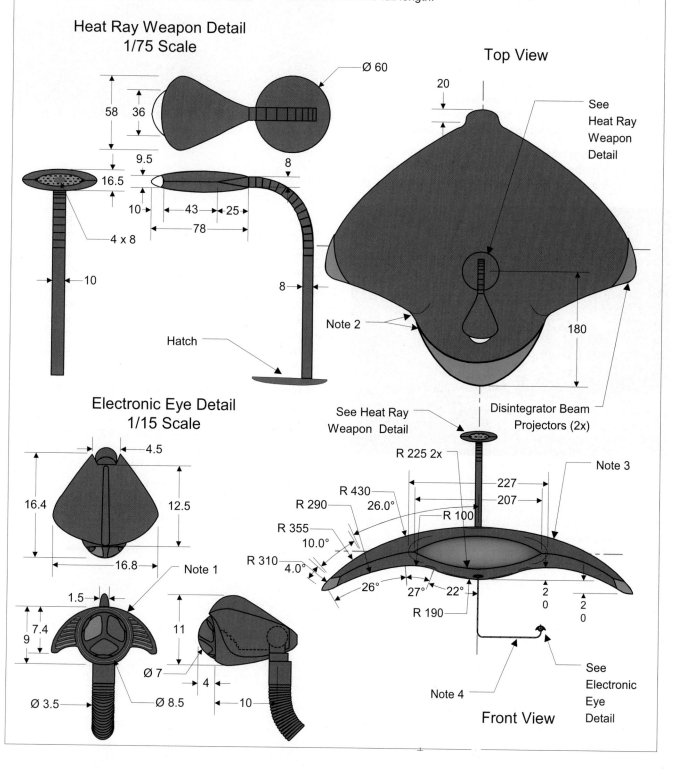

Heat Ray Weapon Detail
1/75 Scale

Ø 60

58   36

9.5

16.5

10   43   25

78

8

4 x 8

10

Hatch

8

Top View

20

See Heat Ray Weapon Detail

Note 2

180

Disintegrator Beam Projectors (2x)

Electronic Eye Detail
1/15 Scale

4.5

16.4   12.5

16.8   Note 1

1.5

7.4

9

Ø 7

4

Ø 3.5   Ø 8.5   10

11

See Heat Ray Weapon Detail

R 225 2x

R 430

R 290   26.0°

R 355   R 100

10.0°

R 310   4.0°

227

207

Note 3

26°   27°   22°

R 190

2 0   2 0

Note 4

See Electronic Eye Detail

Front View

111

The Saucer Fleet

lined shape, meaning that there were only a few items to detail.

In Figure 5 you can see the three ray projectors that supported them, as well as the one opening for the Electronic Eye probe.

Including the opening for the telescopic Heat Ray projector on its top, these were about the only openings visible in the otherwise featureless skin of the craft.

As I have noted in my drawing, the umbilical for the electronic eye is curious because it seems to have a very indeterminate length. It appears to be about 8 to 10 feet long when it comes through the farmhouse window. However, this does not show how long it would have to be to extend clear back and connect to hardware inside the hovering War Machine. I have estimated it would have taken about 40 feet of umbilical to achieve this task, but this must taken as an estimate only.

The final observation about the details of the War Machine comes in the scene where the Martians succumb to our normal earth bacteria. As the fearful earth people approach the fallen War Machine we can see a small hatch opening in the underside of the machine (Figure 6).

©CBS Paramount Network Television

Figure 5

©CBS Paramount Network Television

Figure 6

It is from this hatch a single Martian arm will extend to be felt by Dr. Forrester who will pronounce the Martian dead, thus bringing about the final outcome of the film.[2]

However, the hatch itself is interesting. If you will observe in Figure 6, the hatch is offset from the machine's centerline. Of course this is convenient for the sake of the plot as it allows the hatch to open after the crash. However, plot convenience would not be the normal reason to design such a hatch. Its real use would, more likely, be for entry and exit for the Martians. (Remember the Martian in the farmhouse?)

Since this hatch is placed on the port side, there would be good reason to place an identical hatch on the starboard side for reasons of redundancy, safety, and security. A good designer --especially from an advanced species-- would never allow himself only one means of entering the machine unless absolutely necessary. This is why I have included another, speculated, hatch in the data drawing.

Of course, there is no evidence seen in the movie to indicate that it was really there, it is just likely. There may have possibly even been a third hatch somewhere else in the machine because, as one of the scientists said, "Isn't it curious, how everything about them seems to come in threes!"

---

[2] In this scene, the "magenta line running off to the left suggesting another opening" as stated in the caption on page 100, is not confirmed. What that feature is, remains a mystery.

Modelers' Note

The Sci-Fi model broker "Monsters in Motion" has several versions of the Martian war machines available, ranging from a detailed (well, as detailed as they can be since a featureless surface was part of the appeal) stand-alone version to several mini- and micro-dioramas.

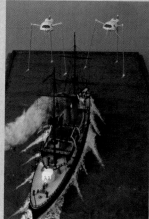

Scratchbuilders should applaud a remarkable diorama depicting a scene from the book created by Dan Thompson. Dan modeled the battle between the Martians and HMS *Thunder Child*. Rather than basing it on Warwick Goble's rather vague illustration (left), he used Michael Trim's painting from the cover of Jeff Wayne's musical album (described in the next section). This meant that the war machines are depicted, incorrectly, as standing completely out of the water (the book clearly describes them as "standing so far out to sea that their tripod supports were almost entirely submerged."[96]). However, you must admit that this version makes for a much more dramatic presentation. But Thompson didn't care for the depiction of the ship itself in either Goble's drawing (what you could see of it) or Trim's painting, and instead did considerable research on British warships of the 1890's. He based his *Thunder Child* on features from, appropriately, HMS *Mars* with its distinctive parallel funnels, and other ships of the period.

©Dan Thompson. Used with permission

Epilog

George Pal was an enigma, a philosophical contradiction. On the one hand, he was a technophile of the first order. He loved gadgets and tried to put the latest technological marvels in his films, from computers (even if they were mechanical) to the latest aircraft (helicopters and flying wings) to atomic bombs. On the other hand, he was deeply religious. A life-long Roman Catholic, he worked religious references into nearly all of his films from the subtle (e.g. in *Destination Moon* Barnes calls out "You got a prayer in your pocket?" as he desperately maneuvers the *Luna* to a hard landing) to the overt retelling of the story of Noah and the Ark in *When Worlds Collide*.

Both sides of his personality were evident in *The War of the Worlds*. His optimism in technology let him create a scientist-hero as the male lead, but this is tempered by the female lead who is the niece of a Protestant minister. Pal created a technological priesthood with the Pacific Tech

scientists who are able to expound on any technical subject in detail. But when Bilderbeck relates his calculation that the Martians will take over the world in six days, Sylvia responds with "The same number of days it took to create it," referring, of course, to the Biblical story of Genesis. At that we are shown not one, but four double takes by the scientists who react to the incredible profundity of what she said. What point was he trying to make? Was this supposed to be the scientists realizing a greater truth that had escaped them due to the earth-bound limitations of their thinking?

Actually, that is the entire philosophical point of the movie, something that many reviewers, especially modern ones, have problems with. The message is quite clear at the end as the Martian machines come crashing down at the entrance to a church when God strikes them dead just as they were about to destroy His house. Cedric Hardwicke brings it home in the dénouement: "The Martians had no resistance to the bacteria in our atmosphere to which we had long since become immune. After all that men could do, the Martians were destroyed and humanity was saved by the littlest creatures that God, in His wisdom, had placed on the Earth," followed by a great "Amen!" sung by a choir. While

---

[96] *The War of the Worlds*, pg. 106.

this is close to a direct quote from the book ("…the Martians [were] slain by disease bacteria against which their systems were unprepared; slain, after all man's devices had failed, by the humblest things that God, in his wisdom, has put upon this earth."[97]), it puts a completely different slant on things than Wells had intended.

H.G. Wells was a renowned atheist. His point in mentioning God in that sentence, which Pal and Lyndon either didn't get, or deliberately ignored, was to give it an ironic twist. Far from being the concluding moral of the book (the sentence appears more than ten pages from the end), he goes on to describe how bacteria had been killing people since our earliest ancestors, and that billions of human deaths over the millennia had developed our immunity to most of them. The true purpose of that statement was an attack on mankind's hubris in thinking of itself as the master of the world; the endpoint, in fact the entire point, of evolution. Wells wanted to take us down a notch or two, to show that the human brain and abstract thinking, of which we are so proud, were completely helpless against this foe, which had no problem completely undoing 10,000 years of civilization in only a few weeks. We were eventually saved by animals at the far opposite end of the evolutionary scale.

Probably the clearest illustration of the difference between Wells' and Pal's philosophies is in their depiction of the clergy. Both the book and the film have a main religious character. Pal's Pastor Collins is intelligent, rational and well meaning; an extremely sympathetic character. His pleading with Colonel Heffner not to start shooting is completely logical considering that they are intelligent beings with whom "no real attempt has been made to communicate."[98]

Wells' Curate, on the other hand, is completely irrational and delusional. A weak and self-absorbed individual, he gets no strength or solace from his faith. He interprets the events he sees around him as God's retribution for the sins of mankind. Towards the end, he becomes completely deranged and is convinced that the invasion is his fault because he hasn't prayed hard enough to lead people out of sin. At that point he charges out of his hiding place (where he had been safely hiding for a week) where the Martians could see him, and drag him to his fate. This delusion is

©CBS Paramount Network Television

barely hinted at in Pastor Collins' march, full of the courage of his convictions, into the face of the advancing war machine while reciting the 23rd Psalm.

If Wells' book was a pointed look at 19th Century British Imperialism, Pal's movie was equally successful in reflecting the Cold War paranoia so prevalent in the early '50s. Sociologists wasted little time in pointing out that these "alien invasion" movies were simply social metaphors for our dread of this new type of enemy. Whether they be the Red Menace, the Yellow Peril, or little green men from Mars (although, in this film they were little brown men) they played on our insecurities in this new age of propaganda wars and atomic bombs.

The War of the Worlds holds the distinction of being the granddaddy of the genre, the very first alien invasion film. True, *The Day the Earth Stood Still* also involved aliens landing (and threatening) to take over, but that was more like an ambassador's visit than an invasion. This was the first film to put it all right up front. They want our planet, and consider us a minor inconvenience on their way to taking it.

It was not only the first, but many consider it the best. It's hard to argue that any of the wave of films that followed could measure up to the standards set by this movie. Two of them, *This Island Earth* and *Earth vs. the Flying Saucers*, plus the later TV series *The Invaders*, are included in this book. As you will see in their chapters, they are entertaining in their own right, but none conveys quite the sense of menace and immediacy of danger as *The War of the Worlds*.

The powerful experience provided by *The War of the Worlds* in all of its forms has motivated generations of film, television and music artists, inspiring remakes, sequels and imitations for decades. There have been almost continuous references to this movie or the radio broadcast over the years in other movies and TV shows, far too numerous to detail here, but we can give you the highlights.

Even before Pal's film was made, the Howard Koch radio play was translated into Spanish and adapted to the local cities and terrain in Ecuador. It was broadcast in 1948 and the effect was exactly the same as in New York and New Jersey ten years earlier. Hundreds of listeners fled from the coastal cities to the presumed safety of the mountains.[99]

---

[97] *The War of the Worlds*, pg. 162.

[98] In the book, this sentiment is expressed by Ogilvy, the astronomer, which is why he was one of the men carrying the white flag to the first encounter.

---

[99] Enrique Tovar, *La Nación*, 3 March 1996.

A TV movie about the radio broadcast called *The Night That Panicked America*, based on the Howard Koch book, *The Panic Broadcast*, was produced in 1975. Written by Nicholas Meyer (who would take on Wells once again four years later in *Time after Time*) it starred Paul Shenar as Orson Welles, and is still well respected today by fans.

In 1976, Jeff Wayne started work on a rock musical, based on the original book, which was released in 1978. While fans consider it a period classic, equivalent to the early rock operas like *Tommy* or *Superstar*, it is not well known to the general public. As of this writing, the original recording (with the main character played by Sir Richard Burton) is still available on

©Columbia Records/CBS

CD,[100] and the lyrics and dialog can be found online. While the musical style is certainly dated today, the composition is a very respectful and reasonably complete, although abbreviated, telling of the Wells story (with an extra character or two). At 95 minutes, it is actually ten minutes longer than the Pal film. Of far greater interest than originally intended (at least to modern day audiences) is an amazing historical coincidence. To update the story some, Wayne added a brief epilogue set in the near future (the near future of 1978, that is), depicting NASA landing two spacecraft on Mars. Controlled from Pasadena, they started returning images of "a remarkable landscape littered with different kinds of rocks [and] dune fields."[101] Since its initial release on vinyl, the album has sold more than 12 million copies.

In 1988 PBS Radio (now NPR) produced a new version of Koch's radio play for the 50th anniversary of the original broadcast. Directed by David Ossman (an original member of the Firesign Theater comedy troupe) it starred Jason Robards as professor Pierson. It also had comedian Steve Allen, in a serious role, playing the reporter who watches the machines cross the

Steve Allen records his role on a New York rooftop

Hudson and eventually succumbs to the "black smoke" poison gas. Allen had heard the original broadcast as a 17-year-old and, along with his mother, had been among the thousands that had panicked.[102] Sound production was by Skywalker Sound designer Randy Thom. Broadcast on 30 October 1988, the exact date of the anniversary, many listeners found the show dull and unengaging despite the stellar cast and expert production; perhaps due to PBS's efforts to avoid another panic situation by keeping the tone very reserved.

Also conveniently timed to coincide with the 50th anniversary of the radio broadcast, a syndicated television series premiered in the fall of 1988. Lasting two seasons, it was a sequel to both the radio show and the movie, ambitiously trying to reconcile both versions into a single plot line. In the backstory to this series, the 1938 landings were a scouting expedition that was beaten back not by bacteria, but by the New Jersey militia. The Orson Welles broadcast was just a cover to give a rationale for the ensuing panic (which really was fleeing the Martians). The 1953 movie de-

©Paramount Pictures
Series Advertising Logo (above) and opening title

picted the main attack. The force fields protecting the war machines had been developed to defend against the weaponry the first landing encountered. Finally, the Martians weren't actually killed by the bacteria, but rather slipped into a state of suspended animation. The bodies were unceremoniously stuffed into nuclear disposal casks and buried in a nuclear waste site. The aliens, by the way, were never called "Martians" in the series and towards the end of the first season they were described as being from the planet Mortax, four light years away.

After the radiation from the site had killed the bacteria, the aliens revived and emerged to continue their war of conquest by hijacking humans in a "Body Snatchers" sort of way. This rather clichéd plot device probably comes from a misunderstanding of a line from the book. Wells was describing the sophistication of the Martian's machines, saying that they essentially wore their machines as clothing: "They have

©Paramount Pictures
An alien emerging

[100] Columbia C2K 35290

[101] Ibid., disc 2, track 7.

[102] *Herald & News Showcase*, 28 October 1988, ppg. 3-4.

[103] *The War of the Worlds*, pg. 124.

become practically mere brains, wearing different bodies according to their needs just as men wear suits of clothes."[103] He was referring to bodies metaphorically, not suggesting anything supernatural.

While that, and other, plot devices seem a little strained, the series got high marks from fans for its writing; especially the first season, where, again, the humans don't always win against this new wave of invaders. In the second season, though, the show got a complete overhaul and did not fare well. Two of the four main characters that had been fighting the aliens for a whole season were killed off, and the aliens were shown to be only mercenaries for an even nastier alien race from the planet "Morthrai." The suspenseful writing of the first season was replaced by a "battle of the week" series of stories in a post-apocalyptic setting. When the cancellation notice came, the last show had not yet been written, so they were able to create an actual series resolution.[104]

Then there are the bright young minds at "Best Brains Inc." In 1988, when they were putting together their satire/social commentary show *Mystery Science Theater 3000* (MST3K), they had to come up with names of the main characters. Since his character is the one shot into space to be force-fed bad movies, series creator Joel Hodgson hybridized his own name to create "Joel Robinson," as he was equally "lost in space." However, for the chief mad scientist responsible for putting him there, Trace Beaulieu, who played the role, asked himself, "Who is the most famous science-guy in all of filmed science fiction?" The name he picked, complete and unabridged, was "Dr. Clayton Forrester." To enhance the "tribute," he even had his Dr. Forrester wear bright green glasses, spoofing the reddish tortoise shell glasses worn by Gene Barry in the film. This choice of name for the character is all the more remarkable when you consider that every single member of the show's cast had been born after the film was released; some nearly 20 years!

©CBS Paramount Network Television ©Best Brains

The Doctors Clayton Forrester

One of the more direct copies of the Pal film, not actually based on the Wells book, was *Independence Day*

(1996), even though the mechanics of the plot were more closely aligned with *Earth vs. the Flying Saucers* (the aliens come in saucers and destroy most of Washington DC). On the other hand, their technology is completely unstoppable as they lay waste to every major city while we throw everything we have at them, including a nuclear bomb carried by the Northrop B2 "Flying Wing."

In the late '90s, Martin and Robin Wells (H.G.'s grandchildren), the trustees of the Wells estate, started negotiating with Hallmark Productions to produce a TV miniseries based on the novel for the book's centennial, but they discovered that Paramount still owned the rights based on the 1951 license signed by their father and George Pal. They sued Paramount, trying to draw a distinction between "motion pictures" and "television miniseries." The judge ruled against them saying that there was functionally no difference between them since movies are broadcast on TV all the time, thus the project was cancelled.

At well over the 100-year mark, the story still has the power to entrance filmmakers. The critically acclaimed, but logically confused 2002 film *Signs* is not overtly a version of the Wells book, but the arc of the story is unmistakable. Invading aliens show up first as mysterious lights in the sky before starting on their war of destruction and conquest that takes only a week or so. While the larger story goes on unseen, the main characters eventually wind up hiding from the aliens in the basement of a house. The aliens, who kill using poison gas, are eventually vanquished by one of the most common and abundant substances on the Earth's surface.

The 2003 British production of *Peter Pan* even contains a reference to this story. The director wanted to establish that aunt Millicent (who was watching the Darling children while their parents were out) was getting quite nervous, thus jumpy, at the noises coming from upstairs. To do this, he simply showed her reading a huge, leather-bound copy of *The War of the Worlds*.

In Ray Harryhausen's autobiography, *An Animated Life*, the last chapter is a compilation of the dozens of projects that he started, but was never able to finish. He ends the book thus:

> Looking back over the years, I have been lucky to have been involved in so many exciting projects, the best of which did grow into full-length feature films. Of those [that were never finished], there are not many I really regret not making, although if pressed, I think I would have liked to have realized H.G. Wells' *War of the Worlds*. Even now, at the beginning of the 21st century, I believe the story still has great power and so much to offer a modern audience. George Pal made an excellent adaptation of the story in 1953, but it surprises me that no enterprising producer has taken the original story in its original Victorian setting and made a modern interpretation of the conflict between two worlds.[105]

Nothing better illustrates the power of this story and its never-ending popularity than to have the great Ray Harryhausen single out not making it as the sole regret of his long and varied career (see test photos, page 80).

[104] *Starlog*, #137, December 1988, ppg. 37-44.

[105] *Ray Harryhausen, An Animated Life*, pg. 298.

Harryhausen actually got his wish to see the story done in its original setting the year after his book came out. In June 2005, three new movie versions of *The War of the Worlds* were released within three weeks of each other (see comparison sidebar on the next page). The Steven Spielberg/Tom Cruise production is the one everyone knows about. A very lavish film with a nine-figure budget, it is another update of the book that again places it in the present day (this time on the eastern seaboard of the US). Given Spielberg's affection for the subject matter, it was not surprising that nearly every scene contained a reference or homage to the Pal film. Here are but two examples of the dozens that can be found if you look for them:

The tethered scanner makes an appearance with its business end updated to look appropriately high tech, almost like a camera from a research submarine with its built-in lights and fish-eye window, but we

©Dreamworks/Paramount Pictures

©CBS Paramount Network Television

can see that, like Pal's, these Martians patterned the window after their own featureless, bulging eyes. The "Mars helmet" shape of the original, though, is echoed more strongly in the Martians themselves.

At the end of the film the machines come crashing down, of course, but not with the remnants of humanity cowering in a church. This time it was helped along by the army shooting shoulder-launched missiles into it. The hatch opens not with a sterile hiss and flood of violet light, but more like the hiss of a sewer line

©Dreamworks/Paramount Pictures

©Dreamworks/Paramount Pictures

bursting and a flood of noxious pink and yellow liquid. The spindly arm with three suction-cup tipped fingers comes crawling down the hatch and dies, this time much more dynamically by

©CBS Paramount Network Television

falling limp and swaying, rather than desiccating before our eyes as in the Pal version.

There are plenty of other references in the dialog. For instance General Mann's line, "once they begin to move, no more information comes out of that area," is repeated verbatim, only this time by a news producer in a remote broadcast truck. The greatest tribute, though, was snagging the two leads from Pal's film, Gene Barry and Ann Robinson, to play a five second, non-speaking cameo as Tom Cruise's in-laws.

©Dreamworks/Paramount Pictures

The film was slickly produced with flawless special effects, but it will be interesting to see how it plays with fans fifty years hence. It's doubtful that any future *War of the Worlds* producer will be looking to place Tom Cruise in a cameo decades from now.

Even with that much talent and budget, it did not fill Harryhausen's desire to see the story filmed in its original form. One of the other two projects meets this requirement. Tiny Pendragon Pictures, in England, started their own film in the summer of 2001, slightly before the Spielberg film. At first, it too was set in the present day, but after the events of September 11 that year, they almost cancelled the project. Director Timothy Hines explains:

> After an initial two-week hiatus, we saw the light in adapting a dead-on accurate version of The War Of The Worlds from the original source material, thanks to the influence and advice of people such as Charles Keller, the director of the H.G. Wells Society and tens of thousands of fans who wrote us.[106]

With this decision, Hines became the first filmmaker to attempt a film the way C.B. DeMille wanted to, placing the story in its original late Victorian setting.

---

[106] Interview, www.sfcrowsnest.com, September 2004.

The Wars of the Worlds

The movie industry often, by coincidence or design, brings out two movies with the same subject in the same season. Past examples include *Deep Impact* and *Armageddon* (1998) both about planetary impacts threatening the Earth, and *Dr. Strangelove* and *Fail Safe* (1964), both about the failure of military controls to prevent an unauthorized nuclear attack. In 2005 we had not two, but three versions of *The War of the Worlds* released within a few weeks of each other in June. These ranged from the megabuck/mega-hit Steven Spielberg production to the miniscule budgeted labor of love by Timothy Hines released by Pendragon Pictures. In between was a horror film treatment of the story by David Latt of The Asylum Home Entertainment. Let's take a brief look at how these three productions approached the rendering of the Martians and their machines.

©Pendragon Pictures

©The Asylum Home Entertainment

©CBS Paramount Network Television

With the full power of Spielberg and Industrial Light and Magic (ILM) behind it, the aliens in the Paramount remake are the most naturalistic of the three (left). The only problem is that they are a little too recognizable as an Earth-style animal with two-segment jointed limbs ending in a hand with fingers. The head is in the "right" place on the body, and the face is a recognizable distortion of a human. There are two eyes for stereo vision and a human mouth with teeth. There's no nose, but there is a place for one. While it is repulsive, it lacks the "alien-ness" of the Pal version.

Pendragon's rendition (center) is very faithful to the description in the book, as would be expected since that was the goal of the production. With their tiny budget, the CGI creations are only on the screen for mere seconds, so we never get a good appreciation for just how well they are depicted with the two large eyes (adapted to the dimmer light on Mars), eight tentacles sprouting from either side of the body/head, and the "V" shaped structure at the bottom that served as the mouth.

The Asylum's film's aliens are on the screen even less. We catch them in tiny snippets throughout the film in extremely under lit scenes where we briefly see the end of a mouth-like appendage or something like a 3-fingered claw. At least Latt gives us a look at the whole alien at the very end. That's it silhouetted in the right image above. It's hard to get a handle on just what it is, or how big, but at least it scores big points for "alien-ness"!

The depictions of the fighting machines are likewise rendered. With the megabucks available to fuel ILM's creative engines, there's no wonder that the machines are the most realistic ever produced. Not only could they realistically place them in a daytime shot (a lot of effects shots are placed at night to help hide the matte lines as much as for dramatic effect), but they put plenty of atmospheric effects in front of them giving a true sense of distance. The first time the machine rises up to its full height (near right frame) it seems to be 1,000 feet (300 meters) tall rather than 100. The motions are slow and majestic, as would be expected from a device this big, and it's depicted in daytime, nighttime, long shot, medium shot, close up, and right into the "harvesting basket" hanging from the back, all with total believability. It has the correct tripod stance and tentacle manipulators from the book, but the rest of it is definitely 21st Century style technology.

©CBS Paramount Network Television

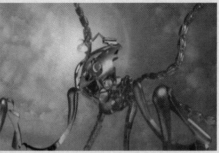

©Pendragon Pictures

Hines and Pendragon again showed what could be done on a very limited budget. The film was produced to not only take place in the late 1890's, but to look like it was actually shot in that period. Most of the scenes, therefore, have a heavy color overlay (amber here) to look like it was printed on color stock. Once again, the images are so fleeting that you pretty much have to freeze them to get a good look. After the initial reveal of the machines in a night scene (done that way, in this case, because that's how they're introduced in the book) we next see them as they storm across Shepperton. The vastly tall legs (three of them, naturally) have a very delicate, almost fragile look to them, which belies the tremendous "thumps" each makes as it walks. The body atop the legs, though, is a surprising departure for the movie that otherwise took such pains to be absolutely authentic. Rather than a simple ovoid body as depicted by Warwick Goble (such as on page 102, which looks for all the world like a kettle BBQ), they went for an insect reference, specifically an ant. The joints where the legs and tentacle manipulators attach to the body are obviously insectoid, and the control hood is very much an ant head. The small oval just above the head that is emitting the bright light is the reflector of the Heat-Ray that dangles from the bent aiming arm.

Finally, the machine in the Latt production even abandons the tripod stance for a fully insectoid six-legged walker. The body of this one is more like two horseshoe crabs stacked on top of each other, the upper one being the swiveling hood where the Heat Ray (which is closer to a green laser) is mounted. The frame on the far right shows an homage (whether intentional or not) to the Pal film with the machine walking down a city street setting fire to the buildings. This gives us a good sense of it's size.

©The Asylum Home Entertainment

Though hampered by a miniscule budget, many Wells fans consider this the definitive film version of the story. CGI (Computer Generated Imagery) techniques made it possible to realize the original-concept war machines with their tall tripod stance and tentacled manipulators, and the huge-eyed squid-like Martians. It's icing on the cake that an English film company produced it.

Completely out of left field was yet another film, this one produced by The Asylum Home Entertainment, normally a purveyor of horror films. The movie, written and directed by David Michael Latt, definitely reflects its roots. This one is set in northern Virginia with the climax in a devastated Washington DC. The aliens show up right away and by 10 minutes into the film they're busy inciner-

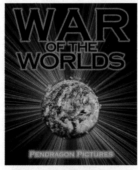

©Pendragon Pictures

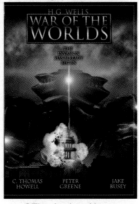

©The Asylum Home Entertainment

ating people. Even though the makeup and effects are more horrific than the other films, it is actually quite faithful to the story, allegorically. The reason it took fans by surprise (most fans didn't know about it until it was released) is that it was filmed under the working title of "Invasion;" presumably to let them work things out with the Paramount lawyers.

If that's not enough "War of the Worlds" for you, Jeff Wayne is producing a CGI animated version of his musical. Originally scheduled for release in 2007, the most recent information at press time tells us that the project is still alive and should be out sometime in 2009.

This story is so deeply entrenched in our culture that there will probably be versions of it being made for the next 100 years as well.

[Note: I started this chapter with an exciting, upbeat observation on current events, and must close it with an equally sad one. While right in the middle of writing it, I was shocked to learn of the passing of both Albert Nozaki and Les Tremayne. Both men died late in 2003 within a month of each other (Nozaki in November and Tremayne in December). I was shocked because I didn't realize either was still alive (Nozaki was 91 and Tremayne was 90). Had I known, I would certainly have approached them for their remembrances and to critique this chapter – JH]

This Island Earth

Evolution is an amazing concept. In the animal world, it sometimes works sequentially, such as how *Eohippus* evolved into the modern horse. Sometimes it works in parallel, where two animals from widely different starting points wind up very similar due to their occupying similar ecological niche (such as birds and bats).

The same thing happens in the evolution of story ideas. Ridley Scott's *Alien*, which went on to spawn an entire franchise, can trace its roots linearly to the 1958 low budget wonder, *It! The Terror From Beyond Space!*,[1] making it an example of sequential evolution. Conversely, one of the most intelligent Sci-Fi movies of the past decade, *Contact* (based on the Carl Sagan novel of the same name), is an example of parallel story evolution. The basic premise of *Contact* did not originate with Sagan, but can be seen over 30 years before the book (and 40 years before the film) in the Universal Studios production *This Island Earth* from 1955.

Universal has a long and accomplished history dating back to the silent era. Founded by Carl Laemmle in 1912, it is, in fact, the oldest continuously run studio in the US. Like Louis B. Mayer at MGM and Darryl Zanuck at Fox who guided their studios for decades, Laemmle was the central guiding manager, developing a distinctive style for the studio's films and ensuring a continuity of corporate culture.

An early peak of Universal's fortunes came at the dawn of the sound era. A series of now-classic horror films from directors like James Whale (*Frankenstein, Bride of Frankenstein, The Invisible Man*), Tod Browning (*Dracula*), and Karl Freund (*The Mummy*) firmly established Universal's style in the mind of the public. But the Laemmle family sold the company in 1936, and without a strong central leader, the studio floundered. After WWII, they were in serious financial trouble, and in 1946 merged with International Pictures, which produced mostly exploitative "B" pictures. This expanded the company's facilities and talent pool, but did little to help their cash flow situation.

©Universal Pictures

Their financial savior came from an unlikely source. The comedy team of Abbot and Costello were the most bankable stars of the late '40s and early '50s. Their long string of theme comedies, including *Abbot and Costello go to Mars* (1953),[2] rescued the studio and set the stage for the studio's second "golden age."

The new Universal-International, or "U-I" as it was known, lasted from the late '40s until the early '60s when the name was changed again back to simply "Universal." Even though the U-I era lasted some 16 years, the movies that the studio is most famous for, including *It Came from Outer Space* (1953), *Creature from the Black Lagoon* (1954) and its sequels, *The Incredible Shrinking Man* (1957), and, of course, this chapter's subject, *This Island Earth* (1955), were all made in the '50s. As Blake Lucas explains:

> U-I may essentially be thought of as a 1950s studio, excelling especially in genres widely considered to have thrived in the period, especially the Western, the melodrama, and the science-fiction/horror film, which was the '50s equivalent to the earlier horror films. U-I's cycle of Sci-Fi/horror films falls neatly in the middle of the U-I years, 1953 – 1958. So strong is the studio's identity with the genre and so good the reputation of this cycle among devotees, who will know all the films, that it may surprise anyone coming to it to know that they were not just churned out.
>
> However apparently modest in artistic ambition and economically budgeted, they tend to be beautifully, if resourcefully, produced, with great care lavished on their overall look and texture. The studio had under contract first-rank cinematographers, wonderful art directors and set decorators in a highly imaginative art department, excellent costume designers, a superior sound department and a number of gifted composers, fine editors and makeup people, and for the Sci-Fi cycle, brilliant special effects work.[3]

---

[1] See *Spaceship Handbook* for a brief entry on this film.

[2] See *Spaceship Handbook* for an entry on this film. Even though *Abbot and Costello go to Mars* did include flying saucers in a minor role as Venusian patrol ships, it was too incidental to rate an entry in this book. However, the film did have some influence on the production of *This Island Earth* as we will see.

[3] *The Science Fiction Film Reader*, Limelight Editions, 2004, pp. 70-71.

Meanwhile, in the print entertainment media during roughly this same period (late '30s to mid '50s), the SF pulp magazines had already reached their peak and were declining. Hugo Gernsback, who was largely responsible for SF pulps, had combined his various titles into *Wonder Stories* in 1930, which he ran successfully until 1936 when he sold it. The magazine became *Thrilling Wonder Stories* and went through a succession of owners until its final issue in 1955. For several years, the new owners focused on the juvenile SF market with Bug Eyed Monsters threatening scantily covered ladies. When Sam Merwin became editor in 1945 he dropped the wild covers and set about attracting higher class SF writers like James Blish, Leigh Brackett, Ray Bradbury, and A.E. Van Vogt.[4] Then, in 1949, he received a submission from Raymond F. Jones.

Jones was an electrical engineer who had worked for Bendix Radio during WWII. In the postwar era he supplemented his income by writing stories for the SF pulps. Like missile engineer G. Harry Stine (who wrote fiction under the pen name Lee Correy), his stories reflected an insider's knowledge of the technologies he used in the plots. In a 1992 interview he recalled the genesis of the idea that became *This Island Earth*:

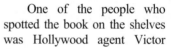

Practically every allied bomber [during the war] carried a Bendix radio compass, and one of its most important components was the instruction and maintenance manual. Afterwards, I wondered, "What if someone encountered an instruction manual that told how to build a gadget that was totally alien to all current technology?"[5]

A WWII-era Bendix radio compass.

This was the premise of the short story that Jones submitted to Merwin. He not only bought it, but encouraged Jones to expand it into a novella. It was published as *The Alien Machine*

in the June 1949 issue, taking up 52 pages. Even though it didn't even rate a cover mention (the cover being dominated by Leigh Brackett's *The Sea Kings of Mars*, the first appearance of what was to become a classic action/adventure SF story), it was very well received. So much so, that Jones wrote two sequels, *The Shroud of Secrecy* and *The Greater Conflict*. His agent, the equally legendary Forrest J. "Forrey" Ackerman, encouraged him to combine the three stories into a single novel-length book. Jones did this, but wasn't happy with the result:

The story was not really planned as a novel, it just kind of grew. The stories all dealt with the same characters and were just chained together. The book had some major defects which would not have been there if it had been planned as a novel from the beginning. Because of these defects, no one has ever been willing to publish it in paperback, in spite of the popularity of the movie.[6]

Jones titled the combined trilogy *This Island Earth*, and Ackerman sold it to Shasta Books, who published it in early 1953. Incidentally, Jones' assertion that "no one has been willing to publish it in paperback" is no longer true. In 1999, Forrey Ackerman contracted with Pulpless.com to do just that as part of his "Forrest J. Ackerman Presents" series.[7]

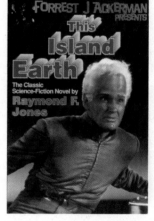

One of the people who spotted the book on the shelves was Hollywood agent Victor Orsatti. Agents are always on the lookout for "properties" that could be turned into films, but Orsatti had a special interest this time. One of his clients was Joseph Newman, a successful, if not especially famous director who worked in a wide number of genres. He had recently been talking with Orsatti about starting an independent production company:

I thought Westerns have been done from here to there, and I'd already made a series of melodramatic crime pictures. Going down the line at the time-1950-the only thing that hadn't been done to almost absolute saturation was science fiction. I'd never been associated with SF, but I'd been interested in it as a kid. Jules Verne was probably my greatest influence. So I told Vic (Orsatti) that I'd like to find something with a science fiction flavor as my first independent production.[8]

So Orsatti read the book with keen interest, and was favorably impressed:

It was really ahead of everything else at the time, so far above and beyond what was going on then, with all the gadgets, space travel and so on.[9]

---

[4] *Science Fiction Encyclopedia*, 1st Ed., Dolphin Books, 1979

[5] Robert Skotak, *This Island Earth*, *Filmfax* #33, June/July 1992, pg. 52.

[6] *Filmfax* #33, pg. 53.

[7] As of this writing, it is still available, ISBN: 978-1584450511

[8] *Filmfax* #33, pg. 53.

[9] Ibid.

The Saucer Fleet

Raymond Jones (inset) with some
This Island Earth memorabilia.

When he gave the book to Newman to read, the director was equally impressed:

> I think the overwhelming thing that came into my mind when [reading] Jones' novel was to illustrate, as the title suggests, that this planet is, in reality, a small island in an extremely vast and unknown universe. Our concerns might be, in the not-too-distant-future, forces and elements beyond the present known universe. I felt we couldn't be the only form of life this advanced mentally. There must be endless planetary systems.[10]

Orsatti and Newman formed their company and purchased the film rights to the book. In August 1953 they announced their intentions on starting production in the fall. Newman hired screenwriter George Callaghan to adapt the novel for the screen.

Up to this time, Callaghan was best known for writing the "Charlie Chan" comedy/mysteries and other low budget genre thrillers. He possessed a vivid imagination, though, which helped him through some early problems with the screenplay. In the book, the aliens communicated and manipulated things by means of thought waves. As Callaghan recalls:

> It had an intriguing start, the mystery of the gadgets and blueprints the two boys receive…but when it came to the point where they were just thinking, transmitting their thoughts and battling with the power of thought waves, there was no action![11]

Callaghan kept up with the latest scientific theories. He had heard of the neutrino, a chargeless, massless subatomic particle that had just been recently theorized by physicists.[12] The fact that the neutrino (whose existence would not be proven for another quarter of a century) could travel through solid rock like it wasn't even there impressed him as the perfect medium for a communications device. Since more than 99% of all the neutrinos striking the Earth pass completely through it without even slowing down, they could travel between the stars without being absorbed by other planets and stars along the way.[13] The concept had a secondary benefit. With such a powerful communications beam, it could be used destructively at close range, a type of "death ray."

Other contributions by Callaghan were moving the battle to Metaluna[14] and, in fact, changing it to a battle that audiences would recognize (in the book, the Llanna, as the aliens call themselves, conduct the war telepathically mind-to-mind with the enemy, which doesn't allow for much action). A physical battle also meant creating the surface and artifacts of Metaluna, plus all of the trappings of the journey (e.g. the heat barrier) and the hardware of the saucer (the green tractor beam, "conversion tubes," and bridge layout to name a few).

With the script underway, Orsatti needed someone to create the visual part of the package for him to sell to a studio for production. He picked the Dutch commercial artist Frans van Lamsweerde to convert Callaghan's incredible concepts to pre-production art. Van Lamsweerde (whose name, by the way, means "of lamb and sword," the symbols on his family coat of arms) had only immigrated to the US from his native Holland the year before to take a job at Disney Studios. In the Disney art department he was working on some of their classic animated features of the period (*Peter Pan* and *Alice in Wonderland*), but was not happy in the organization. What he really wanted to do was start his own motion picture design company. Orsatti's offer for a freelance job was just the thing to get him started.

Although well versed in the arts, van Lamsweerde had no experience with science fiction. He created the pre-production package working only from a partial copy of Callaghan's script and Jones' novel. He moonlighted the project in the evenings, working while his fiancé read the novel to him.[15]

For three months Newman and Orsatti shopped the package around to different investors to try and get backing for the project. They were trying to find a studio willing to act as the production house while they (Newman and Orsatti) maintained creative and financial control. The studio would be paid for their production costs, plus a percentage of the profits. Unfortunately, this approach wasn't working. Most investors still felt that Sci-Fi was too risky, and studios wanted to have more control over what went out with their name on it. Republic Pictures and United Artists were mildly interested, but far from

---

[10] Ibid.

[11] Ibid.

[12] We now know that neutrinos are not completely massless, but do, in fact, have a very tiny mass.

[13] Of course, this begs the question, if neutrinos pass right through everything, how do you detect them when you want to capture the signal?

[14] In the book, the battlefront stretched over several star systems, but their unnamed home world was not directly involved in the conflict.

[15] *Filmfax* #33, pg. 54.

One of Frans van Lamsweerde's wonderful pre-production pastels (this example of which, unfortunately, could only be found in black and white)

a deal. Then, in November 1953, Newman walked into the office of Jim Pratt at Universal.

Pratt had already had a long career in films. One of his first jobs was assisting Willis O'Brien on *The Lost World* in 1925. After spending more than a year on that landmark production, he worked his way up the studio ladder, and by the early '50s, was a production manager at U-I. After Newman and Orsatti made their pitch to Pratt, he surprised them by offering to buy it outright. With no other offers in the works, the two men agreed. As Pratt described later:

> By 1953 Universal had had its run on *It Came form Outer Space*. We'd done very well and had a lot of fun with it, so we thought it'd be damn good business to make another picture in this genre. When Joe and Vic came in with *This Island Earth* under their arms, we decided to go with it. Bill Alland was assigned [to produce] because he'd been associated with these kinds of pictures at the time.[16]

U-I was a very efficient studio. It had to be, since its low-end production budgets were always very tight, meaning they couldn't waste time in long negotiations. Within two weeks after Newman walked into Pratt's office, they had signed a contract giving U-I ownership of the production, but hiring Newman to direct. They held the first production meeting on 2 December 1953 with Pratt, Newman, screenwriter Callaghan, producer Alland and the studio heads from the production and budget departments. Naturally, the thing that Newman feared most, loss of creative control, started right away.

Bill Alland proved to be a formidable taskmaster. A

---

¹⁶ *Filmfax #33*, pg. 56.

¹⁷ Alland was part of Orson Welles' Mercury Theater troupe, and even performed several incidental parts in the famous *War of the Worlds* broadcast. When Welles moved to film with his legendary *Citizen Kane*, Alland played the part of the reporter to whom the Kane story is told.

former radio and movie actor,[17] he moved into production when he came to Universal in 1952 and quickly became famous as their lead genre producer for horror and Sci-Fi.

George Callaghan had been hired because Alland had reservations about the script, and wanted him on hand for re-writes.[18] Of course, the main thing they wanted to do was trim the story to fit the budget. Alland felt that the script had too many parallel storylines, and described too many expensive location sets and effects. Newman fought to retain as much of the story as possible, but in the end he had no real authority, and Callaghan wound up deleting entire subplots and simplifying what remained.[19] Alland even wanted to do away with what became the film's signature character:

> Alland argued strongly against the Mutant, insisting it was not only unnecessary to the plot, but cheapened the whole concept of the film. Management disagreed. Monsters had proven to be box office draws thought most of Universal's history.[20]

There were more reasons than just limited money driving the frantic pace. The studio production department had scheduled the film for a summer 1954 release (only eight months after the contract was signed!). Of course, they had no idea how much work this ambitious story, even when trimmed down, was going to take. The 1953 holiday season

William Alland (right) rehearsing stuntman Regis Parton on how to do a "monster walk" without being able to see his feet.

---

¹⁸ *Filmfax #33*, pg. 56.

¹⁹ *Filmfax #33*, pg. 60.

²⁰ Ibid.

was already upon them as they started in on the special effects groundwork, and on 29 December 1953, Callaghan turned in his final version of the script.

Alland assigned Richard Reidel as art director who started reviewing Frans van Lamsweerde's preproduction art. Reidel's boss, Alexander Golitzen, thought that van Lamsweerde's renderings were "too Disney-like," and told Reidel to simplify them to something that could be done on their limited budget. Newman managed to guide Reidel in ways of keeping the most critical concepts of the original art while still making it simple enough to produce.

Parallel tasks were started for the elaborate prosthetic costume for the Mutant, and the complex miniature sets for the spaceship and Metaluna scenes. Fred Knoth was in charge of the mechanical end of the special effects. He had to create the explosions, blast clouds, fiery spaceships, etc. He worked with Stan Horsley, the f/x photographer whose job it was to capture Knoth's effects on film.

Horsley had cut his professional teeth on such early epics as *The Hunchback of Notre Dame* with Lon Chaney (1923), and *Ben Hur* (1925). In a later interview with Philip Riley, he describes what it was like to work under the financial restrictions at U-I:

> I had caught hell by going over budget on an Abbott and Costello picture, so they were keeping an eye on me. I would do my experimenting at home since they told me I couldn't have any new equipment. I never paid them any attention though, and went ahead and built what I needed. I'd call it something else on the reports and they never caught on until the end.[21]

While Knoth and Horsley were developing their photographic effects (which wouldn't be shot for weeks until after the principle photography was finished), makeup department supervisor Bud Westmore was getting his crew started on the Mutant. Westmore had replaced the legendary Jack Pierce[22] at the studio in the mid '40s. His team had recently finished the grueling task of creating the Gill-Man costume for *The Creature from the Black Lagoon*, a much more difficult job than the Metaluna Mutant since the stuntman had to be able to swim underwater while wearing it.

Development on the Mutant started with a brain storming session the first week of January 1954. Director Newman and producer Alland were there along with Westmore and his expert latex casting technician, Jack Kevan. They reviewed dozens of alien illustrations from science fiction books and

Jack Kevan (left) and Bud Westmore (right) show the original Gill Man suit to Ben Chapman (who played the Creature in all of the non-beach land scenes)

magazines, and concluded that they wanted an insect-like creature, but for the head they used a discarded concept from *It Came from Outer Space*.[23] This was the origin of the skull-less brain and bulging eyes.

Kevan, along with his assistant Beau Hickman and sculptor John Kraus spent most of a week developing a dozen different concept models for review. A version of the one we eventually saw on the screen was chosen, except that there wasn't enough in the budget to produce the whole costume. As Beau Hickman recalled later:

> All of these models showed the entire body. Well, they thought it'd cost too much and take too long to go all the way down, so we decided to put pants on it and just have these crazy clawed feet come out underneath![24]

Stuntman Regis Parton (see photo, page 123) was assigned to play the Mutant. Parton had played the Gill-Man in *Black Lagoon* for all the non-swimming scenes on the beach, but found this more difficult:

> I had tunnel vision [when wearing the costume]. I couldn't see down and couldn't look down. If I took a tumble in this outfit, well then, I'd have likely wrecked it. And me too![25]

Even though Alland had given in to management on leaving the Mutant in, he still didn't like the script as delivered by

---

[21] *Filmfax* #33, pg. 60.

[22] Pierce was most famous for creating the makeup for all the classic Universal monsters: Frankenstein's monster, the Mummy, Wolfman, etc.

[23] *Filmfax* #33, pg. 57.

[24] Ibid.

[25] *Filmfax* #33, pg. 58.

[26] *Filmfax* #33, pg. 61.

Jeff Morrow, in his "Exeter" makeup, waits out a camera setup under some immense hot lights

looks, and booming baritone voice, he was the perfect mid-50's image of the scientist-hero in the mold of Gene Barry's Clayton Forrester from *War of the Worlds* (who was also a nuclear physicist, a pilot, a baritone and a…well, you get the idea).

U-I hired Faith Domergue to play the role of the beautiful, brainy and somewhat mysterious Dr. Ruth Adams. Domergue was probably the most famous member of the cast, having been "discovered" by Howard Hughes some five years earlier.

Russell Johnson clowns around with Faith Domergue between takes.

Supporting roles went to Russell Johnson, who will forever be "the Professor" from *Gilligan's Island*, as fellow scientist Steve Carlson. Lance Fuller was cast as Exeter's assistant, Britt (it was "Warner" in the book, but was finally changed to "Brack"

Callaghan. Production manager Jim Pratt agreed that the dialog needed work, and brought in staff writer Franklin Coen to massage it. "The first script was virtually unshootable," Franklin recalled, "it was not a practical script. Callaghan's work certainly had value, but it was not filmable."[26]

With these three things (the visual effects, the Mutant costume and the script re-write) now working in parallel, Alland started the casting.

The first major role cast was that of the lead alien, Exeter (a major improvement for the screen from "Jorgasnovara", the character's name in the book). Jeff Morrow was the studio's first choice and he jumped at the chance.

Universal contract actor Rex Reason was assigned to play the main human character, the handsome nuclear physicist and jet pilot, Cal Meacham. This was Reason's first leading role; up until then he had done only bit parts in westerns and costume dramas. With his chiseled good

before shooting began). Among the unnamed bit players was Richard Deacon (who gained fame as Mel Coolie on the *Dick Van Dyke Show*) placed at the helm of the Metaluna ship as pilot.

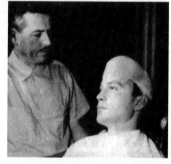

Lance Fuller partway through a multi-hour makeup session with Jack Kevan

Principle photography began in Stage 16 on 31 January 1954, a Saturday. They were able to start so soon by reusing some sets from *Abbot and Costello Go to Mars*[27] while the other sets were being built on Stage 28. They wrapped the principle photography (the parts that just used actors without special effects) just over a month later on 3 March. At that point the real work started: the special effects.

Special effects were done in Stage 12, which Universal called the "Process Stage." It was huge. At 200 feet long and 100 feet wide with a 50-foot ceiling (60 x 30 x 15 meters), it was the largest stage on the lot.[28] Designed to house any conceivable special effect, it had a heavy-duty

Charlie Baker sets up a glass matte painting using a storyboard for reference

Joseph Newman (back to camera) goes over some script points with Rex Reason and Faith Domergue .

[27] *Filmfax* #33, pg. 79.

[28] And it still is. In fact, since this movie was made it was expanded to 200 x 150 feet.

The Saucer Fleet

Image courtesy Robert Skotak

Stan Horsley (standing) and Charlie Baker (in crater) carefully measure the setup for the landing scene where the ship passes through the surface to the subterranean city. Note the Zagon ships to the left, one of which is passing over technician Ed Baldwin (far left) and in front of tech Eddie Stine.

overhead structure to support effects rigging. In the mid '50s the stage was the domain of effects supervisor Charlie Baker, whose career spanned the period from *The Invisible Man* (1933) to *Airport '77* (1977). It was his job to integrate Fred Knoth's miniatures into sets and process shots by Stan Horsley.

Back in this period, before motion-controlled cameras, a moving spaceship in a special effect shot had to really move. It was either suspended on wires from a trolley overhead, or supported from beneath on a dolly. Often times they needed to use both. A dolly track was built stretching the full 200-foot length of the stage. Props for either the Metaluna ship or the attacking Zagon cruisers (which were mostly invisible) had plenty of time to interact along this huge effects setup.

Elsewhere on the stage, the Metaluna globe (used in all the process shots showing the planet from space) was being built. This was actually the same six-foot (two meter) diameter Earth globe used in the Universal logo shot seen at the beginning of all their movies. Charlie Baker had

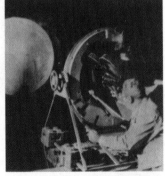

Stan Horsley photographs the "departing Earth" scene prior to converting the globe

built the globe in the late '30s, and now he was supervising re-skinning it with an alien surface.

Technician Ralph Howe in the process of re-skinning the globe to make Metaluna

Meanwhile, Frank Tipper was working on the optical effects in the animation department. These ranged from fairly simple (the neutrino beams from the Interocitor) to the film's signature effect, Meacham's airplane being drawn up into the saucer with the green tractor beam. While that may have become the signature effect over the years, it wasn't the most memorable. That title belongs to the "conversion tube" sequence when Meacham and Adams are prepared for the atmospheric pressure of Metaluna (see image sequence page 138). Stan Horsley describes the process:

> Frank did the layouts for the transformation scene. We got some anatomy books showing the skeleton, the muscles, the arteries, veins and so on, and Frank drew the muscles, as well as all the other portions that make up our composite anatomy, with chalk on black cards. These were photographed in black and white giving us our mattes. I had shot pieces [of film] of the people in the tubes and the people not in the tubes, just as we had done with the bottle scene in *The Bride of Frankenstein* twenty years earlier. This way we kept the reflections.[29]

> I laid out all of these dissolves with the different parts of the anatomy, with the pieces that were to be put together. I gave Ross Hoffman a set of dissolves all keyed in such a way that I would have different color records. As these black and white items would appear, they'd be in different hues and relationships, and the hues would change as they dissolved in and out. So that gave us the final effect, which was fully appreciated by the audience and yet was done at a relatively low cost.[30]

Another very effective special effect was Exeter's ship glowing red hot from the friction as it passes through the thermal barrier on its way to Metaluna, and later when it reenters the Earth's atmosphere. This effect was so real, with the ship glowing and vigorously spewing flame and smoke that the viewer had to believe it was really happening. The reason is because it really was! The saucer miniature (built by Fred Knoth's crew) was made from steel to survive very high temperatures. A special gas furnace was built on the stage that would heat the model up to about 1,500°F (800°C). A special boom dolly (much like a microphone boom) was built to hold the hot model from underneath and move it from the furnace into the camera's field of view. To get the shots, the camera was set up with a large wind machine to one side. When the saucer model was at temperature, the wind machine was started and two grips (stage hands) pulled the glowing model out of

---

[29] Note that this is exactly how George Pal's crew put the Martian Machines inside of the force field domes in *War of the Worlds*.

[30] *Filmfax* #33, pg. 80.

126

the furnace with blacksmith tongs. They placed it in the three-fingered support on the dolly that was quickly moved into range of the camera where technician Ed Baldwin sprayed a light oil mist into the wind machine. The oil flashed instantly when it hit the hot saucer, flaring into a huge flamethrower. The model stayed hot long enough to get the footage they needed.

©Universal Pictures
The oil droplets can clearly be seen hitting the red-hot model in this frame from the film's final scene

For other saucer scenes, such as the approach and landing on Metaluna, the model was flown on wires. With the long ceiling-mounted track available, a high degree or realism was achieved.

With a last few effects shots (explosions, etc.), the film wrapped all of its stage work on 14 December 1954, just over a year from the start of production.

Once the f/x photography was completed, the film moved into post-production. For the optical end, this was Roswell Hoffman's responsibility. His job was to take Joseph Newman's live action shots, Frank Tipper's animation, Charlie Baker's miniature work, and Stan Horsley's painting/process shots and combine them into a single negative. Well, that's not quite accurate. Since they were using the three strip Technicolor™ process, he had to produce *three* negatives that all line up in perfect synchronization. To do this he used a special multi-head optical printer that he had helped develop in the '30s (Hoffman's career ran from the early talkie *King of Jazz* in 1929 to *Earthquake* in 1974). The studio kept their special project under tight control. "Everything was very secretive," said Hoffman in a later interview. "The studio made me keep the doors to the optical department closed and locked at all times, and they wouldn't allow anyone in there to see what I was doing."[31]

While Hoffman wrestled with the optical printing, sound department supervisor Phil Scott was assigned the audio effects. He and his three man crew (Pat McCormick, Bob Bratton and Edwin Lucky) spent the most of the month of January 1955 analyzing and designing the sound effects. In typical Universal style, they came up with very effective results in a very economic manner.

The sound of the neutrino beams that come out of the three corners of the Interocitor was actually an acetylene cutting torch in full blast mode against a thick piece of steel. This was enhanced by a further air-blast sound. For the remote Interocitor beam (like the one that takes out the car that Russell Johnson's character is driving in the escape scene), the basic neutrino ray sound was augmented by the sound of airplane tire screeches (played backwards) from the Universal sound library.[32]

For the sound of Exeter's ship traveling through space, Bratton started with the sound of a jet plane (itself very fu-

The huge Metaluna surface set. Under the bright, flat lighting of this camera setup production still (left), the model really looks like a model. Compare that to the lighting test (right). This emphasizes the importance of lighting and atmospheric effects when doing "live" model work.

---

[31] *Filmfax* #33, pg. 82.

[32] *Filmfax* #33, pg. 94.

turistic sounding in the mid '50s), and filtered it to knock off the edge of familiarity. This created the archetypical sound for spaceships traveling in deep space (even though, as we all know, there are no sounds in space). Everything from *Star Trek*, in all of its incarnations, up through the *Star Wars* double trilogy have used rumbling/roaring spaceships as they pass by.

Some of the sound effects were startling in both their simplicity and effectiveness. To create the eerie phase-shifting sound of the Mystery Plane in flight, Edwin Lucky recorded a carrier wave (of the sort used by teletype machines in the 1950's and '60s) directly off of his ham radio receiver. On the other hand, some of the sound effects were exactly what they sounded like. For the attacking Zagon meteors, they used recordings from the sound library of artillery shells exploding that had been made in the 1930's.

After the sound effects, the only thing left was the music score. That job was assigned to composer/arranger Herman Stein. He was assisted by two staff members, the veteran composer Hans Salter and the youngest member of the Universal music department, Henry Mancini. (Yes, that's the same Henry Mancini who went on to write "Moon River" and "The Pink Panther." Mancini left the studio only two years later to work on the TV series, *Peter Gunn*.) Salter, though, was well past the middle of his career, and had a well-earned reputation for composing for horror films. Over the previous twenty years he had done the scores for many of the Universal horror flicks such as *The Wolf Man*, *The Mummy's Hand*, and *The Ghost of Frankenstein*.

Most of the score used a conventional orchestra to create the atmospheric moods, but for the sequences on Metaluna they trotted out that old Sci-Fi standby, the theremin. Also, the title music was performed on a Novachord, an early synthesizer-like keyboard instrument.

With the score complete, the production formally ended on 3 March 1955 when the copyright prints were filed. With the production almost a year behind its original schedule, the studio wasted no time in releasing it. The reviews were uniformly enthusiastic. All of the big publications, *The Hollywood Reporter*, *Variety*, etc., praised the film for its literate script and eye-popping visuals.

The Story

We know from the beginning that this isn't your typical mid-50's science fiction movie since the opening logo shot is in color. The spinning Universal globe is introduced with a brass fanfare but as that fades out to a moving star field the music changes to a haunting solo on the Novachord. The credits are in bright red, contrasting the black of space, but are unmatted (i.e. see-through), rendering them a bit difficult to read. We fade to the movie proper with an establishing shot of the mall in Washington DC as seen from the Washington Monument. In case we didn't recognize it, a title superimposed on the screen identifies it as such (and it is not see-through, so the effect in the credits was deliberate).

©Universal Pictures

We cut directly to an airport tarmac. A gaggle of reporters are interviewing Dr. Calvin Meacham while standing next to a Lockheed T-33 *Shooting Star*. They shoot questions at him as he gets into his flight suit (right over his business suit!). They are asking about a conference on atomic energy that he has just attended. He tells them that the government is starting a huge project to team atomic energy with electronics to create the true "pushbutton age."

One of the reporters quips about his flying the jet. "One of the boys at Lockheed handed me this one. I hope you taxpayers don't mind!" They all enjoy a laugh as Cal climbs into the cockpit.

©Universal Pictures

He takes off (from an airfield that looks more like southern California than Virginia or Maryland) and flies all the way to California in a montage scene of sweeping vistas.[33]

We dissolve to a close-up of a sign on a gate informing us that this is the private airfield of Ryberg electronics in Los Angeles (Meacham's employer). A guard opens the

---

[33] They leave out a couple of implied landings along the way since the T-33 has a maximum range of less than 1,500 miles (2,400 Km), even with the two wingtip tanks.

gate to let through a blue jeep that parks at the base of the control tower. A somewhat dumpy man hops out and bounds up the stairs. Meanwhile, up in the tower, we hear Cal calling in for landing clearance. As the man (Joe Wilson, Meacham's lab assistant) enters the tiny room he jokes to the controller that his boss is, "the only guy in the world who can travel by jet and still be late." Cal overhears this on the radio and has a joke of his own ready as he buzzes the tower at close to the speed of sound. He pulls up into a climbing roll, but as he levels off, the engine flames out.

The plane plunges earthward out of control, but just seconds before impact, a strange green light envelopes the craft and lifts it back into the air. Accompanying the light are an extremely high-pitched whistle, a weird pulsating oscillation and a rhythmic thumping almost like a heartbeat. It's clear that Cal is no longer in control of the plane, but whatever is, calmly lowers the landing

©Universal Pictures

gear and brings it in for a perfect dead-stick landing. As soon as the plane rolls to a stop, the green light disappears and the sounds stop.

The strange sounds are heard on the ground, too, as we can hear them while Joe bounds down the stairs from the control tower and drives out to meet the plane." How'd you bring it in?" asks Joe excitedly. "I didn't" Cal replies calmly while pulling off his flight gloves. "The controls went out, no power. I should be dead." Joe then describes the lights and sounds he witnessed. Cal warns Joe not to say anything until they figure out what happened.

©Universal Pictures

Joe Wilson watches his boss's plane in trouble, but auto buffs might be more interested in the (now rare) "Henry J." parked behind him.

We dissolve to the two of them walking down the hallway to Meacham's lab. Cal seems completely unruffled by his near brush with death (and his suit seems unrumpled despite being stuffed under a flight suit for the 10 hour flight). They enter the lab and Cal goes straight to a complicated viewing apparatus. His "Little Giant" experiment is all set up and ready to go as if he'd never been away. Not even pausing to don lab coats, Cal and Joe start working the controls that lower a nondescript slab into a complex base. The slab starts glowing while the instruments ping alarmingly. Rather than focusing on his work (obviously something critical is going on) Joe starts making small talk about Cal's trip. Seconds later the device in the test chamber explodes. With alarms sounding and red lights flashing, Joe proclaims, "The 'XC' condenser must have shorted out again." "Get the spare," orders Meacham. "Can't," replies Joe, "it shorted out yesterday."[34] He shows Cal a large can with two high voltage insulators coming out the top and goopy black residue running down the sides.

Joe goes on to explain that he ordered some spares and their supplier sent them some "bead" condensers (small red spheres about ½" (1 cm) in diameter with tiny contacts coming out either side). He tested them and found that they performed just like the big condenser. "If that's true," Cal muses, "we could build a generator

©Universal Pictures

Cal looks unbelievingly at the small bead capacitor that is supposed to replace the large "XC" condenser on the table

to power a whole factory that would fit in a matchbox" (that's quite a conceptual leap from just a small, high-capacity condenser).

Like any good scientist, Cal doesn't just take Joe's word, but tests the bead himself. When it performs even better than in Joe's test, he calls the supplier to ask where they came from. Instead, he learns that the supplier hasn't sent them anything in weeks. Joe then shows Cal a copy of the order he sent by teletype[35] and the receipt that came with the beads. The letterhead has no return address, and identifies the seller as simply "Unit 16." Meacham decides to do more tests on the beads to find out what they're made of.

---

[34] What? He was running this experiment while Cal was in Washington? He obviously needs more supervision.

[35] An amazingly advanced concept for 1955! It resonates especially true today in this age of Internet Commerce.

The Saucer Fleet

©Universal Pictures

Cal performing some destructive tests on the strange bead capacitor

We dissolve to Cal, now in his shirtsleeves and no tie, trying to drill into the bead with a diamond drill. Naturally, it doesn't even make a scratch. As he ponders this, the phone rings. It's the mechanic at the aircraft hanger. A complete inspection could find nothing wrong with the jet.[36]

Moving to the next day, we see a mailman deliver a large catalog to Cal's laboratory. It's from "Unit 16" again, the same place that the magic capacitor beads came from. Meacham thumbs through it, reading about amazing machines called "Interocitors." "What's an 'Interocitor'?" he muses out loud. Joe, showing why he'll always be an assistant and never a researcher, responds, "I don't know, and I don't want to know." Meacham, however, is intrigued and tells Joe to order one using the teletype and hope that the order is intercepted the way the previous one was.

The next scene shows the laboratory packed high with wooden crates. "Let's start unpacking!" Cal orders as he takes off his jacket, ready to get to work. We segue to the two of them up to their armpits in exotic-looking components spread over the entire floor. A quick

montage scene shows them assembling parts, and when done they have a large black console with an atom symbol on the left, a colored disk on the right, a huge triangular screen mounted to the top by a thin pedestal and a smaller triangular screen inset into the console's center.

Plug it in, Joe," says an impatient Dr. Meacham. Joe attaches a thick power cable low down on one end of the console. A glow starts emanating from behind some grilles in the base like large power tubes heating up (no power switch required!) but nothing else seems to happen. "Now what do I do?" asks Cal rhetorically. "Clear your screen, please!" says a voice from the console. The small screen is now a swirl of colors, and lights on the atom symbol flash in synchronization with the voice. It goes on to give Cal some technical instructions to activate the main screen. Soon it, too, is flashing colors which eventually resolve into the image of a man with an enormously tall forehead and completely white hair (even his eyebrows). His deeply tanned, almost orange skin makes the contrast with his hair even starker.

©Universal Pictures

©Universal Pictures

Build your own Interocitor. Some assembly required.

"You have successfully assembled an 'Interocitor,' Dr. Meacham, a feat of which few men are capable."[37] "Who are you?" asks Cal, completely at ease with this very unusual situation. "I'm called 'Exeter.' I'm a scientist like yourself." He goes on to explain that he's part of a group seeking exceptional scientists and that the aptitude test is the construction of an Interocitor.[38] While they chat, Joe sneaks a camera from a drawer and starts taking pictures of the screen. "I'm sorry, Mr. Wilson," says Exeter interrupting his own conversation, "but your camera will pick up nothing but black fog. Images on the Interocitor don't register on film."

---

[36] This is confusing. Does he mean he couldn't find any reason for the flameout or for the way the jet behaved afterwards? We know the reason for the latter (in hindsight), but if there's nothing wrong with the plane, then that means Exeter deliberately caused the flameout so he could demonstrate their powers in rescuing Cal. But that doesn't make sense since the Metalunans didn't want to reveal themselves until after Cal passed their test. It seems more likely that the flameout was caused by Cal's stunt and the Metalunans were hovering nearby to protect their prospective recruit.

[37] The man says this while looking straight at Cal. Obviously this device is two-way.

[38] This makes one wonder about all of the scientists who failed the test. Does this mean there are partially completed Interocitors cluttering up laboratories around the world?

Continuing with Cal, Exeter appeals to his scientific curiosity (not to mention ego) saying that their plane will land at the company airfield at 5:00 AM the next Wednesday. Meacham will have five minutes to get on board before it leaves again. Obviously done with his sales pitch, Exeter instructs Cal to place the large "Unit 16" catalog on the table in front of the screen (which also has the Interocitor blueprints on it).

Working an unseen control, Exeter causes red rays to emanate from the knobs on the three corners of the screen. The catalog and blueprints explode into flame and are instantly consumed.

©Universal Pictures

Cal and Joe step forward to investigate, but Exeter warns them "stay back or you may be harmed!" With a supercilious smirk on his face, Exeter starts some function of the Interocitor that not only turns the entire lab red, but does so at Exeter's end, too.

©Universal Pictures

Cal doesn't care for this at all, and before we can discover what Exeter's intensions are, Meacham dives for the power cord, yanking it violently out of the console. In a huge display of flame and smoke, the Interocitor self-destructs. Checking the remaining pile of glowing slag with a Geiger counter, Cal announces, "There's no reading now." (What? There was a reading before and you stood right next to it?)

We cut directly to a scene of Joe driving his jeep though an impossibly dense fog while Cal operates the manual windshield wipers. Joe is pleading with Cal to reconsider, but he (Cal) is almost delusional as he says he's expendable and that Joe can carry on the experiments with out him. Just as Joe finishes his opinion that no plane could possibly land in this fog, we hear the sound of engines approaching. A DC-3 lands and taxis right up to them.

A door in the windowless fuselage opens automatically and a small stair ramp lowers. Cal climbs aboard and we see that the interior is completely padded with only one seat,

©Universal Pictures

right in the middle. A glance forward to the cockpit shows a fairly ordinary set of DC-3 controls (control yokes and throttle/flap/landing gear levers in the center) but no instrument panel, save for a few minor gauges and switches in two overheads. Even stranger, there are crew seats, but no pilots! Additionally, the cockpit windows are blanked out

©Universal Pictures

with a pair of flat panels, and centered between these, another of the atom symbols with the flashing colored lights. Directly below the atom is a small triangular Interocitor screen. Adding to the unreal nature, everything (except the black control wheels and lever knobs) is colored the same silvery-white.

The Interocitor screen starts flashing and Exeter's voice welcomes Meacham aboard. As Joe is asked to clear the stairway, he pleads desperately for Cal to get off the plane, but it's too late. The stairs retract, the door closes and the interior of the plane is bathed in a green light accompanied by the same oscillating sounds as when Cal's jet was rescued at the beginning.

©Universal Pictures

Outside we hear the sounds of the big Pratt & Whitney's starting up as Cal's seatback abruptly lowers, catching him by surprise. As the plane taxis out, Joe stands there in the fog being blasted by the prop wash, wondering if he'll ever see his boss again.

The Saucer Fleet

We next see the plane flying serenely through the sky. It's now daylight, so it's been flying for some time. Inside, the same green glow bathes the interior, but it's joined by a brilliant orange and red emanating from the furiously flashing Interocitor screen that turn the cockpit crimson. Cal's seatback jerks upright just as quickly as it went down, jolting him out of a sound sleep. "Good morning, Dr. Meacham!" announces a cheery, disembodied Exeter. "Hope you slept well. We'll be landing shortly." Shortly is right. About five seconds later the plane shudders as we hear the chirp of tires on pavement.

©Universal Pictures

From the outside we see the plane taxiing down a dirt runway (which shouldn't have "chirped" the wheels) in a bucolic setting of rolling green hills with a large mansion atop the highest one. As the plane comes to a stop, we can see the blanked out cockpit windows and snazzy black and white paint job. Inside, the green glow and oscillating hum fade out and white light returns as the door opens.

As Cal steps out and surveys his surroundings, a brown Ford "woody" station wagon pulls up. "Where am I?" asks Cal of the pert brunette getting out of the car. "Georgia," she answers.[39] She introduces herself as Dr. Ruth Adams, a nuclear scientist that Cal immediately recognizes as someone he lectured with three years earlier, and had more than a professions interest in. She at first seems to partially recall him, but then claims they had never met.

Cal is a little miffed as she drives them up to the mansion. When they walk up to the door, Cal notices the plane that brought him flying off. For better or worse, he's stuck there.

©Universal Pictures

Ruth gives Meacham an overview of the sumptuously appointed main floor and points out many of the top level scientists milling about. She also identifies another tall-headed, white haired man as "Brack", Exeter's assistant. While Cal is taking this all in, Exeter emerges from his office. "Good Morning!" he offers cheerfully[40] and invites Cal and Ruth into his office. Cal goes in, but Ruth is delayed by an associate who walks up to ask her about a formula they'd been working on. He does it in such a stilted manner that it's obvi-

©Universal Pictures

ous that he's really asking her something else. In fact, he abruptly breaks off the conversation when another scientist happens by.

©Universal Pictures
What are they really saying?

Ruth joins Exeter and Meacham in the office. It's richly decorated in wood paneling, leather chairs and a large oak desk. Exeter gets right to the point: "[You want to know] who we are and what we're doing here," he directs at Meacham. He goes on to "explain" that he represents a group of scientists who are working to end war and they have gathered scientists of exceptional ability to this remote location for that purpose.[41] They are developing "exciting

---

[39] By the sun angles in this scene, it appears to be just about noon. If the plane left California at 5 AM, that means the flight time was only four hours (taking the three time zones he crossed into account) for an average speed of ~750 MPH (1,200 Km/hr). This is four times a regular DC-3's maximum cruising speed (190 MPH/310 Km/hr). Exeter's "green beam" was not only remotely flying the plane, but bumped its speed up to Mach 1! It also doubled its range. A normal DC-3 would have to make at least one fueling stop crossing the country.

[40] It's still morning? That DC-3 was even faster than we thought!

[41] The parallel to the Manhattan Project is probably intentional. In fact, the book version was even closer with Ryberg Instrument Corp. (the company's name in the book) located on the east coast and the secret alien facility in Arizona. This makes sense since Jones lived in the east, so he set the company there. The filmmakers, though, were from California, so switched the story around to opposite ends of the country.

new techniques" to "leap years ahead of the others." "There you have it," he concludes. It sounds like he's explained things, but it only takes a moment's reflection to realize that he hasn't given a single specific. He never said who they are, where they come from and why they look so funny. What techniques are they developing? For what purpose? Who are "the others?"

This doesn't seem to bother Cal whose only comment is, "it all sounds great," but then, "what if I decide I can't go along with you?" Anticipating this, Exeter replies, "then you're free to leave, but let's continue your tour from your chair." He waives his hand over a small box on the table. The bookcase along the wall spins around to reveal an Interocitor, probably the one he used to contact Cal with earlier. Exeter's face glows green as the screen springs to life. He shows Meacham his laboratory, which is not quite ready yet,

©Universal Pictures
Exeter operates his office Interocitor using a non-contact control, exactly like Klaatu with his communication system

but figuring that one peek had done the trick he asks, "Any reason you can't start in the morning?"

Before Cal can answer definitively, the Interocitor starts "ringing" with an incoming "call." Exeter shoos Ruth and Cal out of the office then answers the call. Although we can't see who he's talking to, it's clear that Exeter is being reprimanded by a superior for his lack of progress. Exeter protests that he is being forced by "the council" to use techniques that are "not effective here." He wants to discuss it with the council, but is instead ordered to continue on with the plan (that isn't working). He then reports that Meacham completes the roster of scientists for the project, and, looking slightly stricken, agrees to continue as planned.

We dissolve directly to dinner. With Mozart's "Nachtmusik" playing in the background, we see a very elegant, but stuffy, meal coming to an end. After telling Exeter that the meal was more than promised, Meacham invites Ruth and Steve Carlson (the colleague that Ruth was talking to earlier) to "get some fresh air." They don't get any further than the main hallway before they stop and begin noting all of the other people leaving the dining room and their specialties. Cal notices that they're all scientists involved in the production of nuclear energy, but that there is no one in the support fields to create the practical applications. He then talks Adams and Carlson into showing him his lab now instead of waiting for morning as planned.

They seem hesitant, but agree. Brack sees them go into the elevator leading to the lab areas. Once in the lab, Cal is impressed that it has the same equipment he was working with at Ryberg. Pulling a large, suspended lead slab between them and the Interocitor (everyone's got one, it seems), he turns to the other two and the conversation becomes confrontational. He demands to know what's going on...now! Feeling free to talk, Ruth admits that Cal was right about knowing her from before. They were just being cautious that Meacham hadn't been given the "sunlamp treatment." It's a mind control device that lobotomizes the areas of the brain controlling free will. As near as they can tell, they, and possibly Professor Engelborg from Germany, are the only ones to be spared the "treatment," although they don't know why. Cal then wants to know who Exeter works for. They don't know; only that they're trying to come up with new sources of nuclear power very quickly.

Back upstairs, we see Exeter and Brack marching in perfect step to the office. Exeter is berating Brack for not telling him sooner about the scientists being alone to converse unsupervised. "If you're so concerned," says Brack, "why don't you let me use the transformer?" Exeter protests that the use of the transformer (the official name of the "sunlamp") is "morally abhorrent" and makes the subjects set up mental blocks that cause the scientists to be ineffective, and thus useless to them.[42]

By now the Interocitor has warmed up and they can see the three scientists talking in Meacham's lab. The view is vignetted since they are looking thorough the lead slab that Meacham thought would block the prying eye of the Interocitor, but as we can see, Exeter and Brack have no trouble hearing and seeing everything. Meacham, Adams and Carlson are apparently discussing escape. Suddenly, Exe-

©Universal Pictures
Exeter's own little United Nations of scientists

---

[42] Now you know the "technique" that Exeter was protesting in his conversation with his superior earlier.

The Saucer Fleet

ter's cat, Neutron[43] (who likes to frequent Meacham's lab) lets out an angry, frightened yowl as he hops up on the lead slab. Meacham immediately understands that they are being watched and shifts the conversation to some platitudes about what an honor it is to work for Exeter. "Our little Neutron gave us away, he felt the impulses," says Exeter. Brack is upset (at both the cat and Exeter's refusal to follow the council's orders) but they turn off the Interocitor.

©Universal Pictures

The next morning we see Cal in his lab copying down a formula being dictated remotely by Exeter from his office. He then asks Meacham to slide the lead plate in front of the Interocitor and stand back. He wants to acquaint Meacham with "another of our accomplishments." He pushes a switch and, like in Cal's lab back in California, red beams issue forth from the three knobs on the corners of the Interocitor screen melting a perfect hole in the lead.[44] "Why does a communication device need a destructive ray?" he asks. "Television waves can't penetrate mountains, but neutrino rays can. You've just observed one in action." He goes on to say that the rays can be used for other purposes. "One was used recently to save your life." "The green light?" asks Cal rhetorically, solving an earlier mystery. Exeter then admonishes Meacham not to meet with any of the others outside of normal channels.

In the next scene we see him doing exactly that as he meets again with Adams and Carlson behind the perforated lead slab. They keep Neutron around to detect if they're being watched by Exeter. Carlson shows him some sketches he's

©Universal Pictures

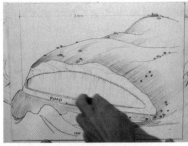
©Universal Pictures

made of Exeter, Brack and a hollowed-out hillside covered by "an acre of canvas," but he doesn't know what's inside. Meacham says goodbye to Neutron as they leave, apparently planning to escape.

Meanwhile, Exeter and Brack are communicating with their superior again. This time we can see who they're talking to, another odd character like themselves with white hair and eyebrows, and an extremely high, indented forehead. He has Exeter confirm that they've abandoned "Plan A" and put an alternate plan into effect. Apparently things are going badly back at home base. "Our ionization layer is failing rapidly" (whatever that means). Brack and Exeter are being recalled. Exeter pleads for a little more time because Meacham and Adams are "achieving positive results." He's instructed to bring them along so they can complete their work at headquarters.

While Exeter and his boss are having this little tête-à-tête, the three scientists sneak out the front of the building and drive off in the station wagon. Inside the house, the conversation wraps up with Exeter being instructed to destroy the installation when he leaves. No word as to what will be the fate of the personnel. As Exeter leaves to make preparations, Brack takes over the Interocitor and immediately tunes in on the escaping car as if he knew about it all along. He reaches down and works a hidden control that starts throwing red neutrino bolts at the car, Zeus-like from above (we'll ignore the fact that there is no Interocitor at the far end for the bolts to emit from).

Carlson, who's driving, tries desperately to avoid the attack, but realizes it's hopeless. He tells the others to jump out when he stops the car with the understanding that they'll all run for the lake. But once Cal and Ruth are out and bounding down the hill, Steve drives off to draw the attack to himself. From the water, Cal and Ruth watch horrified as Brack scores a direct hit and the car explodes in a gigantic red, yellow and green fireball. They have barely assimilated

©Universal Pictures

---

[43] In the earlier lab scene, Ruth had said he was named "Neutron" because he's so positive, a surprising science gaff in this otherwise very literate script (obviously his name should be "Proton").

[44] Why does he think it's necessary to demonstrate this to Cal? He'd already seen it earlier when Exeter destroyed the catalog and plans. Then again, drilling through a 4" (10 cm) thick lead plate is a bit more impressive.

what's happened when Professor Engelborg (the only other one not to have gone through the thought transformer) appears speaking excitedly in German. Cal screams at him to get away, but it's too late. With a single shot, Brack scores another bulls-eye and Herr Professor vaporizes in a smaller version of the explosion that killed Carlson.

This takes care of all the non-thought-transformed scientists except for the two they're supposed to take with them, so Brack stops the attack. Cal and Ruth don't know that, though, and they make a break for the airstrip.

We cut back briefly to the outside of the mansion where we hear an oscillating humming sound and see a pulsating green light coming from behind the hill. As the two scientists continue their breakneck run, a large silver saucer rises up from behind the hill, surrounded by a green glow. The glow (which was apparently coming from the ground) stops as the craft glides off and out of sight.

©Universal Pictures

Our two heroes reach the small plane[45] still parked on the edge of the runway. It's not only unlocked, but the keys are in it. Cal fires it up and pushes the cold engine to an immediate take-off. No sooner are they airborne than the mansion blows up in a giant white explosion. The little plane is rocked violently, but stays airborne just above treetop level. Exeter's ship appears above the plane and, in the film's signature shot, pulls it up though a round portal in the base of the saucer using

©Universal Pictures

©Universal Pictures

a green neutrino beam. The accompanying sounds are the same as when Exeter rescued Cal's jet at the beginning. (By now it should be obvious that green neutrinos are good and red neutrinos are bad.)

Once inside the ship, the portal is closed under the plane. Cal and Ruth get out cautiously and look around. Directly above the plane is a large, pulsing, green sphere floating in a yellow cage suspended from the walls by three lightened beams. This, apparently, is the neutrino ray projector. Other fantastic machinery lines the walls, looking very much like the Krell furnace room machinery in *Forbidden Planet* thanks to the similar technique used in creating the scene.

Another tall-headed, white-haired alien (it's obvious now that they're from another planet and not just weird humans) appears to escort them. He's wearing the same gray tunic seen earlier on the leader during the Interocitor call. A cut to the outside shows the ship racing away from Earth at tremendous speed. Back inside, the escort completes his task bringing Cal and Ruth to the control bridge. We can see now that he's also wearing a

©Universal Pictures

strange transparent helmet. Not a large bubble, like so often seen in Sci-Fi movies of this period, but a form-fitting prosthetic appliance that perfectly matches his odd cranium (and, with his white hair makes it appear that you're looking directly at his brain).

©Universal Pictures

The Saucer Fleet

The control center is a large, sparsely appointed room with flying buttresses ringing it to support the cathedral-like ceiling. In the center is a raised platform for the commander. Instead of the ubiquitous atom symbol, there is a large 3-D model of an atom with at least 50 or 60 nucleons. The "electrons" are multi-colored and the whole thing pivots back and forth. In front of the platform is a large round screen with an operator. There are no Interocitors as such, but the screen control panel has the colored wheel control and the flashing atom symbol.

©Universal Pictures

As we enter, Exeter is in the middle of issuing a command to be careful about the temperature since, "our two passengers are very sensitive to heat." All of the bridge personnel are wearing the odd, clear helmets except, strangely, Exeter (although he is wearing the gray tunic uniform). He strides over to greet the two scientists while revealing his true nature for the first time. His salutation is warm, even apologetic, but it's completely rebuffed by Meacham who is furious over the mass killing he has just witnessed. Exeter claims

©Universal Pictures

that was beyond his control and asks for the two of them to reserve judgment since they are starting a "long strange journey."

Their destination is the planet "Metaluna" in a different solar system.[46] To convince them that they're under way, they move over to the "Stellarscope" (the big observing screen). He has the operator bring up an image of Earth (perfectly centered on North America, of course). Before we can be shown anything else, we see the ship from

©Universal Pictures

the outside again, but it slowly changes from silver to light red. Back inside, Cal and Ruth are starting to show the effects of the rapidly rising temperature, although none of the Metalunans are bothered in the slightest. The walls turn a bright red and steam (steam?) starts billowing up from their base. Outside again, we see the ship glowing a dark

©Universal Pictures

red and streaming copious amounts of flame and smoke as they pass through the "thermal barrier."[47] Exeter warns that the temperature will be "unpleasant" for a sort time.

After a minute or so, things return to normal and the two scientists start peppering Exeter with questions on how the ship operates: How did they control the temperature just then? What power source are they using? What's keeping them from floating around? Exeter grins broadly at that last question, like a parent whose children suddenly started seeing things from an adult perspective. He answers only the last one, saying that they generate their own gravity, but ends the Q&A abruptly when

©Universal Pictures

©Universal Pictures

---

[46] The movie does a much better job than the book in getting its astronomy correct. Jones may have been an engineer, and could spin a good tale, but his handling of the cosmos was quite slipshod. He refers to the war as taking place over "several galaxies" when he obviously meant star systems.

---

[47] This is a welcome change from the already cliché meteor storm that Sci-Fi ships of this era always seemed to encounter.

he says, "If we're going to get you to Metaluna alive, there's a little procedure you're going to have to go through."

They walk over to another duty station behind the control platform. It has a control console similar to the Stellarscope, having an Interocitor style color wheel control on both sides and a lighted atom symbol in the center (but no big round screen). The operator sits facing three "conversion tubes" that are currently each occupied by a Metalunan in a gray jumpsuit and clear skullcap helmet. They're surrounded by swirling smoke or mist. Exeter continues, "Metaluna's atmospheric pressure is like that in your greatest oceans. If we entered Metaluna's orbit[48] without conversion we'd be crushed to death." Meacham notes that going the other way their bodies would disintegrate. "Correct," says Exeter, "If we're fortunate enough to return to Earth."[49]

©Universal Pictures

©Universal Pictures

At that point the process completes for the three crewmen in the tubes. The mist clears, the tubes rise and with a sound like crackling static electricity, the "magnetic" handrails release their grip. The one on the right turns out to be Brack, who is instructed to take the two Earthlings out and get them changed into the special Metaluna suits. They don't get the cool clear helmets, though (maybe the ship's quartermaster didn't have any small enough to fit the humans' tiny crania).

As they step onto the platforms, Exeter instructs them to place their hands over the grip rails. The same crackling static is heard and their hands are drawn spasmodically to the rails. "They're magnetized", he explains (which would mean something if their hands were made of iron…). Exeter gestures to the operator who, we see, is a woman, the only female Metalunan seen in the entire film (played by Char-

lotte Lander in her only film appearance). She starts the conversion process by lowering the tubes and filling them with the smoke/mist. Cal and Ruth talk for a few seconds before loosing consciousness. We see the process reduce them to skeletons, and then rebuild them (see sequence photos next page). Before it's done, though, we hear the Master Control announce that they're entering the enemy sector and that he's awaiting instructions.[50] We note that the atom model is swinging back and forth rapidly with the "nucleus" glowing brightly, so something important is up.

Exeter and Brack (who's no longer wearing the helmet) stride onto the bridge. Exeter barks some commands to avoid contact with the enemy and conserve power (immediately the atom dims and slows its oscillations). They then walk over to the conversion tubes that have just finished and are rising. Cal and Ruth stagger off of the platforms and Exeter and Brack help them down the stairs towards the view screen. In sharp contrast to the military style orders he was just giving, Exeter now gently acknowledges the two scientists fatigued conditions and tells them that they won't want to miss the approach to Metaluna.

©Universal Pictures

[48] He probably meant to say "atmosphere" here.

[49] We have to assume that the conditions on the ship are still close to Earth-normal so the question arises as to why the people already converted aren't disintegrating.

[50] Even though "Master Control" sits on the pedestal that, if this were the *Enterprise* bridge, would be the commander's chair; the station is more like the helmsman. Exeter is the commander.

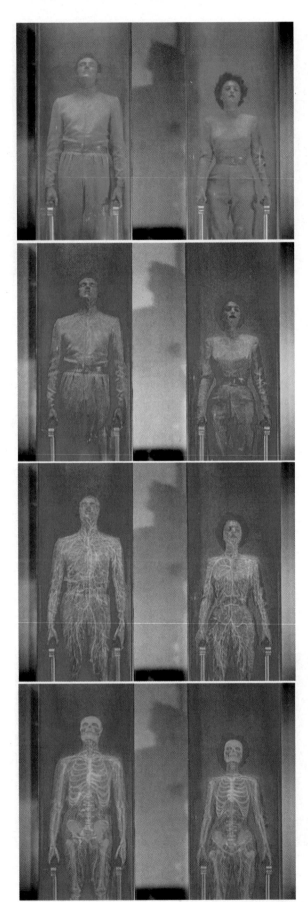

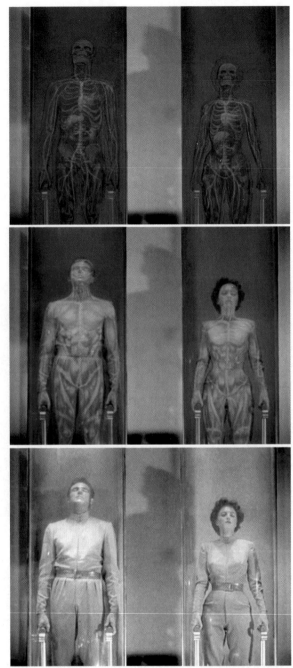

Stan Horsley's brilliant "conversion tube" sequence (see page 126 for a description of how it was done). In the left column from top to bottom we see 1) Cal and Ruth more-or-less normal, but beginning the process, 2) becoming energized with a false-color solarizing effect, 3) exposing the nervous system, and 4) the final reduction to the skeletal structure. In the right column from the top we have them being reconstructed 1) with the addition of internal organs, 2) outside ligaments and musculature, and 3) process complete, back to normal.

We're sure we weren't the only ones trying to sneak a peek at some of Ruth's "internal organs" along the way!

©Universal Pictures

He leads them to the observing chairs[51] and motions to the Stellarscope operator. The screen lights up with a star field and two bright objects traveling in parallel. Meacham says that they appear to have "some intelligence controlling them." Exeter confirms this and explains that they are weapons of the planet "Zagon" with whom Metaluna is at war. "They're going to hit us!" shrieks Ruth, and the two Earthlings recoil in anticipation of impact. The Metalunans, however, watch calmly as the ship's gunners destroy the menace with red neutrino rays. Exeter explains that the Zagons use flaming meteors as weapons, guiding them with their ships up to the release point. They dispatch a couple more "meteors" and the Zagons apparently give up.[52]

The ship travels on until Master Control announces that they are approaching Metaluna. Exeter calls up a series of increasingly magnified views of the planet. It goes from an attractive green globe to a hellish world with a desolate cracked surface and strange red "hair" in bands reaching into space. All over the surface we see flashes. Exeter finally drops the bomb, so to speak, and reveals his true mission. The fuzzy red zone above the planet is the "ionization layer" that forms a shield against the Zagon me-

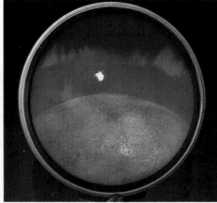

©Universal Pictures

teor weapons. The layer is artificially generated using atomic energy, but they've used up all of their available uranium and the ion shield is failing. As usual, the Earth scientists are one step ahead of the explanation and figure out that's why Exeter was sent to Earth to recruit all of the nuclear scientists for a crash program in generating nuclear material.

We see that the Zagon meteors are now getting through more than they aren't, and Exeter tells the Observer ('scope operator) to go back to the normal, unmagnified view. It doesn't pull back very far as the ship has already made contact with the ionized layer and is beginning its descent to the surface. They break out of the clouds and glide over a desolated, post apocalyptic surface of craters and smoking ruins (it could have been used as the prototype for Venus in *First Spaceship on Venus* three years later).

This is the best effects work in the movie as the gleaming ship travels over the cracked and mangled surface, a stark contrast between the height of Metalunan technology (starships) and its current state on the edge of oblivion. While watching the surface of Metaluna, we notice that are no signs of civilization at all. No remains of buildings, roads or aircraft, just de-

struction. As we pull closer to the mangled surface, we can see that some of the "craters" are actually openings in the crust and we can catch glimpses of buildings below. The whole society has moved below ground!

©Universal Pictures

The ship floats effortlessly through one of the holes, just large enough for it to fit, as meteors continue to explode all about it. Moving underground with it, we can see some of the fantastic architecture of the remaining buildings, as the ship pulls up to a mooring mast.

©Universal Pictures

---

[51] They're really not chairs, but more like a slightly reclining board with armrests and a little "butt shelf" to locate yourself. The viewer is still essentially standing when using it.

[52] Actually, this scene has some pretty outrageous astronomical terminology errors, mixing up "planet", "comet" and "meteor" interchangeably. As with the cat's name earlier, this is surprising for this otherwise accurate script.

The Saucer Fleet

©Universal Pictures

©Universal Pictures

A large square hatch opens in the side of the ship and Exeter steps out onto the saucer's rim with the two Earthlings. Brack, back in the skullcap helmet, stays behind in the doorway as Exeter instructs him to reassign the crew to local duty. Then he ushers the doctors into the elevator in the mooring mast. Meteors now start breaking thorough into this underground complex as the elevator travels down its green beam.

©Universal Pictures

They transfer into a transportation pod[53] that will take them to "The Monitor." On the way, Exeter points out, with great anguish, the shattered remains of his civilization (we presume that he's seeing much of this devastation himself for the first time as things have deteriorated significantly in his absence).

They pull up to the stop for the Monitor's Palace and go in. Everything inside is in eerie false color. Their faces are blue and their gray jumpsuits have a green tinge.[54] Inside we see the familiar (by now) Metaluna technology. The Monitor sits on the same raised platform as the Central Controller onboard the ship, although the "nucleus" of the

"atom" glows purple. Around him are the same sparse control centers: the Stellarscope against the wall where our party entered, and the conversion tube controller (without the tubes).

Cal and Ruth are introduced to "The Monitor", the supreme ruler of Metaluna. It is the same person (if that's the right term for an alien) we saw on the Interocitor in Exeter's office back on Earth. His manner in addressing others is very stiff and mechanical.[55]

Even though the scientists had been brought there to continue working on uranium synthesis, we learn that it is too late. They are expecting Zagon to start a final crushing attack now that their shields are failing. All they can hope for is maintaining the shields locally so that the handful of Metalunans still alive can begin their relocation. "Relocation?" asks Meacham, "To where?" "To your Earth," says the Monitor matter-of-factly. Seeing the shock and anger of the two Earthlings at this little revelation, Exeter steps in and tries to soften the blow. "A peaceful relocation. We hope to live in harmony with the citizens of your Earth." But, "we would be your superiors, naturally," finishes the Monitor.

©Universal Pictures

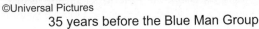
35 years before the Blue Man Group

---

[53] Attentive genre fans will recognize the pod as the Venusian "atom car" from *Abbot and Costello Go to Mars* with its tail fin removed.

[54] Like *War of the Worlds*, this movie was shot in 3-strip Technicolor®, so this effect was quite easy to do by swapping the negatives during the printing process. The swap was different here than in the "epidiascope" scene since that one had green faces and purple clothes. *This Island Earth* was, in fact, one of the last major films to be shot in the 3-strip process. The film *Foxfire* (in production at Universal simultaneously with *This Island Earth*, but released slightly after) was the very last 3-strip feature film made in the US.

[55] This was probably intended to show a pre-Vulcan concept of someone completely emotionless and logical, but it comes across as just bad acting.

This really gets Cal's dander up and he lets the Monitor have it verbally, but his (the Monitor's) only response is to Exeter, "Do you still insist that we can allow any of these Earth creatures to have free minds?" "I do," says Exeter in defense of Earth. But the Monitor is not amused. "You have wasted our time. Take them to the thought transference chamber!"

They head back outside and Exeter leads them nonchalantly down the corridor to the chamber room as meteors strike closer. Ruth balks at going into the chamber and Cal joins her in bolting, but before they can get two steps, a hideous creature emerges to block their escape. Exeter, again apologetically, explains that these insect-like creatures are mutants, bred for generations to do menial work (how can they be mutants if they're selectively bred?). The thing is completely repulsive with what appears to be an exposed brain at the top of the head, two large dead eyes, pincer

©Universal Pictures

claws and some sort of external vascular system outside of its exoskeleton.

Exeter pleads with them to go into the chamber voluntarily and he promises that they won't really be converted. Cal doesn't believe him and decks him with a rabbit punch. As Cal and Ruth try to dodge around the mutant, a meteor scores a direct hit on the Monitor's Palace and the creature is buried in a pile of collapsing rubble. The two scientists run for the transport tube and a quick shot shows us the Monitor lying in the collapsed remains of his control center, dead.

Just as Cal and Ruth reach the tube car, a recovered Exeter sprints towards them. This time he convinces Meacham that he really wants to help, and the three of them hop in and head back to the saucer. When they arrive, they find another mutant guarding the entry hatch. Exeter orders it to stand aside. It does so, but after Cal and Ruth get by, it suddenly attacks Exeter, plunging a pincer into his abdomen. The hatch leads to the cargo hold where the plane is stored, so Cal grabs a fire extinguisher from the cockpit and uses it to beat the mutant savagely on its exposed brain. They pick up Exeter and go inside. Cal closes the hatch remotely so they don't see the injured mutant drag itself into the ship as the hatch closes.

In the control room, Exeter sits at the Master Control station and instructs Ruth to turn on the Stellarscope. From the intensity of the bombardment it's clear that no defenses

©Universal Pictures

are left. In fact, we see, for the first time, the arrowhead-shaped Zagon ships atop each meteor as it's directed to its target. Exeter guides the ship up out of the underground city and, buffeted by explosions, out into space. The Zagons toss two final meteors at them before concentrating all of their efforts back on Metaluna. Exeter dispatches one with a neutrino beam, and the other he simply dodges.

From a safe distance they watch as the impact flashes on the planet fuse into a uniform glowing surface and the whole planet ignites into a star! Exeter's reaction is not what you'd expect from someone who has just watched the annihilation of his home planet. Ever the optimist, he is almost wistful as he philosophizes that maybe this new "sun" will warm some other planet and create life.

©Universal Pictures

Exeter joins Cal and Ruth in a conversion tube. Their cycle has already started (Exeter's has no smoke yet). Note Exeter's abdominal wound from the Mutant.

Once safely underway towards Earth, the trio have to be re-converted back from the Metaluna standards. Cal and Ruth get into the tubes while Exeter sets the controls to automatic cycling before joining them. As the process progresses, Ruth wakes up first and peers out from the smoke-filled tube only to see the injured mutant drag himself into the control room. It heads for her tube (naturally) just as it finishes and starts to open. She tears her hands free of the "magnetic" handgrips and runs around the control room with the mutant in pursuit. Just as it gets Ruth in its clutches, Cal's tube starts to open.[56] Before he can break

---

[56] Why didn't his tube open at the same time as Ruth's? (other than it being a plot device). They started together and previously they finished together.

The Saucer Fleet

free of the hand-grips, the mutant suddenly releases Ruth and collapses. He wasn't converted for the pressure and conveniently evaporates just as Meacham arrives to "save" Ruth.

©Universal Pictures

We dissolve to the ship approaching Earth with our trio still sitting on the floor and/or Master Control. Exeter tells them to get back into the plane down in the cargo hold and he'll drop them off (literally!). "What will you do?" asks Ruth. "I'll explore," says Exeter, "maybe find another Metaluna." Cal isn't buying it. Outside of the fact

©Universal Pictures

that Exeter's injuries are so severe that he can't even sit up straight, he (Cal) has learned something of how the ship works. "You're a liar," he says gently to Exeter. "You've used up all your power bringing us here." "Come with us," Ruth begs in full-on nurturing mode. "You have things to teach us," adds Cal. Exeter ignores him and says only, "It's time for you to go."

Cal and Ruth head for the doorway, stopping only to wave goodbye. Exeter returns a feeble gesture and, as soon as they're through the door, collapses. The two scientists get into the plane and barely get it started before the hatch it's sitting on dissolves and it drops into the air.

©Universal Pictures

The little Stinson stabilizes and Cal and Ruth fly off into the sunset, metaphorically speaking. Exeter, though, has a much less pleasant fate. We see the ship dive for the ocean and begin to glow red as it heats up from the air friction (see image, page 127). Inside, Exeter, struggling to stay upright at the controls, watches the 'scope as the ship plunges towards the ocean. It finally bursts into flame and smashes into the water as a giant fireball, thus making the Metalunan race completely extinct.

With this sad goodbye to the conflicted alien, the final "The End" superimposes over the flames.

©Universal Pictures

The Vehicle

We think we know a lot more about Exeter's ship than we really do. We see throbbing machinery in the bay where Meacham's airplane is brought aboard. We hear Exeter issuing commands to the crew operating the machinery, and even see some of it (the converter tubes and view screen, mostly) working. After passing through the thermal barrier, he starts to answer the two Earth scientists' questions about how their spaceflight technology works, but we don't realize that after answering only the first question, he interrupts himself to put them into the converter tubes, and never goes back to finish.

We do know that it is fission powered; probably using uranium for fuel since that's the material they needed "in gigantic quantities." This may seem a little primitive today, but remember that this movie came out the same year that the *USS Nautilus* was launched, a uranium powered submarine and the absolute pinnacle of high-tech vehicles at the time.

We can speculate on the origins of the design, even though we were not able to confirm this by any production documents. In 1952, there was a famous photo of a "real" UFO making the rounds. Supposedly taken off of the coast of Brazil near Rio de Janeiro, it showed a classic round saucer shape, perhaps slightly elongated, with a smaller second tier stacked on top of it, and a small knob shape on top of

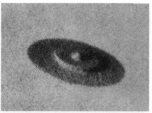

©Universal Pictures

The 1952 Brazil "sighting" (left) with the saucer part of the image enlarged (top right). Compare with the design of the Metaluna ship (bottom right).

©Universal Pictures

that. There is even a hint of some portholes or other structure around the rim of the second tier.

Even though this photo was quickly exposed as a hoax (the light on the trees is coming from the right, but the light on the saucer is coming from the left) it would have been in all of the UFO literature that Stan Horsley and Charlie Baker reviewed when coming up with the design. Along the way, though, they made some definite esthetic improvements. In the Brazil photo, the saucer may or may not be elongated, but if it is, then the right edge, which we presume to be the "front," must be slightly wider than the "back." This was exaggerated in the shape of the Metaluna saucer rim, and the second tier structure on top of it was further streamlined (for atmospheric flight, no doubt) into a teardrop shape. The knob on top was enlarged and moved forward of center to give it a more commanding position as the bridge. It was also given a streamlined shape.

As for the portholes, we simply assume that's what they are. Never, when inside the ship, do we see any sort of windows or other ways of seeing out except for the Stellarscope.

Another interesting feature of the Metaluna ship, seen here in the contrast-stretched shot, is a checkerboard pattern on the rim. Knowing that the prop was made out of steel tells us that these are machining marks, often seen after a part has been cross milled for flatness. The odd thing, though, is that such marks are easy to remove by a light sanding or buffing so they had to have been left on deliberately, but what they are supposed to be (in the context of the story) is anyone's guess.

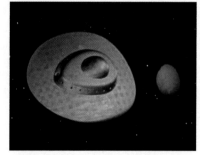

©Universal Pictures

Dodging a comet nucleus

While *Forbidden Planet* is usually the one given credit as being the inspiration for *Star Trek*, there are some more than superficial similarities between Exeter's command bridge and that of the original *Enterprise* that were not seen on the *C57-D*. The Master Control sits on an elevated central platform where all of the other stations can be observed. The "chair" has intercom and other controls built into the armrests (although in this film they are more like "handrests" requiring the occupant to hold his arms awkwardly outwards). In front of the Central Control there is a station for the navigator/helmsman who also operates the large view screen.

Then there is a similar characteristic that is almost certainly a coincidence in that both control bridges are offset with respect to the centerline of the ship. This is undoubtedly for the same reason, i.e. that the people building the interior sets and the exterior models were working independently to meet their respective requirements for filming, and this doesn't always result in completely compatible designs. In fact it's usually fans, trying to build a model or do a drawing that puts the interior into the exterior, that discover the discrepancies.

And if you still don't think that *This Island Earth* influenced *Star Trek* then just check out this Zagon ship!

Some of the technology was even borrowed by Irwin Allen for *Lost in Space*. The converter tubes were the obvious template for the freezer tubes onboard the *Jupiter 2* (see image page 242), although an energetic glow replaced the swirling mist. However, they are functionally closer to the "DC" pads in *Forbidden Planet* in that their job is to protect fragile humans (and other life forms) from lethal consequences of space travel.

©Universal Pictures

After Jon gives us his Archeologist's Report on the Metaluna saucer, we'll be back with some more thoughts on the impact of this film over the years.

Archeological Report on Exeter's Saucer

By Jon Rogers

If you have *ever* seen Universal Pictures' 1955 release, *This Island Earth,* and I mention Exeter's Saucer from the planet, Metaluna, then you immediately know what it looks like! It has a memorable shape, unlike any other saucer of the 50s. But what else do you know about it, really? What are its features? Is it bigger than a breadbox?

This report, based on findings from having closely examined numerous screen shots from the movie and promotional still pictures taken at the time of its release, will give you the answers. I used these sources because no known models or studio drawings of this saucer (our usual source of primary information) have survived.

In studying the saucer, one of the first data points to establish is its overall size. After that, other details can be derived from their relationship to that one overall dimension.

After a careful analysis was completed, it was discovered that this famous saucer was larger than previously believed. I also discovered several interesting details about its interior characteristics. All in all, it proved to be an impressive machine.

I planned to get the size of the saucer by comparing it with another object of known size. The scene where the saucer was "beaming up" the aircraft in which the two Earth scientists were flying, seemed an ideal place to start (Figure 1).

The airplane's position with respect to the saucer would reduce parallax error and should yield a good size comparison. However, there was a problem with this scene. As you can see, the image of the airplane was made indistinct by the green "beam" that surrounded it. At this distance, that fact alone could create a significant error in measurement. Luckily, another similar, but more sharply focused scene was discovered.

©Universal Pictures

Figure 1

In the final part of the movie there is a very short and clear scene where Exeter releases the airplane from beneath the saucer (Figure 2). The airplane is shown facing the opposite way from when it was taken captive due to the film being printed in reverse. It is not known if this was done intentionally. Still, this scene is better for measuring relative sizes .

Using this scene it was possible to accurately compare the diameter of the cargo hatch, the lower portion of the saucer body, the portholes, and eventually the overall size of the ship.

The first thing to do was to identify the make and model of airplane. This turned out to be a more challenging task than I first assumed it would be.

©Universal Pictures

Figure 2

After some research, the airplane was identified as a Stinson Model 108, more commonly known as the "Voyager." However, it turned out that the Stinson Aircraft Company had built several versions of the Voyager between the years 1946 and 1950. Each version had a different length. Which one was it?

Identifying the length of the particular version of Stinson shown in the movie was critical to my analysis. Research revealed that each version of the Voyager was painted a different color scheme. I finally determined that the Voyager model shown in the movie was one of the original Model 108s built in 1946.

Earlier, I had estimated the plane's length to be approximately 20-22 feet. After the correct model was identified, this was discovered to be a significant error. The correct length of the plane seen in the movie was 24 feet 6 inches, some 20% larger than I had earlier thought.

This new data revealed that the saucer was 198 feet long, 41 feet high, and 171 feet wide. It was one of the largest saucers of the era--quite impressive in size and scope. See Figure 3 for the size analysis.

After establishing the size of the saucer, I began reviewing its other details starting with its external features. There are 9 visible portholes on each side. There is a porthole in the front and center of the saucer, but none dead astern. There is also a streamlined dome on the top of what would, otherwise, be a ship with a completely symmetrical top and bottom.

Exeter's saucer had a surprisingly difficult shape to identify because some of its features were ambiguous in different scenes. For example, in the scene where the sau-

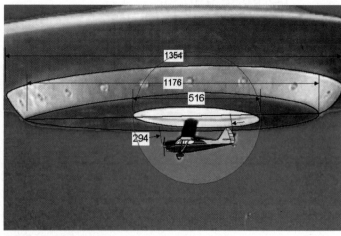

©Universal Pictures

Figure 3

cer is seen flying through space from stage left to stage right, the saucer appears to have a top deck that is slightly shorter and angled more than the bottom deck (Figure 4). Later, this is contradicted by the scenes where it lands on the planet Metaluna and when it delivers the Voyager back to Earth. In these scenes, the underside and top of the saucer appear to be symmetrical.

In the flyby scene, it is likely that the photography and lighting make the rear of the top deck look as though it leans more

©Universal Pictures

Figure 4

forward than it really does. The only other possibility would be for them to have made additional, different shaped models. However, I know of no evidence of this.

The problem of the shape of the top deck being a little more streamlined than the bottom half also become apparent when you compare it to what you see in another scene during the flight. When the saucer veers to miss the comet that flies past it, the top deck and dome are seen from above (see image page 143, bottom). It is observable that the shapes of the deck and the sides are symmetrically proportional and parallel. If the rear of the top deck

were more acutely angled, the edge of the top curve and the edge of the bottom curve would not be parallel as they are in that scene. This is more evidence that the model does not have the kind of curvature we think we see in the flyby. The steeper angle of the top deck is more likely a distortion caused by the motion of the ship and the lighting against a very dark background.

There are a few other external features. One positively identified feature is a hatch on the starboard side, forward. This is the hatch that Exeter, the Earth scientists, and the crew use upon their arrival in Metaluna.

However, another larger hatch is momentarily visible during the flight. You can see its shape and dimensions on Sheet 2 of the Data Drawing. What it's used for remains a mystery. My guess is that it is for either Cargo or major ship refurbishment.

The Saucer Fleet

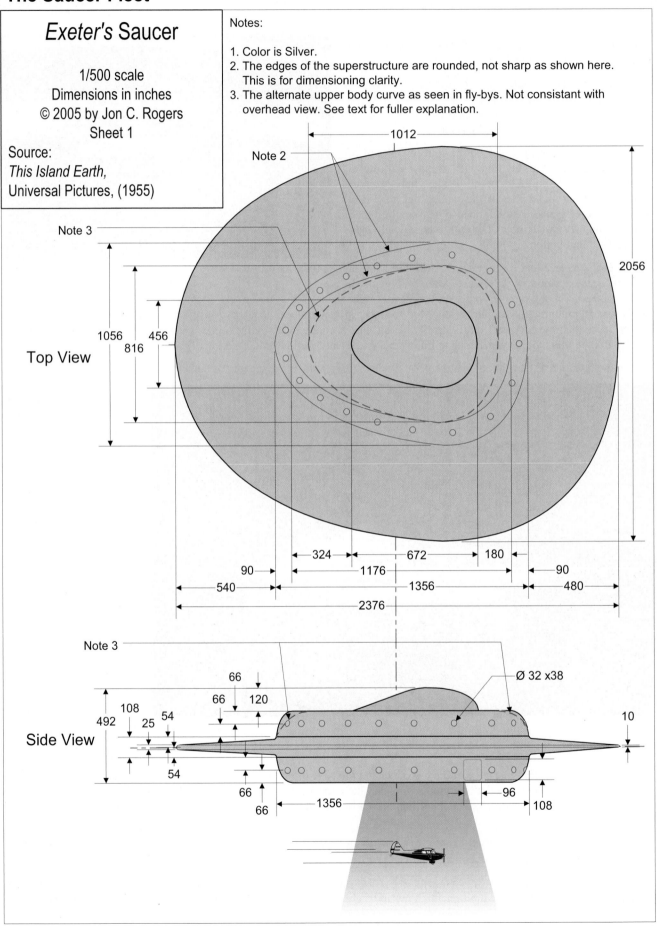

Exeter's Saucer

1/500 scale
Dimensions in inches
© 2005 by Jon C. Rogers
Sheet 1

Source:
This Island Earth,
Universal Pictures, (1955)

Notes:

1. Color is Silver.
2. The edges of the superstructure are rounded, not sharp as shown here. This is for dimensioning clarity.
3. The alternate upper body curve as seen in fly-bys. Not consistant with overhead view. See text for fuller explanation.

Note 2

Note 3

Top View

1012

2056

1056

816

456

324   672   180

90   1176   90

540   1356   480

2376

Side View

Note 3

Ø 32 x38

66

66   120

108

492   25   54

54

66

66

1356

96

108

10

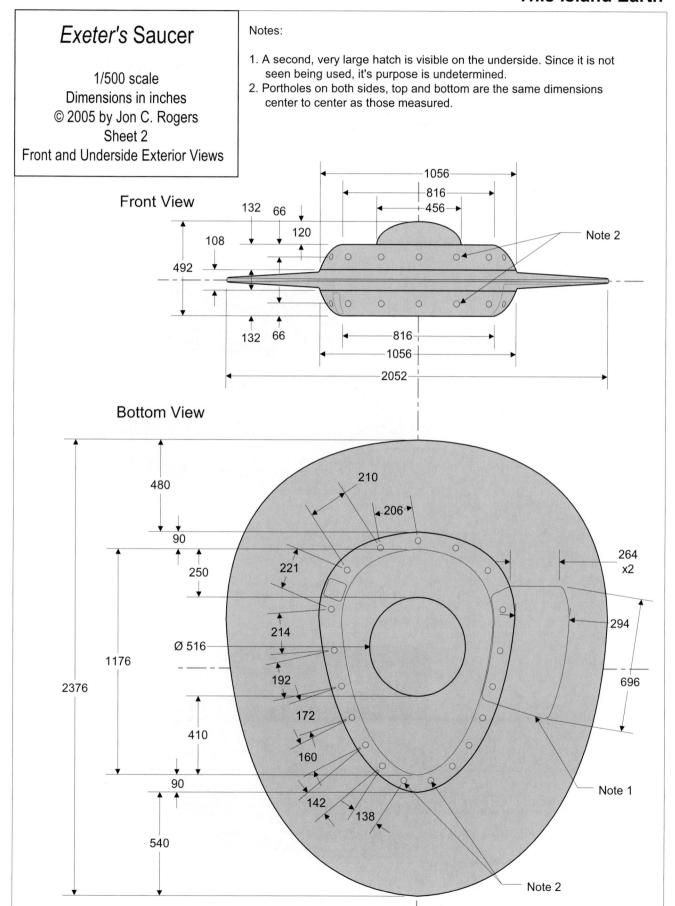

Exeter's Saucer

1/500 scale
Dimensions in inches
© 2005 by Jon C. Rogers
Sheet 2
Front and Underside Exterior Views

Notes:

1. A second, very large hatch is visible on the underside. Since it is not seen being used, it's purpose is undetermined.
2. Portholes on both sides, top and bottom are the same dimensions center to center as those measured.

Front View

Bottom View

Exeter's Saucer

1/250 scale
Dimensions in inches
© 2005 by Jon C. Rogers
Sheet 3
Interior Front and Side Views

Notes:

1. The Interior Set is shown against the ship for comparison. The Interior Set is too high (192 inches) to fit within the Dome (120 inches) height. It is accurate in the plan view only (See sheet 4, also Text).
2. The exterior shape of the Control Bridge is shown by a — — — line.
3. This is the personnel transportation tube seen on the cargo deck.
4. This is the implied walkway that would be required to get from the cargo deck to the bridge. Also see sheet 4.
5. There are approximately 10 each, 48 inch square passageways around the cargo bay. Purpose unknown.

Detail Identification
A. Conversion Tubes Controller
B. Power Monitor
C. Captain/Pilot's Chair
D. Main View Screen
E. Control Room Shield Section
F. Viewscreen Viewing station
G. Pressure "Conversion" Tubes
H. Entrance to Bridge
I. Raised Dias for Pilot
J. Pilot Station Overhead
K. Control Ray Projector
L. Cargo Hatch
M. Bridge Ceiling Struts

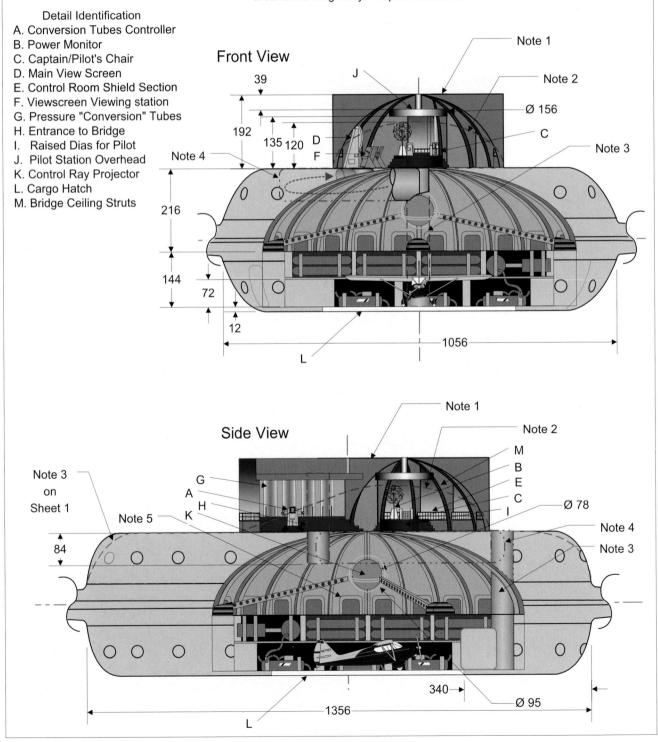

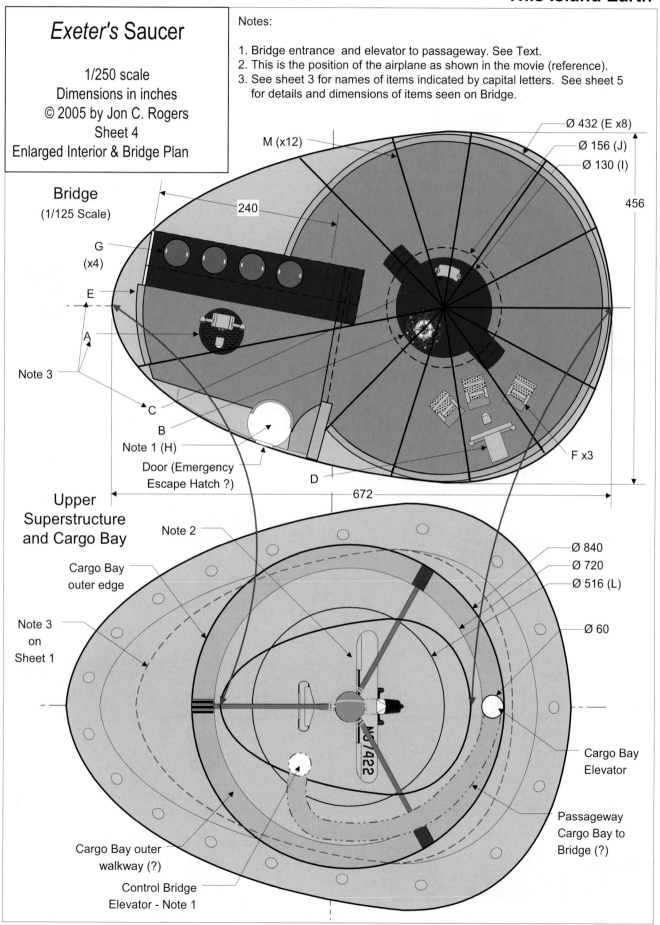

Exeter's Saucer

1/250 scale
Dimensions in inches
© 2005 by Jon C. Rogers
Sheet 4
Enlarged Interior & Bridge Plan

Notes:

1. Bridge entrance and elevator to passageway. See Text.
2. This is the position of the airplane as shown in the movie (reference).
3. See sheet 3 for names of items indicated by capital letters. See sheet 5 for details and dimensions of items seen on Bridge.

Bridge
(1/125 Scale)

Ø 432 (E x8)
Ø 156 (J)
Ø 130 (I)
456

M (x12)
240
G (x4)
E
A
Note 3
C
B
Note 1 (H)
Door (Emergency Escape Hatch ?)
D
F x3

672

Upper Superstructure and Cargo Bay

Note 2
Cargo Bay outer edge
Note 3 on Sheet 1
Cargo Bay outer walkway (?)
Control Bridge Elevator - Note 1

Ø 840
Ø 720
Ø 516 (L)
Ø 60

Cargo Bay Elevator
Passageway Cargo Bay to Bridge (?)

Exeter's Saucer

1/100 scale
Dimensions in inches
© 2005 by Jon C. Rogers
Sheet 5
Bridge Details and Dimensions

Notes:

1. The locations of Bridge Items are shown on sheet 3 & 4.
   Refer to Letter Identifications.
2. Aircraft is shown for size comparison.
3. Registration Number on the aircraft is as shown in film.

Details of Bridge Items

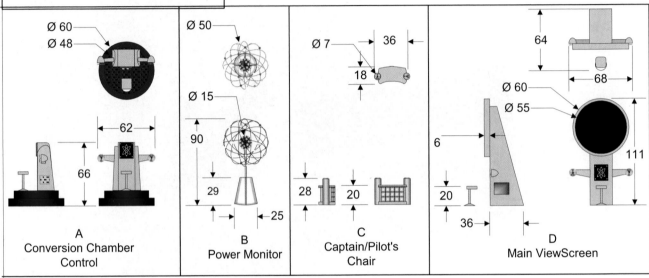

A
Conversion Chamber
Control

B
Power Monitor

C
Captain/Pilot's
Chair

D
Main ViewScreen

E
Control Room Shield Section (8 ea)

F
Viewscreen Viewing
station (3ea)

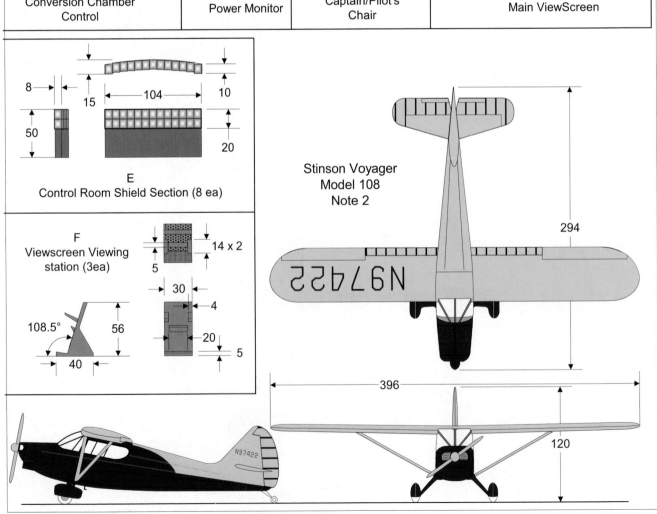

Stinson Voyager
Model 108
Note 2

N97422

The first time we see the interior of the saucer is when the two Earth scientists are beamed in, in their airplane (Figure 5). They arrive in the main cargo hold, which we get a good look at before they are taken to the control room or "Bridge" of the spaceship.

The cargo hold is very detailed. Besides the huge green ball, which is obviously the "Tractor beam generator," there is lots of mysterious machinery, passageways and in the foreground, leading vertically upward, is a transparent, personnel transport tube.

At this point in the story, we do not know what the tube is for, but later, when they arrive on Metaluna, they descend from the landing platform in an angled transparent tube like this one. It is reasonable to conclude that this is the passageway to the bridge, which is located above the cargo hold.

After one of the crew leads the two scientists away from their plane, the next time we see them is when they are about to step out of a passageway into the bridge of the saucer and be greeted by Exeter, himself.

©Universal Pictures

Figure 5

Since the transportation tube in the cargo hold is in the front of the ship and the entrance to the Bridge is in the rear, this leaves us with a question, how did they get from the cargo hold to the entrance of the Bridge?

The layout of the ship as shown on Sheets 3 and 4 of the Data Drawing shows the problem and one possible solution. There must have been some passageway between the two areas. If the passageway was designed as I have shown, it implies a second elevator to the passageway entering the bridge. My design is a "best estimate," and it is unknown if it was actually designed this way.

©Universal Pictures

Figure 6

The entrance to the Bridge is also interesting. As you can see in Figure 6, it appears to be a small room with a smaller bulkhead style door in the back wall. How one enters this small room is not shown, presenting one of the first mysteries of the bridge area.

While it may look like it on first glance, the bulkhead door is not the room's entrance for a couple of reasons.[1] First, it is a sealed, airtight door. You can see its position relative to the outside of the Bridge Dome in the Data Drawing, Sheet 4. This is a sealed airtight door for good reason. It leads directly to the outside of the ship. Its placement and design strongly suggest that it is meant to be an emergency escape hatch. Although the placement of this doorway is not the best for this purpose, no other exits to the Bridge exist.

Secondly the door's design suggests that it is not intended for everyday use. Its overall dimensions are too small for use by more than one person at a time and that person must crouch to use it. It is heavily reinforced with large hinges (internal) where other internal doors are of the sliding pocket design. This evidence all points to a secondary purpose, most likely, emergency escape.

Another feature of this entryway is that it is a raised platform with little space to the left or right from this camera angle. While it could have an entranceway hidden from view to the left or right, it's being on a raised platform suggests an elevator platform. This is consistent with the other elevators we see on Metaluna[2] and is a more likely design. For these reasons a sec-

---

[1] Most moviegoers probably assumed it was the entrance when viewing the picture. The fact that it is undersized was overlooked most people.

[2] The elevator on the Saucer Platform and the one entering into the Prime Sector both had raised platforms. They are so similar it is likely that the same set was used for both the Bridge and the Prime sector entrance.

The Saucer Fleet

ond elevator is implied in the design of the Bridge to cargo hold passageway.

The design of the Bridge of Exeter's Saucer presents more mysteries. After analysis of its dimensions, it was determined that, while it did fit in the plan view it did not fit within the confines of the outside dome on the saucer.

This is another example of Hollywood art directors taking liberties with the internal dimensions of spaceships for dramatic, staging, or photographic reasons. It appears that the art director wanted to keep to the design plan for the saucer but, in building the set, the vertical height requirement was ignored.

In the lower, front view of the saucer in Figure 7 (Sheets 3 and 4 of the Data Drawing) the interior set is shown with a dashed red line to indicate the outline of the dome on the top of the saucer. You can easily see that the height of the interior set far exceeds the available space. In comparing the top, or side view in Figure 7 it is easy to tell that, even if the interior set were to be lowered until the circular overhead on the pilot's platform were within the envelope, most of the area by the conversion tubes would still protrude above the sloping rear ceiling of the dome. Essentially, it was a square set that didn't fit in a round dome!

©Jon Rogers

Figure 7

Another problem would arise from lowering the Bridge into the saucer. As large as this saucer is, the cargo hold is also large so the two spaces would come in conflict. There would not be enough vertical space left for the green "tractor beam" globe. The saucer would have to be much bigger than it was to incorporate the dramatic bridge set and cargo hold.

As far off as these vertical dimensions are, the plan view of the bridge and the topside dome of the saucer fit quite well (see the Data Drawing Sheet 3). Since the plan view fits so well, it indicates that the directors were working to an overall design of the saucer's interior when they built these sets. Like most sets, these plans were created to insure the stage area looked correct. The fact that they took liberties with the height of the stage area was not believed to matter. It is a pity that those studio plans are not available; they would have shed much light on what the inside of Exeter's Saucer looked like.

Finally, I believe we have learned a great deal about this important flying saucer from the 1950s. I have been able to identify the size and shape of the saucer and almost the entire interior set's props all from accurately knowing the size of the Stinson Voyager. For this reason, I have included it in as reference, along with the other details on Sheet 5 of the Data Drawing. It is an honor it has earned in helping reveal the secrets of Exeter's Saucer.

Quickspec: Metaluna Saucer	
Vehicle Morphology	Saucer
Year	1955
Medium	Theatrical Film
Designer: Stan Horsley/Charlie Baker	
Length	198 ft (60 m)
Width	171 ft (52 m)
Height	41 ft (12.5 m)

Modelers' Note

Unlike *Forbidden Planet* or *Lost in Space*, fans of this film haven't exactly beaten a path to model companies demanding kits of Exeter's saucer. There is one low volume, resin cast diorama kit of the saucer, the Stinson (airplane) and a bit of ground below them. This expensive (~$100) kit is lovingly hand made one at a time, and is available from the Sci-Fi specialists at Monsters in Motion

(www.monstersinmotion.com). MiM also has a reproduction of the old Aurora Metaluna Mutant kit box. [Note that's just the box, not the kit.]

Some fans have gone the custom route, supplying other fans on a commission basis such as with this gorgeous interpretation (it's hard to say "scale model") of a Zagon ship found a while ago on eBay.

Epilog

With the remake/sequel mentality that has gripped Hollywood for the past couple of decades, it's amazing that *This Island Earth* has not been remade. From this creativity-barrel-scraping we have movies made from TV sitcoms, cartoon shows, video games and even theme park rides. Why should a literate, well-made film like this not be remade?

Well, perhaps it has. In a review of this movie on amazon.com, Arthur McVarish, writes:

> *This Island Earth* is a visual wonder and thematically solid. Released a year before the genre masterpiece, *Forbidden Planet*, some regard *This Island Earth* as predecessor to *Star Wars*, the way *Forbidden Planet* paved the way for the Star Trek phenomenon. *This Island Earth* is space opera at its finest.

That's an interesting insight. The idea of an interstellar civilization on the verge of destruction, reaching out for help to citizens of another planet ("Help me, Calvin Meacham, you're my only hope!") certainly has a familiar ring to it. In fact, George Lucas gave a nod to *This Island Earth* in the famous *Star Wars* "cantina scene" where the Metaluna Mutant was the model for the musicians (but without the exposed brain).

On the other hand, we don't know anything of the war between Metaluna and Zagon. Which one was the aggressor? From the movie it would seem that Zagon was the bad guy, but we only saw the very end of a war that had been going on for some time. Could it be that Zagon was more like *Star Wars*' rebel alliance that had finally turned the tide against their Metalunan oppressors? The Monitor's attitude

when interviewing Meacham and Adams certainly leaves this possibility open; especially his cavalier attitude when ordering the destruction of all the Earth scientists at the complex when recalling Exeter. If someone were to land in Germany or Japan in the spring of 1945, it would certainly seem that the inhabitants were to be pitied for the pummeling they were taking at the hands of the Allies.

None of the cast from *This Island Earth* went on to especially successful movie careers, although they seemed to have no problem working in TV. Prior to this film, Rex Reason had acted in a few "B" movies, mostly westerns and war films. The studio had bigger plans for him, though. After the critical and box-office success of *This Island Earth* they had other leading roles in mind. They even changed his screen name to "Bart Roberts" for two pictures before he convinced them to change it back. "They were hoping to make a Tony Curtis or Rock Hudson out of me," he said in a later inter-

Reason in *The Rawhide Trail* (1958)

view. Even so, most of his vehicles were still westerns and war movies. He did do at least one more Sci-Fi genre film, namely the last installment of the "creature" trilogy, *The Creature Walks Among Us* (1956). He also did a fair amount of television work starting the same year as *This Island Earth* (1955), including appearing as a regular on two TV series: *Man Without a Gun* (1957-59); and *The Roaring Twenties*. (1960-61).

The Saucer Fleet

Strangely, he left acting just as it seemed his career was taking off. After a 1963 appearance on *Wagon Train*, he abruptly quit acting to get into real estate investment. He still has an active fan base and occasionally shows up at conventions, as we'll see in a bit.

Faith Domergue (which, by the way, is pronounced "Dah-mure," not "Dommer-gue") has a much sadder story. Arguably the most compe-

tent actor in *This Island Earth*, she was already over 30 (well past her "prime" by the Hollywood standards of the day). She did two other films released in 1955; the Harryhausen film *It Came From Beneath the Sea* where she played (surprise!) a woman scientist, and then as a seductive cult leader in *Cult of the Cobra*. After that, though, the leading roles dried up and, like Rex Reason, she disappeared into television for the next 15 years.

Early career glamour shot

She did do one further science fiction role ten years after *This Island Earth*. In 1965, director Curtis Harrington was re-cutting the stilted Soviet film *Planeta Burg* (Planet of Storms) into the completely embarrassing *Voyage to the Prehistoric Planet*. As part of this less-than-stellar effort, he re-shot the portion of the film where the female cosmonaut

Voyage to the Prehistoric Planet

stays alone in orbit around Venus while the men are down on the surface exploring. Through contract negotiations lost to history, Faith wound up in the role where her entire part consisted of sitting in a cheap set talking into a microphone. Accompanying her in a similar situation was the fading Basil Rathbone, in one of his last films, also acting mostly solo as a ground controller on the moon, supposedly in charge of the mission.

After television, her last acting was in very low budget horror films (mostly made in Italy) alongside such genre stalwarts as John Carradine. One of them, *Legacy of Blood* (1971) even reunited her with Jeff Morrow. She gave up acting after 1974's *The House of Seven Corpses* and married Paolo Cossa, a former assistant director with whom she had worked during her Italian period. They moved to Switzerland where they lived until his death in 1996. She returned to America and lived in Santa Barbara, California until passing away on 4 April 1999.

One historical footnote; due to the circumstances of her being "discovered" by Howard Hughes, Domergue appeared as a character in Martin Scorsese's 2004 biopic *The Aviator*. She was portrayed by actress Kelli Garner in some highly fictionalized scenes of her audition before Hughes, accompanying him to a nightclub and later ramming into his car out of jealousy.

Jeff Morrow was already well into his career when he made *This Island Earth*. He had started acting on stage in 1927, doing a mixture of Broadway and Shakespeare. After World War II he moved onto radio, including the title role in the *Dick Tracy* series. He transitioned naturally into television starting with *The Philco Television Playhouse* in 1950. His first film was the big-budget costume epic *The Robe* (1953) but most of his films were "B" westerns until after *This Island Earth*, when he did a few more Sci-Fi films. Among them was *The Creature Walks Among Us* where he worked again with Rex Reason. However, after

the underrated *Kronos* (1957) he worked almost exclusively in television with only the very occasional film role, such as *Legacy of Blood* mentioned above. While much of his TV work was on top-rated series like *Bonanza* and

©Universal Pictures

Reason and Morrow in *The Creature Walks Among Us*

Perry Mason, it was always as a guest actor. He was never the star, nor did he ever work as a regular in a series. He came out of retirement to do one last television role in 1986, with a guest appearance on the second *Twilight Zone* series. He passed away on the day after Christmas, 1993, just three weeks short of his 87th birthday.

Russell Johnson will be forever typecast as "The Professor" from *Gilligan's Island*. That role probably started with *This Island Earth* and the typecasting was only en-

hanced by a couple of appearances on *The Twilight Zone*. In "Execution" (airdate 1 April 1960) he plays a scientist who invents a time machine and inadvertently snatches a criminal from the old west who is in the process of being hanged. The next year in "Back There" (13 January 1961) he

Johnson as you-know-who

played a time-traveling engineer who tries to prevent the assassination of Abraham Lincoln.

Johnson had a few minor roles in TV and film in the early 50's, but he cut his Sci-Fi teeth on U-I's very first science fiction film, *It Came from Outer Space*. He wasn't a scientist in that one, but rather a telephone lineman who is the first "victim" replicated by the aliens.

He didn't just do science fiction, though. There were some low budget action-adventure films and even a comedy, *Ma and Pa Kettle at Waikiki* (1955). This experience (scientist roles, adventure roles and a Pacific island comedy) prepared him for the audition with producer Sherwood Schwartz when he read for the part of Roy Hinckley, the Professor's real name that was rarely, if ever, used after the pilot episode. Before accepting the role, though, Johnson made Schwartz promise that when he made scientific statements that they would be accurate, a promise that Schwartz kept as much as could be expected for a show this detached from reality.

After *Gilligan's Island*, he appeared in a few movies; most notably a minor (and uncredited) role in the Robert Redford thriller, *Three Days of the Condor* (1975). But most of his acting career was on television, including a recurring character on the soap opera *Santa Barbara*. He managed a self referential joke with a guest appearance on *Newhart*. He played a member of a men's organization watching a *Gilligan's Island* marathon on TV. When they are evicted from the room, Johnson's character, protests, "I want to see how it ends!" He is assured that the castaways don't get off the island.

Johnson in 2007

After his son, David, was diagnosed with AIDS in the early 1990's, he quit acting to work as a full-time volunteer for AIDS research fundraising, a commitment that has only intensified after his son's death in 1994.

Joseph Newman (whose screen credit usually read Joseph M. Newman) never did start his own production company, however the Academy Award nominated director[57] did work for ten more years before retiring from film making. Unlike the actors in *This Island Earth*, he managed to work mostly in theatrical films, although he did do some TV including four episodes of the last season of *The Twilight Zone*: the universally acclaimed "In Praise of Pip" (airdate 27 September 1963), "The Last Night of a Jockey" (25 October 1963), "Black Leather Jackets" (31 January 1964), and the very last episode, "The Bewitchin' Pool" (16 June 1964).

---

[57] Assistant director *David Copperfield* (1937) and *San Francisco* (1938).

Newman in 2005

After directing a few episodes of *The Alfred Hitchcock Hour*, Newman did a single episode of *The Big Valley* in 1965 then quietly retired. His career included an impressive total of more than 50 directing jobs and another 20 as assistant director. He had plenty of time to enjoy retirement as he lived another 40-plus years, passing away on 23 January 2006 at the age of 96.

* * * * *

While not as pervasive in our culture as *War of the Worlds*, *This Island Earth* has made its mark in the Science Fiction community. Here are a few high-profile appearances in the genre.

In 1962, the Topps bubble-gum company released a gruesome series of story cards known as "Mars Attacks". They were created by Len Brown and Woody Gelman, and painted by comic book artist, Norm Saunders. The design of the Martians was heavily influenced by the Metaluna mutant with the exposed brain, although the insect-like features were replace by the face of a human skull.

©Topps Co.

The story was a straightforward re-telling of the H.G. Wells tale and was, of course, the basis for the 1996 Tim Burton film of the same name.

A more direct reference is made in Steven Spielberg's 1982 blockbuster *E.T. the Extraterrestrial*. As E.T. explores Elliot's house, he comes across a TV playing *This Island Earth*. He watches the brief scene of the plane being pulled into the saucer and it's hard to tell from the look of surprise on his face whether it's a shock of recognition or confusion (this author goes with the former since his mission was a collection expedition, after all).

Finally, we come to what is either the greatest compliment or greatest insult a Sci-Fi movie can receive, being featured on *Mystery Science Theater 3000*. This show (known as "MST3K" by its fans) is a cult favorite in the US, but doesn't play that well outside the 'states. For those unfamiliar with the show, the premise is that a "test subject" has been launched into space against his will by mad scientists. There he is experimented upon by subjecting him to low grade movies (usually, but not always science fiction) in an effort to break his will. To preserve his sanity, the subject resorts to making fun of the movies along with two robot companions that he built to keep him company.

Okay, that sounds pretty silly if you've never seen the show (well it **is** pretty silly, actually), but it formed the basis for some of the most brilliant comedy writing and social commentary on TV. The show lasted for ten seasons from 1988 to 1999 on two different cable networks (first on Comedy Central followed by the Sci-Fi Network). However, for every fan there was an anti-fan. While most people don't mind them making fun of such dreck as *Manos, the Hands of Fate*, (show #424) many fans took offense at their "riffing" genre classics like *Rocketship X-M* (show #201).

©Universal Pictures/Best Brains

Opening shot from the "experiment" with the signature "Shadowrama" of Mike and the 'bots at the bottom

The show had become a big cult hit by the end of the 1994 season so the production company, Best Brains, Inc., started talking about doing a theatrical version. As mentioned in the film's Wikipedia entry, some fans were upset that BBI picked *This Island Earth* as the "experiment" for the feature film:

> The choice of *This Island Earth* for the film to be riffed in *MST3K, The Movie* was questioned by some. Sci-Fi buffs have a soft spot for [this film], and its visual quality is considerably higher than virtually all of the films seen in the series. This was intentional: Best Brains and [distributor] Gramercy agreed that average movie-goers would not sit through the sort of dreary, black & white fare normally featured on the TV series, and that the concept would be more successful if the film that was featured was visually engaging and action-packed. [Writer] Mary Jo Pehl also described it as "the only one we could get!" and not Best Brains' first choice.

Ironically, the feature film was actually shorter than any of the TV episodes. The regular show ran two hours with commercials, and the actual run time (with commercials removed) varied from 97 to 103 minutes. Ninety to one hundred minutes is a typical run time for a theatrical feature, so that was the length of the MST3K movie as originally edited. However, before going into general distribution, nearly one fourth of it was cut. Again from Wikipedia:

> The movie's brevity came about as Gramercy officials, worried by test-audience reactions indicating that impatience with the concept grew steadily at about the 75-minute mark, pressed Best Brains to cut the film down.

When finally released, the film had been trimmed down to 73 minutes. Not only is that a full half hour less than a typical TV episode, it is 13 minutes less than *This Island Earth* itself. After you subtract the run time for the framing story, credits, etc., there's barely half of the original movie left. Suffice it to say that if you've only seen the MST3K version, you haven't seen *This Island Earth*.

In some quarters, the strategy to use an actual good movie as the subject, backfired. After the movie was released on home video in 1998, critic Richard Scheib observed:

> The whole making-fun-of-bad-movies [premise] that the show bases itself on is really an exercise in fashionable sarcasm. [W]ith the target made of *This Island Earth*, it is an undiscriminating sarcasm that ignores any virtues or artistic merits a film may have and is founded solely around deriving humour out of the less-sophisticated effects and corniness that dialogue has or may appear to have when isolated out of context. The sad truth that *Mystery Science Theatre 3000: The Movie* seems to miss is that *This Island Earth* is not a bad SF film [like] *Robot Monster* (1953) or *Plan 9 from Outer Space* (1959). In fact, in what is rather offensive to one's intelligence, *This Island Earth* is actually a far better film than *Mystery Science Theatre 3000: The Movie* is.

Still, the fans loved it. The film was released just before the second (and, as it turns out, last) MST3K fan convention in Minneapolis. As a tie-in, the con producers managed to get both Russell Johnson and Rex Reason as panelists. Johnson was invited for his role *This Island Earth*, of course, but spent most of the time answering questions about *Gilligan's Island*. For most, the high point of the panel discussion was when Johnson tried to explain to Reason what a "bong" was (that's what the wisecracking robots called the "Conversion Tubes" when they fill with smoke).

©Universal Pictures/Best Brains

Exeter seems to be pleading silently with the audience

This Island Earth today doesn't seem to get the notice that it deserves. Here is a film that was widely praised for its production quality and the intelligence of its story, but today it is largely unknown outside of Sci-Fi fandom. The film has never been remade (directly), so most people today, if they know of it at all, associate it with a humor-satire subject from cable TV. Perhaps our humble efforts here will help people remember it as it was intended.

Forbidden Planet

A good litmus test for determining what "generation" a Sci-Fi fan belongs to is to ask them to pick the movie that started the modern age of science fiction film. Many people will pick *Star Wars*, the 1977 George Lucas opus that, admittedly, did change the nature of SF motion pictures forever. A slightly older, or more analytical fan might pick the Kubrick/Clarke masterpiece *2001: A Space Odyssey* from ten years earlier. A surprising number of people, though, would reach back yet another decade to this film, the 1956 MGM effort *Forbidden Planet*.

More than fifty years after its release, *Forbidden Planet* holds up remarkably well. Its fans are adamant in considering it the greatest Sci-Fi film ever made. While that may be a bit overstated, it did produce the strongest social icons for spaceships and robots for the entire decade of the '50s, and its influence was strongly felt through most of the '60s as well. What is it about this film that resonates so strongly with fans and gives it such *gravitas*? Why do they consider it not just significant, but a major, defining work?

It was not the first Sci-Fi movie made in color; *Destination Moon* had seen to that some six years earlier.[1] It was not, as others have stated, the first post war SF film to have an "A movie" budget. Fox's *The Day the Earth Stood Still* was produced for $1.2 million, over four times the typical "B movie" budget at the time, and on a par with the $1.9 million that *Forbidden Planet* cost. What it represented, though, was the first time since the disastrous *Just Imagine*[2] of 1930 that a major studio was convinced to put its full support behind a science fiction film, using all of its state-of-the-art tricks and tools. In addition to color, it used the new

ultra-wide screen Cinemascope® process, stereo sound, pioneering special visual effects, and groundbreaking original music that created the first totally electronic film score using no traditional instruments at all.[3] Further, it was a completely original story, not derived from a proven work of science fiction. It even claimed to be based on Shakespeare's play *The Tempest*. To what degree this is true, we will discuss as we go on.

The thing that really made this film stand out, though, was the fact that it was being made by MGM, a studio famous for making lavish, high quality "big" pictures, and one that had eschewed the science fiction genre its entire existence.

Allen Adler was a struggling young writer in the new field of television. By 1953 he had noticed that science fiction was undergoing quite a boom in the movies and thought he'd try his hand at it. Since 1950 there had been well-made SF films in the action-adventure style (*Rocketship X-M*), the scientifically accurate style (*Destination Moon*), the social commentary style (*The Day the Earth Stood Still*) and even comedy (*Abbot and Costello Go to Mars*). George Pal had made a cottage industry out of doomsday style Sci-Fi (*When Worlds Collide* and *War of the Worlds*). There were also lots of less well-made films, some good and some very bad. But one thing that Adler noticed was that they were all set close to home (meaning within this solar system).

Adler was a friend of Irving Block, a painter and special effects technician who had his own effects business, Septa Productions, partnered with Jack Rabin. Septa had produced many of the effects for *Rocketship X-M*, but most of their assignments were on very low budget, exploitative SF projects. Block had also done some writing, so in early 1954, he and Adler decided to collaborate on a spec script for a science fiction film. It would be set well out into space, and incorporate some intelligent concepts and situations to try and get away from the "BEM" (Bug Eyed Monster) image the genre was saddled with. Adler had lots of

©Warner Bros.

---

[1] In fact, all of Pal's SF films were in color. In addition to *Destination Moon; When Worlds Collide* (1951), *War of the Worlds* (1953) and *Conquest of Space* (1955) all predate *Forbidden Planet*. The *War of the Worlds* section starts on page 76. The other three films can be found in *Spaceship Handbook*.

[2] See *Spaceship Handbook* for a summary of this reasonably significant film.

[3] Dare we say "music without instrumentalities?"

The Saucer Fleet

Irving Block

concepts that he wanted to work into the story, and what Block brought to the party was his love of the classics and mythology. His favorite play was Shakespeare's *The Tempest*. They decided to use it as a template for the major characters and the basic setup for the story, but the resulting film is, as we shall see, far from being a modern retelling of the story in the manner that *West Side Story* is a remake of *Romeo and Juliet*.

Incidentally, most film historians doubt the legitimacy of the *Tempest/Forbidden Planet* connection. None of the original press releases or publicity material mentions it, nor do any of the contemporary reviews. The first documented reference connecting the two came from a lecture by film scholar Kingsley Amis in 1964,[4] nearly a decade after the film was released. Irving Block doesn't mention it in interviews until the late '60s. While he very well may have had *The Tempest* in mind while writing the screenplay, he never admitted it publicly until after it became "common knowledge." Still, as a service to those who want to decide for themselves, we'll continue with the parallels.

To briefly recap *The Tempest*, the main character is the magician Prospero who is the Duke of Milan. He has been deposed by his brother and exiled to an enchanted island with his daughter, Miranda. There, he has the help of Caliban, a part human, part fish servant; and the spirit, Ariel. To try and get off the island, he has Ariel create a storm (a tempest, get it?) to force onto his island a passing ship that has several Italian noblemen on board including Ferdinand, the son of the Italian king. The ship would not only get him off the island, but Prospero wants to marry his daughter to Ferdinand, which will help him regain his throne in Milan. That's enough of the plot so that you can compare it to the story of *Forbidden Planet* as we get into it further down.

Block's basic outline follows this by creating the characters of Morbius (Prospero), his daughter Altaira (Miranda) and isolates them on a planet with an almost magical secret (the enchanted island). A non-human servant, Robby, a combination of Caliban and Ariel, rounds out the cast of planetary inhabitants. An interstellar cruiser carries Commander John J. Adams (Ferdinand) to the planet. Incidentally, Block named Morbius after the German mathematician August Möbius; best know as the creator of the Möbius strip.[5]

After a couple of months work, their screenplay, titled *Fatal Planet*, was ready to sell. They were planning on approaching Allied Artists, a studio that was producing a large number of science fiction films, albeit all very low budget. Their agent, however, encouraged them to "think big" and try MGM first.[6]

Nicholas Nayfack

In early spring of 1954, Block and Adler managed to get an audience with MGM line producer Nicholas Nayfack. When they offered him the story to read, he refused saying "You tell me what the story is about."[7] Block began to act out the story, but came to one potential sticking point, the invisible monster. After convincing Nayfack that an invisible monster would be more frightening **because** it was invisible, the producer was sold; especially since he wouldn't have to spend any special effects budget on something that's not seen on the screen.

The next hurdle was for Nayfack to sell the idea to his boss, MGM studio head Dore Schary. This would be

Dore Schary

MGM's first science fiction film, so it wasn't an easy sell. Schary was the top man at the studio, but he still had to report to the Loew Company (the New York based holding company that owned MGM at the time). Even though there was a big boom on in science fiction films, he had rejected dozens of projects since the beginning of the '50s without approving a single one. People were beginning to think that there was a systematic rule against Sci-Fi at MGM. Apparently that wasn't the case. In a 1979 interview Schary recalls:

> I liked the idea of the Id forces and its effect on Morbius. It was an imaginative concept and I felt it was the type of idea that could transcend the average space adventure story–the type of

---

[4] Correspondence with Shakespeare authority Mike Jensen, 15 May 2003.

[5] *Cinefantastique*, V4, N1, spring 1975, pg. 7.

[6] Ibid.

[7] Ibid.

picture that was then being mass-produced by everyone else. Up until that time, I had seen virtually nothing even close to what we thought could be the studio's first science fiction entry. There was no total anti-science fiction sentiment among the executives. That would have been stupid. We were simply waiting for the right project.[8]

After approving the project, Schary did insist on a name change to *Forbidden Planet*, feeling that it was less negative and more marketable than *Fatal Planet*. Also, it was not, at first, going to be an "A movie" (despite MGM's reputation of making only "big" movies, they had a "B movie" production unit just like every other studio). The initial budget for *Forbidden Planet* was between $500,000 and $700,000. That's high for a "B" picture, but barely halfway to "A movie" territory, which started at an even million at the time.[9]

The first hire Nayfack made on the film was Cyril Hume, a screenwriter who was an expert at writing scripts for adventure movies set in exotic locales, having worked on several Tarzan films. Irving Block was also hired, due to his previous special effects work, as a pre-production artist to produce concept art. Nayfack picked Fred McLeod Wilcox as director, an odd choice since Wilcox's main claim to fame was a string of successful "Lassie" pictures a decade earlier.[10]

Arthur Lonergan was assigned to be art director, the one responsible for the overall appearance of the film. Under Lonergan's direction the sets and props took on a distinctive "Norman Bel Geddes" look.[11] Robby's "jeep" and, in fact,

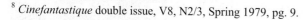

Arthur Lonergan

Robby himself, are perfect examples. Lonergan picked the legendary Arnold "Buddy" Gillespie to supervise the special effects (Gillespie also had a long and varied career at MGM, including supervising the effects on *The Wizard of Oz*). Almost right away Lonergan and Gillespie recognized the opportunity that this picture gave to their department. Since it

"Buddy" Gillespie

was the studio's first Sci-Fi film, they were free to do pretty much anything they wanted, with no precedent to drag them down. But the low budget was a problem, as Lonergan recalls:

> MGM wanted to make a cheap picture on this. Buddy [Gillespie] and I got together and we decided we'd go ahead and design the picture the way it should be done, regardless of the damn budget.[12]

Nayfack, to their surprise, went along with their little conspiracy, seeing it as a chance to break out of the confines of a low budget in other areas. This, naturally, led to some confrontations with the studio higher-ups, as we shall see later.

The first thing that Gillespie turned his attention to was Robby the Robot, since he was, by far, the single most complex prop in the film. He produced this sketch of Robby, which isn't too far from how he eventually wound up (although the "food intake doors" seem unfortunately located) for the prop department to build from. It was turned over to production illustrator Mentor Huebner who refined it, getting rid of the cast rubber legs and replacing them with the familiar ball

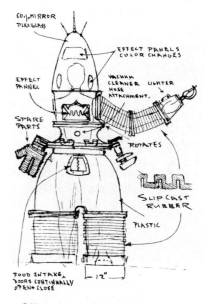

Gillespie's original conceptual design sketch for Robby

[8] *Cinefantastique* double issue, V8, N2/3, Spring 1979, pg. 9.

[9] Ibid., pg. 16.

[10] Wilcox spent his entire, very long career with MGM, but it was an unusual one. Starting in 1926, he worked as an assistant and second unit director but didn't get his first directorial assignment, *Lassie Come Home*, until 1943. It was a commercial and critical success, as were its sequels, *The Courage of Lassie* (1946) and *Hills of Home* (1948). Despite this impressive start, he received very few assignments, and directed only ten films. Probably his best film (outside of *Forbidden Planet*, of course!) is the excellent children's fantasy *The Secret Garden* (1949). He directed only one more film after *Forbidden Planet*, and died in 1964 at the age of 59.

[11] Norman Bel Geddes (1893-1958) was an industrial designer and architect. His form-follows-function designs incorporated streamlining, teardrop shapes and lots of fins. He is probably best known for designing the General Motors "Futurama" exhibit at the 1939 World's Fair.

[12] *Cinefantastique*, V8, N2/3, pg. 20.

The Saucer Fleet

Mentor Huebner's pre-production painting of the "Chariot" version of Robby and the vehicle.

joints, and smoothing it into the final form. Irving Block picked up on perhaps a subliminal aspect to the design. "I saw the robot as looking like [producer] Nicholas Nayfack. He was shortish, with stubby legs, somewhat bald, and a very sweet guy"[13] (see photo page 158).

The design was then given to mechanical designer Bob Kinoshita who was responsible for turning Huebner's paintings into fabrication drawings that the prop department could use to build what we saw on the screen. Kinoshita, of course, served as art director on Irwin Allen's *Lost in Space* ten years later, where he designed that show's enormously popular Robot (thus explaining the strong family resemblance to Robby. See page 248 for a detailed comparison of Robby vs. the Robot). In 1979 he recalled some of his design constraints for Robby:

©Mike Clark
Bob Kinoshita on his 90th birthday in 2004 with the *original* Robby. He (Kinoshita) is still active in *Forbidden Planet* and *Lost in Space* fandom.

One of the first things you do when you design a robot is to try to confuse the audience as to where you put the guy inside. It's difficult to completely fool an audience because they know there is someone inside. But if you make an effort to confuse them, it can work in your favor and make the whole creation more believable.[14]

Kinoshita's drawings were given to Jack Gaylord, head

of MGM's Prop Shop for fabrication. Gaylord's crew worked out the final engineering to make everything work, especially the complicated head assembly and chest panel. One of the electrical techs working for Gaylord was Max

Jack Gaylord

Gebinger, who was also a glass blower. Gebinger made all of the neon "voice tubes" that flashed when Robby spoke.[15] The tubes were connected to a special voice activation circuit built by the sound department. They were controlled directly by the operator's voice as he spoke the lines (the sharp-eyed viewer will notice that not all of Robby's voice tubes activate all the time, but give some shading and nuance to his words when speaking).

[Note: Those who don't want to spoil the illusion of Robby's personality may want to skip the next four paragraphs]

The voice tubes were spaced with gaps between them so that the operator inside could see out through the gaps. It's impossible for the camera to see in when the tubes are on because the gaps are small and the light is bright. To prevent any possibility of catching a glimpse of the operator when the voice tubes were off, he wore black pancake makeup on his entire face. The operator during development of the costume was Eddie Fisher. No, not the singer and father of Carrie "Princess Leia" Fisher, this Eddie Fisher was an effects technician who worked for Jack Gaylord in the Prop Shop. Even though Robby was 6' 11" (211 cm) tall, the costume was so full of wires and mechanism that there was very little room left for the operator. Fisher

Frankie Darro suits up

was picked because, at 5' 6" (168 cm), he was the perfect "fit" for the job. Before filming started, however, the Screen Actors Guild (SAG) stepped in. They said that since Robby spoke lines in the film, the operator inside had to be an actor with a SAG card.

---

[13] Ibid.

[14] Ibid., pg. 26.

[15] "Neon" is being used here as the generic name for electric discharge light tubes. Since the voice tubes glowed blue, the gas was not neon, which glows orange-red. These tubes were filled with an argon-mercury mixture, which is highly toxic and posed a health hazard for the operator when they occasionally broke.

During filming, Fisher was replaced by card-carrying actors Frankie Carpenter and Frankie Darro (although Fisher went on to perform Robby later on TV appearances). This is all the more ironic because it's neither Carpenter's nor Darro's voice that you hear in the final film. All of Robby's lines were dubbed in post-production by

Eddie Fisher gets ready for an appearance on *The Gale Storm Show* (inset) to promote the movie

Marvin Miller (also a SAG member). Miller was chosen for his voice quality and delivery, but even if one of the performing actors' voices had been appropriate, they would

Marvin Miller

have had to post-dub it anyway. The number of motors and mechanisms in the costume made it quite loud; described by some observers as sounding like a chorus of vacuum cleaners. This is also why it took two actors to perform Robby. All of those electrical devices made the suit quite hot inside, especially under studio lights, and with little or no ventilation, they could only work a short while. This was compounded by their having to carry most of the 100 lb (45 Kg) weight of the costume on their shoulders.

For Robby's effects that had to be performed on cue, such as the flipping of his "thinking" relays before he spoke, there was a control box operated by Prop Shop electrician Jack McMasters. The box was attached through a cable that could be connected to Robby at either heel. The connection also carried power for all of the other motors and lights. For scenes where Robby was walking, or even just standing with his feet visible, the mechanisms were powered by a battery pack strapped to the operator's waist adding another 40 or 50 lbs (20 Kg) to his burden.

For all their hard work, none of the actors and support people who collectively "performed" Robby got public recognition. The screen credit reads simply "and Introducing Robby, The Robot."

Much of Robby's enduring charm comes from Hume's dialog for the character. Together with Miller's flawless delivery, the robot managed to be both superior and servile at the same time. One of the core aspects of his personality, included at the very beginning by Allen Adler, is his adherence to the "Three Laws of Robotics." These famous laws were formulated by prodigious author Isaac Asimov[16] for his series of robot stories that ran for 50 years. They started with the 1940 short story *Robbie* (our movie character was probably not coincidentally so named), and running until the 1990 anthology *Robot Visions*.

The three laws state:

1. A robot may never harm a human being, or, through inaction, allow a human being to come to harm.

2. A robot must obey the orders given it by human beings except where such orders would conflict with the First Law.

3. A robot must protect its own existence as long as such protection does not conflict with the First or Second Law.

We see the first two laws in action when Morbius tells Robby to shoot Commander Adams "right between the

Robby and friends Cliff Grant (left), Andy Thatcher (kneeling) and Jack McMasters.

---

[16] According to *Filmfax* Publisher Michael Stein in an August 2005 interview, *Astounding SF* editor John Campbell helped Asimov complete the three laws, and first published them in their final form in *Runaround* in 1941.

The Saucer Fleet

eyes" with a blaster and, of course, he can't.[17] The third law is never explored directly, but is hinted at in two scenes. First, when Morbius tells Robby to put his arm into the disintegrator beam (second law, obedience, overrides third law, self preservation). Second, when Alta "beams and beams" for him, but he is slow to appear (apparently breaking the second law). "Sorry, miss," he says when he finally enters the room. "I was giving myself an oil job," (thus helping protect and perpetuate his own existence).

Lonergan then turned his attention to the sets for Morbius' house and the Krell Lab. His work there, while equally fascinating, is beyond the scope of this book. He simultaneously turned Gillespie loose on the design of the *C57-D* Star Cruiser. According to Steve Rubin:

> Influenced by the rash of UFO sightings bursting out across the country at the time, United Planets Cruiser *C57-D* became a flying saucer. Its only description in the screenplay calls it an "object of polished metal, shaped along the general lines of the planet Saturn."[18]

This seems a little disingenuous, since the "rash of UFO sightings" had been "bursting out" since 1947, as we saw in Chapter 1, nearly a decade by the time the screenplay was written. A better rationale would be that in the mid '50s there were only two choices for a spaceship: a tube with a pointy front end and with fins and fire at the back; or a saucer. Well read SF fans like Adler and Block knew that you weren't going to make it to another star system using the chemical or even nuclear reaction motors usually found in the tube shape. Besides, rockets, even big ones, were already getting a little old fashioned by this time, having been in the public eye for over a decade since the end of WW II. The saucer shape was just so much cleaner, more efficient and modern looking. Even more importantly, in all of the Sci-Fi movies up to this time, the saucers belonged to the aliens, thus the product of an incredibly advanced technology. This time, the saucer would be one of ours, but it still carried the cachet of highly advanced technology.

To get a reality check on their preliminary designs, Gillespie and his team grabbed their sketchpads and notebooks and headed to nearby Cal Tech in Pasadena. Gillespie recalls: "The scientists were extremely helpful and gave us a lot of ideas. They looked over some of our sketches and approved of the designs for the crew's quarters, the navigation devices and the control panels."[19]

Since the *C57-D* is, of course, the main subject of this chapter, the description of the "real" one (in the context of the story) will be dealt with further down in the "Vehicle" section. For our discussion here, we'll just be describing the props used in filming it.

The C57-D interior set

There were three miniatures of the Cruiser made: 20-inch (51 cm), 44-inch (112 cm) and 88-inch (224 cm) in diameter. There was also a full sized set approximately 60 feet (18 meters) across of the lower part of the ship, including the three landing legs/staircases, the central support pylon, lower drive nacelle, and bottom surface of the disc from the center to the edge on the left side of the screen. The set only went about ten feet to the right of center, but it gave the impression of the full saucer about 100 feet (30 meters) across.

The smallest miniature was used for long shots, such as the ship banking over Altair IV in preparation for atmosphere entry. The other two were used together to extend the range of the painted backdrops using forced perspective. This can be seen most clearly during the narrated portion of the opening sequence as the ship moves in from a great distance and flies past the camera. It was shot by starting with the 44-inch miniature as far back as possible against the star field backdrop. The model was moved towards the camera about halfway until its image in the camera was the same size as the 88-inch model placed at the back of the set. The larger model was then placed in the same relative position

---

[17] Actually, a true Asimovian robot would not go into a self-destructive loop here since the laws are hierarchical and the first law automatically takes precedent over the second. It would simply say something like "I'm sorry, sir. That order is impossible for me to execute since it would harm Commander Adams." This would, of course, completely destroy the drama of the scene, and there would be no setup for the similar scene at the film's climax.

[18] *Cinefantastique*, V4, N1, pg. 9.

[19] *Cinefantastique*, V8, N2/3, pg. 25.

©Warner Bros.

The 20-inch model banks over a swirling Altair IV

in the shot and "flown" all the way past the camera. The same technique was used for the landing sequence as the ship flies in from the distance to land directly in front of the camera. Only the 88-inch model had the three retractable legs and the landing support pylon.

The rotating orange light in the lower nacelle, which indicated the main drive in operation, was accomplished different ways in the different models. In the 20-inch and 44-inch props, it was a simple rotating inner dome marked with black segments (all three had clear fixed outer domes). A simple lamp with a red-orange filter up in the saucer body provided the light. The big 88-inch saucer model was a different story. The lights were neon tubes provided by Max Gebinger (the same guy who made Robby's "voice tubes")

that really were neon this time. The tubes were arranged in a rotating fixture that was powered by high voltage electricity supplied through two of the support wires that suspended the ship over the set. The voltage was so high, though, that it would arc from the wires over to the metallic paint on the ship's surface, requiring frequent touch-ups.[20]

With all of this detail attention being lavished on the sets and special effects, the film ran into budget problems almost immediately. Jack Gaylord recalls "We ran out of money on this picture twice before we had the thing finished."[21] Arthur Lonergan was more specific, "The construction department exceeded their budget on every set. Those sets were impossible to estimate. It was a new kind of picture for them and they didn't have any precedent as to how much it would cost."[22]

When Joseph Cohn, an executive production manager (that's studio-speak for bean-counter) complained about the overruns to studio chief Dore Schary, he got no sympathy as Schary had, by then, taken a parental pride in the studio's big adventure into this new field. "Schary wasn't interested in this picture until we started to build sets," recalls Lonergan. "When they were half finished, he became fascinated by it, and was down on the stages every day, watching us build them."[23] Schary concurs: "I was fascinated with what they were doing, just fascinated. I always got protests from the production department whenever a picture was exceeding budget projections, but I had a very close working relationship with Joe Cohn and I just said, 'Oh, come on, we've got something good here, just give the money, transfer funds.' "[24] With this kind of support from the top, the film moved quickly over the $1 million threshold into "A movie" territory. Cohn didn't surrender without a concession,

Lighting test for the "Krell Lab" set, one of the constructions that had studio head Dory Schary "Just Fascinated."

[20] Ibid., pg. 43 (caption).

[21] Ibid., pg. 29.

[22] Ibid.

[23] Ibid.

[24] Ibid.

though, and several planned sets showing other areas of Altair IV and the Krell underground complex were cancelled. Even with these cuts, the sets by themselves wound up costing $1 million, more than the entire original budget of the film.

Principle photography began on Friday, 15 April 1955 and took about five weeks. This may seem short, especially by modern standards, but remember that this was all studio work done on sound stages with no location shooting at all. It is also just the parts with the actors in front of the cameras. There followed nearly a year of post-production work where the special effects were inserted and the groundbreaking music was composed.

Some of the special effects scenes were filmed along with the live photography. Buddy Gillespie had to supervise any of the shots that would be combined later with animation, such as when the crew fights the Id monster at the ship (which, as it turns out, was the very first scene filmed).

Joshua Meador

The animation effects in the film were produced by Joshua Meador, the supervising animator for Disney Studio's Effects Animation unit, who was "loaned" to MGM for this film (Meador, along with John Hench, had just won the Academy Award for special effects for his work on *20,000 Leagues Under the Sea*). The best-known animation sequence in the film is the Id monster battle, of course, but he provided all the other animated effects as well:

- The ship's landing force field
- The forming image of Altaira in the Krell educator
- All of the lighting effects in the Krell machine (the huge lightning bolts in the ventilator shaft and the strobing lights down the furnace corridor)
- Testing of the force-field fence
- The "household disintegrator beam,"
- Robby's "beams" (both his weapons neutralizer and the one he zaps the monkey with in the middle of the story)
- Robby's "shorting out" effect.
- All of the blaster sequences such as when Robby disintegrates the *althaea frutex* in Morbius' garden, Adams vaporizing the attacking tiger, and when he tests the Krell door.

Nothing is so closely identified with *Forbidden Planet* as the unprecedented music score by experimental composers Louis and Bebe Barron. It's hard, then, to realize that

The oh-so-serious Louis and Bebe Barron in their New York studio

originally there was no thought of having anything but a conventional orchestral score for the film.[25] The way the Barrons came to write the world's first all-electronic film score, itself, reads like a Hollywood story.

In the early '50s the Barrons, both contract musicians (he a drummer and she a singer/pianist), started experimenting with making music using the newly available tape recorder and custom electronic oscillator circuits. Working from their Greenwich Village apartment, they had even managed to get their music used in an avant-garde film called *The Bells of Atlantis*. Thinking that film scores could be a new direction for them, they asked their artist friends how to contact the major studios. With great excitement they learned that 1) Dore Schary was the head man at MGM, and 2) Schary's wife, a painter, was just about to open a one-woman show at a local New York gallery. They showed up at the opening night uninvited, worked their way in and located Schary by "look[ing] around for the person who would probably look the least important. We spotted him and sure enough, it was Schary."[26] Using their single film project as an introduction, they told him about their ideas for electronic film music. Schary was intrigued and gave them a standing invitation to visit him at the studio the next time they were on the west coast.

Seeing this as their big break, they immediately left for California taking two weeks to drive cross-country (Schary presumably flew back). True to his word, when they called, he saw them that same afternoon. They brought sample tapes of their music and clips from their one movie. Schary waived off the film but listened to the tapes. He was so impressed that he called Johnny Green in the music department and set up an evening meeting at Green's house with other department members. It was a classic case of being in the right place at the right time as the studio was just starting to score the film and hadn't known how to approach it.

[25] The theatrical trailer, available on the DVD version, has a standard action-adventure sort of orchestral score under it (it's actually pieces of the music to the 1954 film *A Bad Day at Black Rock* which also starred Ann Frances). Very disorienting to anyone who has seen the film many times (and who hasn't?).

[26] *Cinefantastique*, V8, N2/3, pg. 42.

This was the answer they'd been looking for.

They hired Louis and Bebe to produce some music and sound effects to be used in the film but there was one problem. The Barrons didn't want to uproot all of their delicate equipment and ship it clear across country for this three-month job. MGM made the unprecedented concession to let them work in their New York studio, out of the direct day-to-day supervision of the music department. This was the first time the studio had contracted for a music score outside of Hollywood.[27]

MGM sent a workprint of the film to New York for the Barrons to work from. This was an early cut of the film, basically complete except that it had placeholders for the missing special effects composites that were still being worked on.[28] The Barrons spent six weeks composing, recording and editing the music track. For their efforts they were paid a total of $25,000; a considerable sum in the mid '50s, but a tremendous deal for the studio. It could have easily cost ten times that to hire a composer, studio orchestra, recording engineers and sound editors to produce a traditional score. With time (and money) running short, Schary decided that the complete Barron score would be used, and there would be no conventional music in the film at all. One final concern came from the MGM legal department. The original screen credit read "Electronic Music by Louis and Bebe Barron," but the lawyers were worried, remembering the problem with Robby's operators, that the musicians union would object since neither of the Barrons were union members. After some head scratching, Schary came up with the term "Electronic Tonalities" to describe what the Barrons had created.

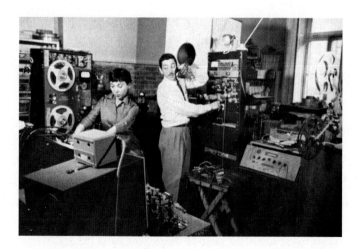

Forbidden Planet went into general release on 30 March 1956, two full years after Adler and Block had pitched the project in Nicholas Nayfack's office. It was not a financial success. The lavish production in Cinemascope®,

Eastmancolor® and stereo sound was well received by both critics and fans, but as Steve Rubin points out:

> Forbidden Planet was definitely a B-picture. The budget overruns caused by Lonergan and Gillespie's elaborate designs for props, sets and special effects had resulted in an anomaly, an expensive B-picture. The final cost was estimated to be about $1,900,000.[29]

The amount the film took in during its original release has been hotly debated, but most sources agree that it was on the order of $1.4 million. It was still playing to a "B movie" crowd and had it been a traditional "B movie" it would have been considered a runaway success. But all that the MGM executives saw, as did the execs at other studios, was that *Forbidden Planet* failed to return even its production costs. Rubin continues:

> Studios wrote off the innovations of *Forbidden Planet* [such as] star travel, robotics, [and] alien civilizations, as money-losers, concepts so costly to achieve in film that they could not be supported by what was perceived to be a limited market for the genre.[30]

So, like *Just Imagine* a quarter century earlier, *Forbidden Planet* caused Hollywood to back away from big-budget science fiction films. It would be another twelve years before, ironically, another MGM film, *2001: A Space Odyssey* proved that the market for good Sci-Fi films was far from limited.

Even though *Forbidden Planet* didn't make money for MGM, it did make their reputation among science fiction enthusiasts as being **the** studio for quality SF. The fans were stunned by the quality and imaginative nature of the film. It immediately went to the top of everyone's "A list" for Sci-Fi films and quickly passed into cult status as the benchmark for other films to meet.

One last thing before we launch into the rather long story synopsis.[31] We mentioned above that the Barron's had received a workprint of the film to compose to. Workprints are made just after the primary photography is completed for use by all of the post-production departments (like music). One of the post-production functions, though, is editing and many times a film loses significant footage in the process.

In the case of *Forbidden Planet*, it was even worse. Film editor Ferris Webster left MGM to work for another studio while the movie was still not completely edited.[32] The editor brought in to complete the project (whose identity is unrecorded) was either not as skillful as Webster or

---

[27] Ibid., pg. 43.

[28] This workprint would later have tremendous importance for film historians, as we shall see.

[29] *Cinefantastique*, V8, N2/3, pp. 62, 63.

[30] Ibid., pg. 65.

[31] At 98 minutes the movie isn't that long, but it is **very** talky. Even with significant trimming in the following synopsis, we are still left with a long summary since so much of the plot is advanced through dialog.

[32] *Cinefantastique*, V4, N1, pg. 12.

simply unfamiliar with the project as many of the scenes have confusing transitions, and several scenes that were setups for later plot points were removed.

In 1977, the Barron's brought out a soundtrack album of their music for the film's 20th anniversary. In preparation for doing the cover art, they got together some artists for a preliminary meeting. They also invited Bill Malone and Steve Rubin, two of the film's biggest fans, as experts to answer questions. As an orientation to some of the artists, who had never seen the film, they had a special screening.[33] Both Malone and the Barrons brought prints to show, and they decided to use the Barrons'. At first there was no sound, so Malone got up to fiddle with the equipment. Steve Rubin continues the story:

> [Lou] Barron suddenly threw up his hands, "Wait a second, Bill, this is that old workprint editor Ferris Webster gave to us. There isn't any music on this print at all." Malone could hardly control his excitement. Feeling like Howard Carter on the verge of discovering King Tut's tomb, he persuaded everyone to leave the print on, and all settled down to watch a strange version of *Forbidden Planet*. It was the first time the workprint had been viewed in twenty-two years.[34]

Rubin then goes on to describe the original workprint in detail. The changes made are not large, and were obviously done to "tighten up" the film. Unfortunately, in the process, they removed some of the explanatory dialog and setups for later scenes as mentioned above.

©Warner Bros.

In the interest of historical completeness, we will be adding some of these missing pieces to the story as we go (and clearly labeling them as such). We won't be doing all of them, just the ones that will make the film more understandable.

---

[33] It may be difficult for younger readers to believe, but back in the mid-'70s, you couldn't just pop a DVD, or even a tape of a movie into a player anytime you wanted to watch it. You had to either catch it on TV and suffer through the commercials (presuming you were lucky enough to be able to watch when it was on); or find out when there would be a "special screening" and go out to see it. Sound movie projectors were very expensive (the equivalent of nearly $10,000 in today's money) and the films themselves, which were really on big reels of film, could be even more if you wanted to buy them.

[34] *Cinefantastique*, V8, N2/3, pg. 50.

The Story

After Leo the lion finishes his trademark roar, a flying saucer swoops in from overhead and recedes rapidly into the distance. Just as it disappears completely, the title comes rushing back, drawing itself across the screen in massive solid-block letters (this was an apparent homage to *The Day the Earth Stood Still*, which did the exact same thing with Klaatu's saucer at the end of the film with "The End" rushing back in a similar, though less exaggerated font). A moving star field tells us we are traveling through space at high speed as we read the names of the cast and crew in some highly stylized credits. Instead of music, our ears are confronted by a strange series of sweeping glissandos, throbbing, pulsing beats, and bubbling arpeggios. It is so completely foreign sounding that it takes some time before we realize that this is, indeed, the music.

The flying saucer returns for a series of passes as a narrator (veteran actor Les Tremayne[35] working uncredited) somberly fills us in on the 300 years of space exploration separating the audience from the time of the story:

> In the final decade of the 21st Century, men and women in rocket ships landed on the moon.[36] By 2200 A.D. they had reached the outer planets of our solar system. Almost at once there followed the discovery of hyper drive through which the speed of light was first attained and later greatly surpassed. And so at last, mankind began the conquest and colonization of deep space.
>
> United Planets Cruiser *C57-D*, now more than a year out from Earth base, on a special mission to the planetary system of the great Main Sequence star, Altair.

I know we just got started in the story, but we need to take time for a little astronomy sidebar.

The term "Main Sequence" is a reference to an astronomer's tool called the Hertzsprung-Russell[37] diagram, or just "H-R diagram." This is a chart that classifies stars by plot-

---

[35] Tremayne is probably best known for portraying General Mann in George Pal's *War of the Worlds* (see page 76).

[36] Screenwriter Cyril Hume is the one responsible for being so outrageously off on the date of the first moon landing. All through the early '50s, books and magazine articles, such as the famous *Collier's* series, were showing that spaceflight, even trips to the moon, were within the grasp of the technology of the day. While everyone's crystal ball is perfect in hindsight, it's still inexcusable for such knowledgeable writers as Hume, Block and Adler being off by 1,000% (the actual first moon landing was only 14 years in the future when this screenplay was written, not 140). Perhaps he wanted to say "the last decade of the 20th Century," which would have been a reasonable assumption in 1955, but it just didn't sound "futuristic" enough.

[37] Named for the Danish astronomer Ejnar Hertzsprung and the American Henry Russell who both independently came up with the concept around 1912.

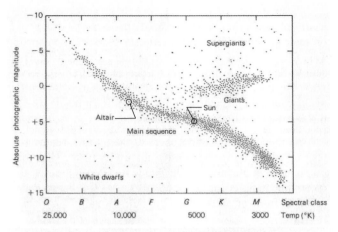

The Hertzsprung-Russell Diagram

ting their temperature against their brightness. When done this way, nearly all stars fall into a narrow band that runs diagonally through the diagram. This band is called the "Main Sequence" since that's where most of the stars are; the exceptions being the giants and dwarfs.

The way the narration above is written, it makes it sound like there is some privilege to being on the Main Sequence, or that there are "greater" or "lesser" stars on it. In truth, there's nothing special about being there. The Sun is a Main Sequence star. Altair is a bright white star, just to the left of the Sun on the diagram (for those astronomy buffs among you, Altair is spectral class A7 with a magnitude of +2.3). It is the primary star in the constellation Aquila, the Eagle, and forms part of the "Summer Triangle" along with two other bright stars Deneb and Vega, the latter of which figures prominently in a later Sci-Fi classic, *Contact*. From most of the northern hemisphere, Altair can be seen almost directly overhead during late summer. Just recently astronomers announced that they had been able to image Altair (i.e. see it as an actual object rather than just a point of light) by linking four telescopes at Mount Wilson observatory in California to work together as one giant 'scope.[38] They discovered that it is a "rapid rotator", turning on its axis every 6½ hours. That's so fast that it bulges some 14 percent wider at the equator than at the poles. Okay, back to the story.

©Warner Bros.

[38] *Astronomy* magazine, November 2007, pg. 26.

Inside the ship we see a small model of the saucer encased in a large astrolabe-device made of transparent spheres that slowly rotate about each other. A crewman works at a control panel at the base of this navigation device. We soon see that the saucer's crew works under a military (specifically Navy) protocol.

"When do I get a DC fix, Jerry?" Commander Adams, the ship's captain, calls out to Farman, the first officer who is working the controls. "Half a minute, skipper," comes the crisp reply. "DC point in less than three minutes!" Adams picks up a wireless mic and announces to the crew that they are about to

©Warner Bros.

decelerate below light speed. "DC Stations, on the double!" snaps the bosun, and the crew starts assembling on what look like Star Trek transporter pads.[39] A protective green energy field surrounds each crewman as the ship starts its trip down through the "light barrier." The equipment screams and shrieks through this tortuous process. The entire cabin is bathed in successive colors of light, from blue to yellow to red as the ship Doppler-shifts its way down the spectrum. The shrieks die down and normal light returns. The green energy fields around

each crewmember dissolve and they stumble off the pads, rubbing their necks and muttering; obviously not a pleasant experience.

©Warner Bros.

The captain sits down at the navigation console, looks into the binocular sight and announces that they are "down to .3896 of light speed." "It's warm in here, skipper," says Dr. Ostrow, the ship's surgeon. Adams looks back into the sight and sees an enormous red star[40] so close that ¼ of it fills half the field of view. Without saying a word, Adams gets up with a worried look on his face and walks over to the main viewplate. "There's Altair right on

[39] This is not a coincidence, as we will see in the Epilog.

[40] Red star? Obviously there must be some filters in place in that viewer. As we just discussed, Altair is a white star, and a brighter white than the Sun.

the nose, skipper," says someone, not on screen. Still silent, the captain stares at the viewplate, his face a mixture of anger and concern. From outside we see the ship passing by while a huge form moves in front of Altair until it is totally covered and its corona starts to show. Back inside, Dr. Ostrow says, "Meanwhile, this ship arranges its own eclipses…" Meanwhile? Meanwhile from what? This doesn't seem to make sense. What was covering Altair? Why does Adams look angry, and why doesn't he say anything? The whole sequence seems disjointed and confusing. This is the first place where some helpful dialog was cut. As originally shot, the scene went like this after Ostrow says, "it's warm in here:"

FARMAN

There's Altair right on the nose, skipper (now we see who says it).

ADAMS

Jerry, you have got to stop cutting them this close!

FARMAN

This is close?

ADAMS

[Angrily] Yeah, there's the hull temperature alarm. Put us in the shadow of the first planet out from the primary.

FARMAN

Aye, aye, sir.

ADAMS

Coolers, Quinn!

QUINN

Aye, sir. [to crewmen] Detail, cooling stations, on the double!

ADAMS

[to Farman] Someday you're going to bring us out right inside some star!

FARMAN

And then you'll probably court martial me for it.

Now the scene makes a lot more sense. First off, it shows some of the danger of travel through hyperspace. If the navigator isn't exactly perfect with his calculations, it could be disastrous. Second, the "arranged eclipse" is now seen to be a deliberate maneuver to cool the ship down, plus we know what it is they're hiding behind. It's Altair I. If you listen closely, in most of the release prints (including the DVD version) you can hear Adams start to say "Jerry, you…" due to the less than completely careful way this scene was edited.

---

[41] In Greek mythology, Bellerophon was the mortal who tamed the flying horse god Pegasus, and rode him in battle. Not an especially ethical character, Bellerophon arrogantly tried to ride Pegasus all the way up to Mount Olympus, the home of the gods. For this, Zeus punished him by pushing him off of Pegasus (while in flight) thus crippling him for life. Some reviewers see this as a metaphor for Morbius' rise on the wings of Krell knowledge after "taming" the brain boosting machine, and his ultimate fatal fall at the end.

Adams orders Farman to "punch in an orbit for the fourth planet," after which he briefs the crew (and thus us) on their mission. They have been sent to locate the expedition from the survey ship *Bellerophon*,[41] which had been sent to Altair IV twenty years earlier, and search for survivors. As they brake into orbit, the ship does a sweeping bank around the planet (see image, page 163). Inside we see the Astrogator mimicking the motion.

As they enter the atmosphere of the planet, Adams and Ostrow study the surface on the main viewplate. There are no signs of civilization at all. Suddenly Chief Quinn at the communications console announces, "we're being radar scanned from an area about 20 miles square."[42] A voice

"We're being radar scanned!"

message comes in from someone identifying himself as Dr. Edward Morbius of the *Bellerophon*. They confirm that he was a crewmember, and then tell him how glad they are that he is alive. He thanks them for their concern, but says he's fine and that they should just leave. Adams' demeanor turns from relief to annoyance as he insists that Morbius give them landing coordinates. He does, but adds, "I wash my hands of any responsibility for what might happen to your ship or its crew." He tries one last plea for them to not land, but Adams angrily cuts him off. "Something's funny down there, skipper," says Ostrow, a master of the obvious.

Adams takes the ship in for landing, barking out orders to Farman who executes and confirms them in a very realistic, military manner:

"Artificial gravity off."

"Grav. Off!"

"Half flux."

"Half flux!"

"Cut primary coils"

"Primaries cut, sir!"

---

[42] Remember that number!

From the ground we see the ship hovering, slowly approaching us for the landing. It wobbles slightly,[43] searching for an appropriate touchdown site. It stops over the chosen area, supported on a blue force field while a large pylon extends downward from the center of the drive nacelle at

©Warner Bros.

The landing sequence. At top the ship approaches from a distance (this is the 44-inch model). Second, the Josh Meador-animated "repulsor" beam supports the ship for touchdown (this and the other two frames are the 88-inch model). Third, the beam concentrates as the central pylon extends. Finally, the legs are lowered to broaden the support footprint.

the bottom. The force field concentrates and kicks up some dust as the ship lowers its sizable bulk gently onto the surface. With the main drive still spinning, three legs are lowered for stability. The drive shuts off just as the legs touch ground, rendering the nacelle opaque. The legs also form gangplank/stairways as the first crewmen come spilling out to secure the area.

©Warner Bros.

"All Secure, Sir!"

With a report of "all secure, sir!" the captain, first officer and doctor come down to assess the situation. "Look at the color of that sky!" exclaims Ostrow, surveying the grey-green horizon. Adams checks the "command mic" on his belt-mounted communicator, then orders the tractor assembled. But before the bosun can even start to carry out the order, a speeding cloud of dust is spotted coming towards them. In an almost comically under-cranked shot, a highly stylized vehicle pulls up, replete with dozens of horizontal strakes, aerodynamic flowing fenders and even itty-bitty tail fins. As the front of the vehicle swings open, we see that the "hood ornament" is actually a large bulbous robot that totters over to the captain, bows elegantly and announces, "Welcome to Altair IV, gentlemen! If you do not speak

©Warner Bros.

English, I am at your disposal with 187 other languages along with their various dialects and sub-tongues."[44] "Colloquial English will be fine, thank you," replies Adams. "Uh, you **are** a robot, aren't you?" he continues, not quite sure how to approach the situation. "That is correct, sir," says the robot diplomatically. "For your convenience, I am monitored to respond to the name 'Robby.'"[45]

[43] Unlike the infamous wobble of the *Space Ark* in Pal's *When Worlds Collide* (see *Spaceship Handbook*) when it starts down its launch ramp, this wobble is deliberate. It was added to make the scene look less static and more familiar, like a helicopter preparing to land.

[44] An early version of the "protocol droid" concept, like C3PO from *Star Wars*.

[45] The word "monitored" would obviously be "programmed" today, but in 1956 there were only a few dozen people in the entire world who were employed to "program" information into thinking machines. It's amazing both that the concept was included at all, and that it was so understandable to mid '50s audiences without explanation.

169

Robby tells them that he has been sent to transport them to the residence where they are expected. Adams makes a quick judgment call and decides that it's okay, but still tells Chief Quinn (who is being left in charge) to track his communicator signal and monitor it for a distress call.

Adams, Farman and Ostrow[46] get into the vehicle. Robby reminds them to fasten their seat belts and they take off at the same immense speed towards Morbius' residence.

©Warner Bros.

Commander Adams wonders if he made the right decision. At least he has a seat, unlike poor Doc Ostrow in the center who has to sit on the vehicle's rear deck.

The final film moves directly to them pulling up in front of the house, but originally, there was a transition scene showing more of Robby's capabilities. A medium shot shows the vehicle streaking down a dirt road while a close-up of the car traveling at tremendous speed shows Adams more than a little concerned for their safety:

ADAMS

Hey Robby, hold the speed down!

ROBBY

(Turning his head completely around to face the officers)
Have no apprehension, sir. My built-in reflexes are infallible.

ADAMS

For Pete's sake, at least watch where you're going.

ROBBY

Not necessary. I do everything by pulsetonic transfiguration.

A Low quality image (from the work print) of the deleted "speed" scene

---

[46] Note: the captain, first officer and doctor. Sound familiar?

This scene was cut not so much for time, but because the model work on the medium shot, and the rear projection of the speeding background in the close-up was considered sub-standard.

They arrive at the residence where Dr. Morbius greets them at the entrance. They exchange pleasantries, and he once again insists that he needs no assistance of any kind, but does invite them to stay for lunch. As they finish up Farman inquires about the robot, since he had prepared the lunch from some of his "synthetics." Morbius demonstrates, in a somewhat offhanded manner, more of Robby's capabilities. As a final, very dramatic demonstration, Morbius asks Adams to give Robby his blaster. After having him

©Warner Bros.

"Aim right between the eyes"

(Robby) disintegrate a tree in the garden, to show that he knows how to use it, Morbius tells Robby to shoot Adams "right between the eyes." The robot just stands there, frozen, while huge electric discharges course through his "head;" demonstrating that it is impossible for him to harm humans, even if directly ordered to do so.

©Warner Bros.

"He's helpless!"

Ostrow, relieved that his commander is still intact, asks Morbius how he came by such a device. "I didn't 'come by' it, Doctor," he replies, "I tinkered him together my first months up here." While this is soaking in, Morbius re-states that he's in no need of any assistance whatsoever, and again tries to shoo them off the planet. "Just as soon as we interview the rest of the *Bellerophon* crew members," inserts Adams, mindful of his orders. Morbius looks stricken for a second then tells them that there are no other members. They are all dead, torn limb from limb by "a devilish planetary force, which never once showed itself." For some reason, he and his wife (who died shortly after of unspecified "natural causes") were immune. The *Bellerophon* itself was "vaporized" as the last three members of the expedition tried to take off. The force has not re-asserted itself in the intervening 19 years "except in my nightmares," he reveals in a nice bit of foreshadowing.

As the stunned space men try to take this all in we hear a female voice inquire, "Father?" A young blonde woman about 19 comes into the room to meet them. The hardened military men turn into love struck schoolboys at the sight.

©Warner Bros.

Farman is more than happy to step
forward and volunteer for this detail!

Morbius introduces her as his daughter, Alta (short for Altaira). Farman starts to put the moves on her over coffee, while Ostrow (still back with the captain and Morbius) says, "We young men have been shut up in hyperspace for well over a year, and right from here the view looks just like heaven!" (And right in front of her father!).

At Morbius' urging, Alta says she's perfectly happy there with her father, Robby and all her friends. "Friends?" asks Adams, who had just been told that there were no other humans on the planet. Alta moves out onto the terrace and blows an ultrasonic whistle. Two deer romp in from the woods, followed shortly by a full-grown tiger. While it appears completely tame, "outside of my daughter's influence, it's still a vicious wild beast," warns Morbius.

Once more he tries to hurry them on their way and off his planet, but Adams says there's now a complication. His orders don't cover the *Bellerophon* fatalities,[47] so he has to call back to Earth for instructions. He describes at some length the amount of work it will take to build an appropri-

ate transmitter, which includes removing the ship's main drive (!) to get enough power. Morbius wants to help in any way he can (just to speed things up, so they will leave) and offers to have Robby run up some lead shielding.

Back at the ship the next day, they're busy removing the reactor core as Robby walks up carrying, on one arm, the shielding. It's not common lead but, "my morning run of Isotope 217," which only weighs 10 tons (9 tonnes)! We see that Alta has ridden

©Warner Bros.

over with Robby to watch. Farman walks over and picks up where he left off the day before. In the background we see them walk off as Chief Quinn asks the captain to inspect the "Klystron transmitter monitor unit" he's building. There's some more great technobabble as Quinn shows the commander what he's up to, with the equipment spread over a large table outside the ship. "Any quantum mechanic in the service would give the rest of his life to fool around with this gadget," he says with considerable pride as he plugs a complicated looking coil device into an equally complicated base.[48]

©Warner Bros.

Chief Quinn seems quite pleased
with his Klystron modulator

---

[47] This is not really true since their mission, as stated earlier, is to search for survivors, not just check on the status of the colony. This implies that they were expecting some fatalities, all the way up to 100%. The problem is that there is only **one** survivor, and Adams regards him with suspicion.

[48] This scene, and this whole concept of having to gut the ship to make contact with Earth, is obviously played for drama and humor, not logic. It is supposed to be analogous to, say, an Antarctic expedition of the early part of the 20th century which has to gut some of their equipment to build a transmitter powerful enough to reach back to home base in the northern hemisphere. But, removing the reactor core? The crew of a nuclear submarine in the Weddell Sea wouldn't yank its reactor onto the beach just to set up a big transmitter. Sure, the high-power transmitter components might have to be located remote from the ship, but the reactor would do much better left right where it is, with all its shielding and controls in place. Besides, if the ship had the equipment on board that could be cobbled together in the field to build such a device, why hasn't the UP just designed that capability into the ships in the first place?

The Saucer Fleet

Cookie, the ship's cook, gets Robby aside and asks him where he can find "some of the real stuff." "Stuff?" inquires Robby. Cookie means genuine ancient rocket bourbon, "just for cookin' purposes, you unnerstan'." Robby takes Cookie's last bottle and drains the remains into his food analyzer port. As Cookie shouts his protests, Robby, not understanding the nature of this particular human weakness, asks, "Would sixty gallons be sufficient?" We leave them with Cookie in drop-jawed amazement and join the other couple, Jerry and Alta to see what they're up to.

©Warner Bros.

Farman is explaining kissing to her in academic terms that he thinks will appeal to her well-developed mind, while taking shameless advantage of her total naiveté regarding, uh, biology. Adams catches them at it, and angrily orders Farman back to the ship. He then tries to explain the situation to Alta, but she is completely obtuse, due to her inexperience, which angers him in a paternal sort of way.

Dissolving back to the residence that evening, Alta is at the tail end of explaining what happened and is busily mimicking the captain's anger. Morbius doesn't seem too upset and turns in. Alta "beams and beams" for Robby, and seems annoyed when he takes all of fifteen seconds to appear ("Sorry, miss, I was giving myself an oil job"). She then tells him she wants another dress. "Again?" he says, coming as close as a robot can to sighing. She says it must cover everything but "fit in all the right places." Despite her outward annoyance at what Adams asked, she is having a dress made that meets his request that she be more conservative.

©Warner Bros.

Another new dress?

Meanwhile, back at the ship, two crewmen, Gray and Strong,[49] are making small talk while on guard duty outside the ship. We float over them and up the boarding stairs as a mysterious electronic beat plays in the background. We move inside the ship and see the hatch raised and the cargo crane move aside by an unseen hand. Cutting directly to the next morning, Adams and Quinn are going over the equipment that was destroyed the previous night. Among the casualties is the coil device. Quinn lists all of the reasons it can't be fixed to which Adams snaps back, "Okay, so it's impossible. How long will it take?" "Well, if I don't stop for breakfast…" Quinn replies, lacking only a Scottish accent to make the scene complete.[50]

©Warner Bros.

Adams and Ostrow get into the tractor to head back to the Morbius residence, but he orders Farman to remain behind. Farman takes this as punishment for the previous day's "kissing" incident, but it seems reasonable, by military standards, to leave the first officer in command while the captain is away.

At the house, Robby is placing flowers in a vase as a mischievous monkey tries to steal fruit from a bowl on the

---

[49] This character name may be a nod to the popular '50s TV and radio show *Tom Corbett, Space Cadet*. Captain Strong was Corbett's commanding officer.

[50] To acknowledge its debt to *Forbidden Planet*, the Star Trek film *First Contact* (1996) contains an homage scene where Lt. Barclay brings up a damaged "warp plasma conduit" to chief engineer Geordi La Forge for inspection. It's a copper coil about 3 feet (1 meter) long whose damage bears an uncanny resemblance to a certain damaged Klystron frequency modulator.

©CBS Paramount Network Television

other side of the room. With "eyes in the back of his head" Robby dispatches the little thief with a low power beam. Adams and Ostrow show up and Robby welcomes them. He says Morbius is in his study, "never to be disturbed while that door is closed." They decide to wait, Ostrow in a chair while Adams takes a walk outside.

Alta is swimming in the pond, apparently nude. Adams politely looks away, but the audience is shown she's actually in a very conservative flesh-colored swimming suit. Some innocent-awkward goofy banter transpires while she changes into the dress she'd had made to please Adams. She asks him to kiss her (using lines she learned from Farman), and this time apparently enjoys it. Her pet tiger shows up on the rock overhang and roars threateningly. Adams draws his blaster but Alta reassures him, "It's okay, he's my friend." Suddenly the tiger leaps at them in a decidedly unfriendly atack and Adams disintegrates it to save them. "I'm sorry I had to do that," he says apologetically. "He didn't recognize me!" she says, astonished. "He would have killed me, why?"[51]

©Warner Bros.

Adams walks back into the house and is informed by Ostrow that Morbius hasn't yet come out of his study. Getting impatient, Adams walks over to the door, which opens for him. They enter the study and find no one there. "The robot lied," says Adams exasperatedly, as if he's discovered a chink in Robby's armor of perfection. Curious as much as angry, they start nosing around the desk. While they're so occupied, the blackboard wall slides aside silently to reveal a rough-hewn stone surface with a large diamond shaped passageway. Morbius emerges from the passageway and is incensed at their intrusion into his private study. His anger

---

[51] This is Irving Block's addition of the legend of the virgin and the unicorn by the Roman historian Pliny the Elder. In mythology, unicorns are not the pastel colored playthings of young girls' fantasies; they are fearsome, deadly creatures (they have that horn for a reason) that can only be tamed in the presence of a virgin. Once the virgin is "compromised," the unicorn reverts to being a deadly beast. This scene is to show Alta's loss of innocence after kissing Adams (and enjoying it). There were other references to this legend, such as Adams and Ostrow back at the ship discussing why they thought the big cat was so tame around her, which were cut from the final print.

quickly subsides and he starts to tell them about the Krell[52] to explain why the equipment on the ship may have been sabotaged. The Krell were a highly advanced civilization that was on this planet thousands of years ago. They were a space-faring people and had even visited Earth in the distant past (hence all of the Earth creatures on this planet). Their massive and benevolent civilization disappeared literally overnight some 2,000 centuries previously. Morbius leads them down the stone passageway into a Krell lab. He shows them a display screen upon which "the entire scientific knowledge of the Krell" can be projected. He first taught himself their alphabet after which he began to learn their technology. "The first result was that robot that you seem to find so remarkable. Childs' play," he says offhandedly.

"What's this device over here?" asks Ostrow, indicating a strange looking piece of equipment with a clear pyramid in its center. "I call it the 'Plastic Educator'," replies Morbius. "They used it to test and train their young." He sits at the console (which has a peculiar chair, very wide with a split back) and draws the 3-armed headset towards him. He places the central arm on his forehead and the outer ones on his temples. He pulls a switch and a large, highly stylized indicator starts floating up a tube nearby. It goes up about halfway. "One of their average children was expected to send that clear to the top!" he says. "Which, by Krell standards, classifies me as a low grade moron, yet I have an official IQ of 183." Pulling another lever he shows them the machines "primary function." A dynamic cloud starts to form under the pyramid in the machine's center. It resolves itself into a perfect 3-D living image of Altaira.

©Warner Bros.

The astounded visitors eagerly accept Morbius' invitation to take the Krell test themselves. "There's something wrong here," says Ostrow, when the indicator rises only a third as far as when Morbius took it. "I have an IQ of 161." Adams tries next and registers a half gradation less than Ostrow. "That's all right, sir," says Morbius patronizingly while casting a sidelong glance at Ostrow. "A commanding officer doesn't need brains, just a good loud voice, eh?" "How do I make an image?" Adams asks, ignoring the remark and reaching for the lever. "Don't! Stop!" shouts Mor-

---

[52] Some sources spell it with only one "l" (Krel) since that looks more alien, but all MGM documents use the double "l" form.

"A commanding officer doesn't need brains…"

bius in a panic, yanking the headset away. "Our *Bellerophon* skipper tried that and it was instantly fatal!" "So you're immune to this, too?" replies Adams, suspiciously. "In my first attempt to make an image there, I lay unconscious for a day and a night," Morbius explains. "But you can imagine my joy when I discovered that the shock had permanently doubled my intellectual capacity!"

Switching subjects he continues the Krell story. "Just before their annihilation, the Krell turned their entire racial energies towards a project that would free them from physical instrumentalities." The doctor senses the import of this statement, but Adams is incredulous. "Dr. Morbius, everything here is new, not a sign of age or wear on any of it." "These devices, self-serviced and maintained, have stood as you see them for 2,000 centuries!" Morbius declares. "2,000 centuries?" Adams returns. "What is their power source?" Morbius points out the gauges wrapping around the room. "They're set in decimal series," he explains, "each gauge registering ten times as much potential as the one before it. 10 x 10 x 10 x 10 x 10… The number 'ten' raised, almost literally, to the power of infinity!"[53] Seeing that his guests are properly stunned, he offers, "Would you like to see more of the Krell wonders?"

They hop into a small shuttle and start down an infinitely long tunnel of lights.[54] "Prepare your minds for a new scale of physical scientific values," warns Morbius.

The next scene is the special effects centerpiece of the film; a vertigo inducing shot of an infinitely deep

shaft. Huge machine pieces slide up and down rails as massive green and yellow discharges arc between two of them. A flat walkway bridges the shaft. Three infinitesimally small figures emerge from a doorway and walk to the center of the walkway. (To make the figures so small, the live action part of this shot was filmed from the roof of the nearby Bekins Van and Storage warehouse with midgets, costumed like the actors, walking in the parking lot below. They did such a good job that none of the test audiences ever saw the actors walk into the scene, so they had Josh Meador animate a blue light flashing onto the walkway to draw attention to them.[55])

We come down to our party's level as they stop in midspan. "Twenty miles!" says Morbius, pointing down the corridor. "Twenty miles!" he repeats, pointing the other way. "This is one of their ventilator shafts. You can feel the

warm air rising. Look down here," he says as he points over the edge. Adams and Ostrow are reluctant to approach the edge of an unguarded walkway that drops off apparently to infinity. Morbius continues somewhat condescendingly, "Look down, gentlemen! Are you afraid?" (actually, that would seem to be a perfectly reasonable reaction!). They peer uneasily over the edge. "Seventy eight hundred levels, and 400 other shafts like this one," continues Morbius as they walk back to the doorway.

They next visit one of the "furnace rooms" where titanic machines hum and flash to some unknown purpose. Morbius continues his soliloquy, "Yes, a single machine, a cube 20 miles on each side. For 2,000 centuries it has been waiting patiently here, tuning and lubricating itself, replacing worn parts…" "But what's it all for?" interrupts Adams. "Uh, sometimes the gauges register a little when the buck deer fight in the au-

[53] Actually, with 18 dual gauge panels, it's the number 10 raised to the 36th power, which, while really big, is a long way from infinity. Morbius may be a super brain-boosted philologist, but he's no mathematician.

[54] Could a young Stanley Kubrick, just starting his film career, have been impressed enough by this effect to use it 12 years later as the concept for the "Stargate" sequence in *2001*? The view out the shuttle window is very similar to that out of Bowman's pod.

[55] *Cinefantastique*, V4, N1, pg. 11.

tumn. Nearly a whole dial became active when your ship first approached from deep space,"[56] offers Morbius hesitantly.

At this point in the final film, they ignore the fact that Morbius has completely avoided answering Adams' question, and move on to the next scene. As originally written, though, the scene's much more revealing.[57] It continues:

ADAMS

I asked you, Dr. Morbius, *what's it for!?*

MORBIUS

(Looking haggard and haunted) *I don't know!* In twenty years I have been able to form absolutely no conception at all…

Perhaps they just didn't want to show Morbius' fallibility just yet, but it would have been good to see his limitations.

"I'll show you one of the power units," he says, continuing the tour. They move off of the walkway into a (comparatively) small room where Morbius instructs them to look only in the mirror[58] as he opens the shields onto one of "9,200 thermonuclear reactors in tandem; the harnessed power of an exploding planetary system!" (More foreshadowing.)

©Warner Bros.

As Morbius closes the shields, we jump back to the *C57-D* where it's night and they have just finished setting up a force-field fence around the ship. Farman tests it by tossing a small branch into the field. It disintegrates in a blue flash.

Cookie shows up and evasively asks permission to leave the perimeter. Farman reluctantly gives in and has the fence turned off long enough for him to leave. Cookie races out so fast that he trips rounding a large rock, and as he gets up, comes face-to-face with "480 pints, total: 60 gallons" of bourbon, "as you requested," reports Robby. The humorous scene turns ominous as Robby detects something. What's

©Warner Bros.

the matter?" asks Cookie, "Something coming this way?" "No sir," replies Robby, "Nothing coming this way," although he continues to scan the area warily.

Back at the ship, the fence "shorts out" in a huge display of blue discharges. The men all rush over to check it out, but rather than joining them, the camera goes right down to ground level. We hear the same "beta beat" as before, but rather than floating over to the access stairway like the last time, something unseen is leaving enormous, single toe footprints in the dirt. "It" continues into the ship up the boarding stairs as they bend under a huge, but unseen, weight. Farman and the bosun are still

©Warner Bros.

discussing the fence shorting out, when a blood-curdling scream is heard in the ship.

We dissolve back to the residence where the trio is back in Morbius' study. A heated argument is in progress. "A scientific find of this magnitude has got to be taken under United Planets' supervision. No one man can be allowed to monopolize it!" demands Adams. "For the past two hours I've been expecting you to make exactly that asinine statement!" returns Morbius. "Such portions of the Krell's science that I deem suitable and safe, I shall dispense to Earth. Other portions I shall withhold!"

The two are interrupted by Adam's communicator beeping, "Skipper, the Chief's been murdered!" reports Farman. "Quinn? Murdered?" How was it done?" stammers Adams. "Done? Skipper, his body is plastered all over the communications room!" As Adams and Ostrow race off, Morbius stares blankly into space. "It's started again," he croaks out, as he sinks back into his chair.

Back at the ship the next day, Ostrow presents a plaster cast made from one of the footprints. He describes it in taxonomical terms, and how it "runs counter to every known law of adaptive evolution. Anywhere in the galaxy, this is a nightmare!" (yet another nightmare reference). They bring Cookie in to discipline him for leaving the perimeter, but it turns into an interrogation. After learning that he was with Robby the whole time, it eliminates the robot as a suspect in the murder. Frustrated by the lack of leads, Ad-

---

[56] Remember being "radar scanned" from an area 20 miles square?

[57] *Cinefantastique* double issue, V8, N2/3, pg. 12.

[58] Whether intentional or not, the staging of this scene is lifted directly from Meador's work on *20,000 Leagues Under the Sea* where Nemo shows Professor Aronnax the power source of the Nautilus.

©Warner Bros.
Adams and Ostrow study the plaster cast of the Id Monster's footprint. Note the similarity to the UP logo, seen here on the side of the tractor, but also present on the uniform caps. Morbius may have subliminally incorporated this into his monster as a way of warning them they were "harming themselves" by staying.

ams suggests "Maybe you and I should drop over to that Krell laboratory and get our IQ's boosted a couple of hundred percent!"

Later, outside, they finish burying Quinn and turn around to discover Morbius and Alta standing near by. "I warned you," Morbius says, "while your ship was still in space. I begged you not to land! The next attack on your party will be more deadly...and general" "How do you know that?" interrupts Adams. "Know? I seem to visualize it," says Morbius, injecting the final foreshadowing.

We dissolve to night where the crew has just finished setting up the large artillery-size blasters. They test them to devastating effect into the surrounding hillside. Then Randall, the radar operator, announces that he has detected a target moving up the arroyo towards them, as the "beta beat" starts again. Adams calls for "automatic control" and the blaster men take their hands off of the controls. They send two heavy bursts down the canyon, which Randall reports were "dead on target,"[59] but that it is still coming towards them. "Are you sure you have a real blip there?" Adams asks, as everyone strains to see in the darkness. "Big as a house, sir!" comes the frantic reply.

Suddenly the fence starts to "short out" again, but this time in the discharges we see a huge creature outlined in the energy beams. A tremendous firefight ensues with the space men using their main batteries, augmented with hand blasters. Most of the blasts hit the creature directly but splash harmlessly over it. Two crewmen near the fence, Strong and

[59] If they have "automatic controls" that are this good, why would they ever use manual?

©Warner Bros.
Two views of the Id Monster. Top: Josh Meador's masterful work showing the energy bolts from the blasters "splashing" over the creature giving it a sense of mass and solidity, even though it's made completely from energy. Bottom: Jerry Farman sacrifices himself for his shipmates.

Gray, are tossed aside like dolls and die instantly. Farman rushes up to make sure his blaster shots will hit square on. The creature picks him up and holds him in the air, the red energy discharges enveloping his body. His lifeless form is tossed aside and the crew resumes fire. We cut back to the Krell lab where Morbius is asleep at the desk, twitching in a

©Jon Rogers
Some people have a hard time seeing the Id Monster since it's never really clearly shown. This was, of course, deliberate. Jon has taken a frame and filled in all the "missing parts" from Meador's character design.

©Warner Bros.

nightmare. Behind him, over half the Krell power indicators are flashing as the machine roars in a manner similar to the creature. We hear Alta scream from outside which startles Morbius awake. A brief cut back to the ship and we see the creature simply disappear (well, it was always invisible when not trapped in the energy beams, but now it ceases to exist). As Morbius gets up, the power indicators quickly drop down to their normal "off" position. Alta runs to meet him saying that she'd had a terrible nightmare about a creature attacking the ship. Morbius says he is powerless to protect them a long as they stay.

Back at the ship, Adams and Ostrow are trying to make sense of what they just saw. "Doc, an invisible being that can't be disintegrated by atomic fission…" begins Adams, gesturing towards the fence. "No, skipper. That is a scientific impossibility," replies Ostrow. "Any organism dense enough to survive three billion electron volts," he continues, tapping one of the main battery blasters, "would have to be made of solid nuclear material. It would sink by its own weight to the center of this planet!" "You saw it yourself standing in those neutron beams!" counters Adams. "Well, there's your answer. It must have been renewing its structure from one microsecond to the next," concludes the doctor.

Adams makes the command decision to return to Morbius' house and "evacuate" them as per regulations, even against their will. To prevent what happened to the *Bellerophon* when it took off, he says that one of them has to take the Krell brain booster (although why they think getting a brain boost would prevent them being vaporized is never made clear).

They arrive at the residence and are met by Robby who is "monitored to admit no one at this hour." They draw their blasters, but he neutralizes them with a pair of beams that emanate from the sides of his head. Alta shows up, overrides Robby's security program, and lets them in. Adams and Alta have a heart-to-heart

©Warner Bros.

about leaving, by force if necessary. "I can't leave without my father," Alta protests. "Darling, I'm not leaving without you!" says Adams, sounding like a lovesick teenager instead of the steely-eyed military commander he's been up 'till now. As may be obvious, there is a scene missing previously that was a setup for Adams being so romantically disposed towards Alta. When she and Morbius visited the ship at Quinn's funeral, Morbius said that he didn't like how she had "attached" herself to the commander, and that she would soon have to make a choice between Adams and himself.

Alta throws a lip-lock on him to shut him up. "Doc? Will you talk some sense to this girl? I'm in over my head," says Adams in full mush-mode. And then alarmed, "Doc?!" suddenly realizing that Ostrow isn't there. He whips around just in time to see Robby carrying Ostrow's limp body out from the Krell lab. Placing him on the sofa, we see there are burn marks at his temples and in the center of his forehead, obviously from the Krell educator's headset.

"So you took the brain boost," says Adams, despairingly. "You should see my new mind!" replies Ostrow, obviously in severe pain. "It's bigger than his now." He continues, "Morbius was too close to the problem. The Krell had completed their project. The big machine…no instrumentalities…true creation! But the Krell forgot one thing. Monsters, John! Monsters from the Id!" Ostrow gives one last convulsion and dies.

©Warner Bros.
"Monsters, John! Monsters from the Id!"

Morbius enters from his bedchamber. "The fool! The meddling idiot!" he explodes. "As though his ape's brain could contain the secrets of the Krell!" Alta, shocked at her father's callousness, addresses him while clinging to Adams, "Morbius, you wanted me to make a choice. Now you've chosen for me."[60] Turning to Adams she says "I'm ready to go with you now, darling," and runs from the room (presumably to pack). Adams restrains Morbius from following her.

Morbius, what is the Id?" he demands. There follows a brief discussion of the primal subconscious. Adams relates what Ostrow had said, and how, as soon as they switched on the machine, the "inner demon" of every Krell was set loose.

---

[60] Referring, of course to the missing scene mentioned earlier. As edited, the audience is left to wonder, "When did he ask her to make a choice?"

The Saucer Fleet

"My poor Krell!" says Morbius finally understanding. "They could hardly have understood what power was destroying them. "But," he continues, "the last Krell died 2,000 centuries ago, and today, as we know, there is a monster loose on this planet." "Your mind refuses to face the conclusion…" starts Adams, who is interrupted by Robby who has entered at the back of the room.

"Morbius," Robby calls out in alarm. "Something is approaching from the southwest. It is now quite close." They rush to the window-walls only to see trees parting and crashing as the invisible thing approaches. Morbius closes

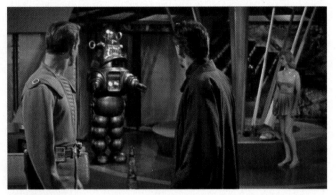

©Warner Bros.

the steel shutters, but the thing starts punching through them like they were sheets of plastic. "You still refuse to face the truth," says Adams. "What truth?" shoots back Morbius. "That thing out there," Adams continues, as if speaking to a child, "it's you!" "You're insane!" Morbius replies.

By then, the creature has broken through. "Stop it, Robby! Kill it!" implores Morbius, but Robby knows the truth. He gets into the same 1st-2nd law conflict loop (knowing that to follow Morbius' command and stop the monster, he would have to kill Morbius), this time all the way to his destruction. As Robby stands there, lifeless, swaying slightly back and forth, the trio runs into the Krell lab and Morbius closes the 4-

©Warner Bros.

part door. The power indicators are flashing, indicating the machine is in operation again.

Adams forces Morbius to sit at the educator. "Here is where your mind was artificially enlarged. Consciously it lacked the power to operate the great machine, but your subconscious had been made strong enough!" "I won't hear you!" shouts Morbius. Adams won't let up, "20 years ago when your comrades voted to return to Earth, you sent your Id out to murder them. When our ship threatened your little egomaniac empire, your subconscious sent its Id monster out again. More deaths, Morbius! More murder!"

"And now this?" Morbius wails, "harm my own daugh-

ter?" "She's defying you, Morbius. Even in you, the loving father, there still exists the mindless primitive!" (This guy's a pretty good psychoanalyst for a starship captain…) "Now you're whistling up your monster again to punish her, and if you don't do something about it soon, it's going to be coming right through that door!" He points to the door, which is already starting to discolor from the intense energy being applied from the other side. "That's solid Krell metal, 26" thick!" Morbius protests. "Look at your gauges!" demands Adams. "That machine is going to supply your monster with whatever amount of power it requires to reach us. Look, now!" The door is glowing a dull read, but it rapidly gets brighter as it heats. "Alta, say you don't believe this!" Morbius implores, but Alta just looks down silently. "Then it must be true," he says in total surrender. He looks to the door that is now white hot and starting to fall away in large chunks. "Guilty! Guilty!" he wails, pronouncing sentence on himself. "My evil self is at that door and I have no power to stop it!' But Adams does. He draws his blaster, preparing to kill Morbius to save his and Alta's lives. The door is now crumbling completely as the Id monster punches through it; every power gauge flashing.

©Warner Bros.

Morbius rushes to the door, "Stop! I deny you! I give you up!" he says to the unseen terror coming through. The machine reaches a shrieking crescendo as Morbius faces the ultimate horror. We only see Adams' and Alta's reactions as the machine peaks and starts to wind down. Quickly it is back to only the "beta beat" as Alta rushes to her mortally wounded father. (Many people consider this scene a "cheat" on the audience since we never see what it is that attacks Morbius. The original screenplay has the monster become

©Warner Bros.

I deny you!

178

visible for the final confrontation, but there was no logical reason for it being so, unlike when it was glowing in the energy beams at the ship. With time and budget running out, it was probably decided to be left invisible for purely practical reasons, although it is very confusing just what is happing.[61])

"Son," croaks Morbius, now suddenly paternal towards Adams, "turn that disk," pointing to a console in the center of the lab. Adams does so and a long rod emerges from the floor. "The switch, throw it!" Adams pushes the rod down until it clicks at the bottom. The rings around it start pulsing red, and an ominous rumbling can be heard. "In 24 hours you must be 100 million miles into space," he gasps, barely audible. "The Krell furnaces…chain reaction…cannot be reversed…" and collapses, dead.

©Warner Bros.

We dissolve to the *C57-D* streaking away into space. Inside, Adams looks at the Astrogator and declares "98 million point 6, we're clear now." He walks, with Alta, to the other side, and there sits Robby! (Wait a minute, didn't he "blow every circuit in his body" during the Id monster attack? Who fixed him? Both miracle-worker Quinn and super-brain Ostrow are dead. Maybe he has a giant reset button somewhere.) "He's quite an Astroga-

©Warner Bros.

[61] *Cinefantastique* double issue, V8, N2/3, pp. 34, 35.

tor," says Adams (and they need one, too, with Farman also dead). "A genuine privilege, commander!" says Robby. "Activate main view plate" Adams commands. "Aye, aye, skipper!" a jovial Robby replies.

They walk over to watch. "Fifteen seconds," someone announces. Adams is holding Alta in his arms as he recites, "Your father, my shipmates, all the stored knowledge of the Krell." He counts down the last few seconds, and exactly on time, Altair IV turns into a hot, blue nova for about five seconds. Alta turns away and buries her face in his shoulder. Adams, very patronizingly, gives his final soliloquy, "In a million years, the human race will have climbed up to where the Krell stood at their great moment of triumph and tragedy, and your father's name will shine again. True, it will be to remind us that we are, after all, not God!"

©Warner Bros.

With that final insult to her father out of the way (well, that's what it is), the *C57-D* continues off into the void. In the final scene, cut from the film shortly after the premier, Adams and Alta stand side-by-side as the bosun officiates at their wedding. The crewmembers look on in amusement, especially Cookie who has fashioned a small wedding cake for what's left of the crew to enjoy. This gave the film the requisite Hollywood happy ending, but makes very little sense. This is one cut that most fans approve of.

©Warner Bros.

The Vehicle

*[Author's Note: Velocities and accelerations in the following discussion are described in straight Newtonian terms. I don't want any complaints from physics majors saying that I'm ignoring the relativistic effects of these speeds. Relativity says you can't travel **at** the speed of light, but doesn't prohibit going faster. Besides, the movie itself ignores the relativistic predictions of faster-than-light (FTL) travel such as time running backwards – JH]*

The *C57-D* is a bit of an enigma. We know from the unedited version of the opening narration that it uses "quanto-gravitetic hyper drive"[62] for propulsion. This has a nice, techie sound to it with an obvious basis in real physics (perhaps it propels the ship by gravity waves at the quantum level) but outside of this single reference, which was cut from the final movie, we have no clue how the main drive works. Is it powered by fission? Fusion? Antimatter? Some other currently unknown subatomic reaction? Again, there's not much to work from.

What we do know is that it moves the ship right along. Altair is 16 light-years from the Sun. The opening narration tells us that they've been traveling "more than a year" on this mission. That's more than 12 months, but probably less than 18 (or the narrator would have said "nearly two years…"). To make the math easy, we'll assume it's 16 months. Traveling 16 light-years in 16 months is one light-year/month or 12c (12 times the speed of light). A potent drive indeed!

©Warner Bros.

They give us a pretty good look at the drive hardware in the scene where they pull the main core when setting up the transmitter to contact Earth. It has the appearance of a small, but massive magnet, which would imply some sort of fusion device as the prime energy source. This is bolstered by the network of copper coils in the "frame" that holds the core. This frame slides out of the central landing pylon, thus giving easy access to the ship's main drive (this ship is well designed for field repairs!). On the other hand, there was an indication, cut from production, that the core was danger-

ously radioactive, which would suggest a fission source:

> The script called for helmets and radiation armor to be worn by the crew as they unshipped the saucer's main core to power communications back to Earth, but this special costume idea was scrapped when the film began to go over budget.[63]

The two-level main cabin is enclosed mostly by the massive upper dome, but it also extends out into the saucer disc. Given the size of Arthur Lonergan's interior set design, we can see that the "full size" exterior set of the lower part of the ship is nothing of the sort. The exterior set, as mentioned earlier, was a 60-foot (18 meter) piece of the ship that would have been about 100 feet (30 meters) in diameter if it were all there. But the interior set is 30 feet (9 meters) from the center of the Astrogator out to the furthest "DC Station" pads, making the dome, by itself, about 60 feet (18 meters) in diameter. With this scaling reference, the overall diameter of the disc is 160 feet (49 meters).

©Warner Bros.

The only "real" part of the *C57-D* is the set enclosed in the white box. Everything else is a matte painting. Compare this to the cutaways in the Data Drawings on the next few pages to see how undersized it is.,

The crew cabin can't extend too far down into the disc, because of the central landing pylon (that supports the ship's weight on landing) needs a place to retract up into. The three landing legs, with their integral boarding stairs, are mainly for stability rather than support (this was a popular Wernher von Braun concept, which he used on his *Collier's* moon landing craft and the later Disney Mars landers). Just prior to touch down, the ship is supported on a blue force field beam, but has the ability to hover without it, as shown in the landing sequence. Coincidentally, the alien ships in *Earth vs. the Flying Saucers* (see page 216) also use the central pylon idea for landing support. Those saucers, though, didn't have legs, only the pylon.

No indication is given as to why the ship is a saucer, other than it looks futuristic. There is some aerodynamic benefit to this particular shape for maneuvering in atmosphere, but any ship that can travel faster than light and balance on force field beams for landing probably doesn't need much help from something as primitive as mechanical lift from gasses flowing over the hull. Neither is there any hint as to what the rest of the volume of the disc is used for. The crew is under the large top dome and the main drive in the smaller lower dome extending up into the ship. Not knowing how quanto-gravitetic hyper

---

[62] Starlog Photo Guidebook *Spaceships*, V1 1979, pg. 87. Also, the unedited opening narration is included as a bonus feature in the 50th Anniversary "Collectors" DVD set.

[63] *Cinefantastique*, V8, N2/3, pg. 28.

drive works, the "Primary Field Coils," mentioned by Adams in the landing sequence, might have to be arranged in a circle some distance from the central drive nacelle. Note that the outmost edge of the disc is a darker color than the rest of it, suggesting some sort of different structure, or at least a different material (such as how the leading edge of the Space Shuttle's wings are a different material because they have to be more heat resistant than other parts). As for the rest of the disc, it's probably consumables storage, space for the three landing legs plus the tractor in a disassembled state (Adams orders the bosun to "assemble the tractor" just prior to Robby's arrival). Jon has some speculations on what this space is used for in his "Archeologists' Report" coming up.

Inside the ship, we see that it is functionally very close to a submarine, although arranged in a circle rather than along a tube. There are no windows or viewports, and the captain usually sees out by peering through a binocular instrument analogous to a periscope. The floor plan is divided up into ten wedges of 36° each, although some are subdivided further into 18° segments. The crewmembers all work at stations along the wall on two levels, and re-

©Warner Bros.

port in, as necessary to the captain in the center. There is a ship's cook, played mostly for comic relief, but we never see a galley, or even the crew's mess. This is strange since this two-level crew area is the only occupied space we see. The command station is in the very center of the lower level (more about that in a second), surrounded by the communications center, a large radar screen, the main viewplate, a ward table and the heavy-duty hatch leading down to the exit area below. In a ring further out there are the "DC" (deceleration) stations and crew bunks. The stations on the upper level are not clearly defined, except for the gunner stations. Just where the beams from these weapons exit the ship is not known. Again, Jon's Archeologists' Report will cover this in more detail.

The command station in the center consists of a large complex navigation device, called the "Astrogator" (see photo page 167). This suggests an advanced level of technology while still being understandable to the general audience. By putting a model of the saucer in the center, we intuitively understand that the surrounding domes somehow represent the universe and its "guiding stars." The de-

©Warner Bros.

vice is a powerful social icon from the past when astronomy was just starting to break from astrology and become a science. Early astronomers sought to capture the beauty and complexity

in the motions of heavenly bodies by making elaborate clockwork mechanisms called "armillary spheres." This device mimics an armillary sphere quite closely, substituting clear etched spheres (the "crystal spheres of the heavens" imagined by those ancient geocentric astronomers) for the elegantly engraved metal, the other difference being that we are now traveling among the stars, rather than simply observing them.

Surrounding the Astrogator is a circular console with the three control panels that directly operate the ship. The main control station has two panels flanking the binocular sight, while an auxiliary station is placed 180° around from it with a single panel. Both control stations have cutouts in the console forming a station for the operator. At 90° around from both control stations are two chart stations for course plotting, patterned after similar navigation stations on contemporary naval ships.

The DC stations also show some advance conceptual thinking. The standard way of dealing with high accelerations in Sci-Fi movies up to this time was a horizontal cot or reclining chair with some straps to hold you in place. That seems pretty primitive when dealing with FTL travel. The deceleration phase, as shown on screen, lasts about 40 seconds. During that time the ship decelerates from 12c (8 billion MPH or 13 billion Km/hr) to "only" .3896c (261 million MPH or 421 million Km/hr), which is a deceleration rate of 9 million gees! Somehow cots and straps seem woefully inadequate, so each DC station envelops its occupant in an energy field that protects him during this unthinkable transition.

©Warner Bros.

There's obviously something else at work here since, energy field or not, this level of acceleration is sufficient to completely collapse any molecular structure into solid atomic material, and maybe even start collapsing the atoms themselves. There's a suggestion that the deceleration event is survivable without the DC station as the bosun snaps, "you wanna bounce through this one?" to a crewmember who is slow getting to his station.

The *C57-D*, despite all of the rewrites and awkward edits, is a major milestone in the evolution of the spaceship. It is the film world's first venture into faster-than-light travel (by humans) and does it in a completely believable way that rings true even today. Its influence was strongly felt throughout the rest of the '50s and most of the '60s, and following Jon's Archeologists' Report we'll be back with the post release history of the film, and its impact on Sci-Fi culture.

Archeological Report on the C57-D

By Jon Rogers

As Jack has already stated, being filmdom's first attempt at creating a believable, human flown, *interstellar* spaceship; the flying saucer *United Planets Cruiser C57-D* was a milestone.[1] Since this was their first on-screen spaceship and because they knew it would be seen by millions of moviegoers, MGM invested a lot of money and talent in its design and development. The amount of thought and detail they lavished on this project was the major reason the *C57-D* became the extremely influential design it did. The *C57-D* helped persuade people that flying saucers were of interstellar origin. It really impressed the future producers of our present science fiction movies who saw it and, because of that; its influence can still be seen on the silver screen today.

Because of its familiarity and the thorough analysis it has already received, you might think there is little more to be learned about this MGM movie creation. On the contrary, there are still many details to be discovered.

What is known about the ship? Studio diagrams, the movie, and other behind the scenes photos are the basis of this study. They reveal that the *C57-D* had a diameter of 160 feet (50 meters) and a dome 60 feet (18 meters) across. These sources also give us a firm understanding of the saucer's overall shape and the size of the internal set that appeared in *Forbidden Planet* as the main deck. Since everything must fit inside the saucer, we start the analysis by comparing its envelope and its internal set.

You can see the features of the main deck set in Figure 1. Comparing the set to the saucer's outside shape reveals that the set does not take up all the room inside the saucer. This is unusual for movie sets of spaceships of the era. According to our sources, there is even sufficient room for another deck (see Data Drawings sheets 4 and 5) below what we see in Figure 1. This has many implications.

Referring to Figure 1, it is evident that the interior set of the *C57-D* is actually divided into two decks. I have designated them "Top Deck" and "Main Deck" for clarity.

The main features of the Top Deck are the two "Gunner's Stations," some hinged panels, the illuminated ceiling above it, and two elevator poles that the crew use for moving between it and the Main deck. There is also a Gyroscope (the large black drum) between the two gun stations, but it is never seen in use.

A review of the space available for the Top Deck (Data Drawings sheets 2, 4, & 5) reveals that there is quite a bit of area between the visible walls and the outer skin of the ship. There is space for a 7 foot (215 cm) wide, circular hallway that would extend completely around the viewed upper deck. The height in this hidden area outside the top deck varies from about 5 feet down to zero. This space could be used for DC Station equipment, to house the ship's weapons, storage or other purposes. What it was actually intended

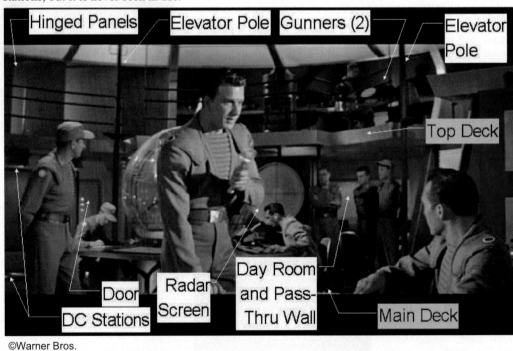

©Warner Bros.

Figure 1

[1] Ironically, TV was there first. *Space Patrol* had already shown interstellar flight with the *Terra-V*. But they had done it on a limited budget with a "Buck Rogers style" spaceship to a Saturday morning children's audience. This was to be MGM's first class, "big buck," serious adult audience approach.

for is unknown, as it was never shown in the movie or identified in surviving studio documents.

The two elevator poles that the crew uses to travel between the Top deck and the Main deck deserve closer scrutiny. To use the elevator pole, the crewman approaches it and steps onto a small platform extending out from one side of the pole. He next grabs onto a small handhold on the pole which is at about chest level. The platform then raises or lowers him to the other deck. When space is at a premium, this is an efficient, elegant solution to the problem of moving personal between decks.

One curious thing about the Top deck is while the men are shown manning the defensive consoles, or ray gun stations; they are never actually shown in use. The fact that they are there implies a need for them in a space battle that is never seen. Who would they be fighting?

Another intriguing feature of the Top Deck is visible directly above both the deceleration stations, and the communications station. It is an area that has, what appear to be, large, movable panels. These panels are clearly identified in the studio plans as "hinged fins." The normal engineering use for "fins" is to dissipate heat. These fins are seen moving during the deceleration phase of the ship. Since the ship is supposed to go through a period of intense heat during this time, we may infer that these are part of some kind of thermal transfer system. Why they would need fins *inside* the ship, we can only guess.[2]

When we start to examine the Main Deck we find that the purpose of some areas are clearly spelled out while others are ambiguous or undefined. One example is in the day room area (Figure 1 and Data Drawings sheet 2).

The "Dayroom"[3] is located behind the large radar screen. It is an area designed to hold over half of the crew at one time. It has a long semicircular table that will seat nine men. There is also seating against the back wall with three viewing instruments in front of it at various places, as though they were for reading. This seating may be movable or storable.

On the wall behind the seating is a sliding panel that, at times, opens up to reveal a pair of library shelves. At other times this panel is closed off. According to the set plan and ship's outline, there is a room behind this wall that is about 9½ feet (3 meters) wide and 40 feet (12 meters) in arc length. There are also two doors on the wall, one at each end of the seating that lead into this room beyond. This is a large room for a ship this size and must have some important purpose. Because of this and the following reasons it is very probable that this is the ship's Galley.

On ships where floor space is at a premium (like submarines) dayrooms serve a double purpose, both as a resting/recreational area (Dayrooms) and a dining area (mess hall). If this area also serves as the crews mess then the opening in the wall could also serve double duty because it connects the "galley" and the mess hall. At different times it could be either a library or the serving station. The two doors on each side of the seating area, leading into the large room directly behind the common wall certainly support this theory. So does the fact that the off duty cook is seen several times near the dayroom area. However, this theory is just a good, reasonable explanation for what is seen. The fact remains that it is not known absolutely where the galley is, or how the men are fed.

Another space that was not identified is the large room located between the Dayroom and the communications station, which has a bright blue, horizontally corrugated, plastic panel wall (Figure 2). A door leads to this area, but you never see it being used. At this writing, no information exists that identifies this area's real purpose. However, an intelligent guess can be made based on what *is* known.

©Warner Bros.

Figure 2

---

[2] Actually, their position makes little sense. Fins to remove heat would be on the outside of the ship. Fins to bring in heat would be on the inner side of the outer wall, which is in the outer hallway, not visible here.

[3] This area is not named this in the studio plans. It is my designation taken from the apparent purpose of the room. Its design duplicates areas on board all naval vessels called Wardrooms or Dayrooms.

The Saucer Fleet

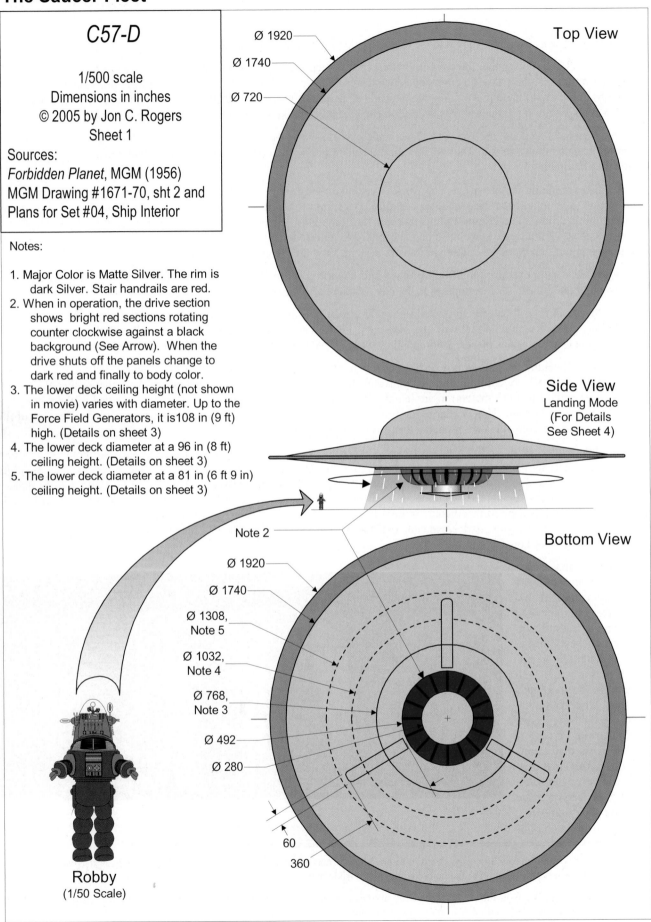

C57-D

1/500 scale
Dimensions in inches
© 2005 by Jon C. Rogers
Sheet 1

Sources:
Forbidden Planet, MGM (1956)
MGM Drawing #1671-70, sht 2 and
Plans for Set #04, Ship Interior

Notes:

1. Major Color is Matte Silver. The rim is dark Silver. Stair handrails are red.
2. When in operation, the drive section shows bright red sections rotating counter clockwise against a black background (See Arrow). When the drive shuts off the panels change to dark red and finally to body color.
3. The lower deck ceiling height (not shown in movie) varies with diameter. Up to the Force Field Generators, it is 108 in (9 ft) high. (Details on sheet 3)
4. The lower deck diameter at a 96 in (8 ft) ceiling height. (Details on sheet 3)
5. The lower deck diameter at a 81 in (6 ft 9 in) ceiling height. (Details on sheet 3)

Top View

Ø 1920
Ø 1740
Ø 720

Side View
Landing Mode
(For Details
See Sheet 4)

Note 2

Bottom View

Ø 1920
Ø 1740
Ø 1308, Note 5
Ø 1032, Note 4
Ø 768, Note 3
Ø 492
Ø 280

60
360

Robby
(1/50 Scale)

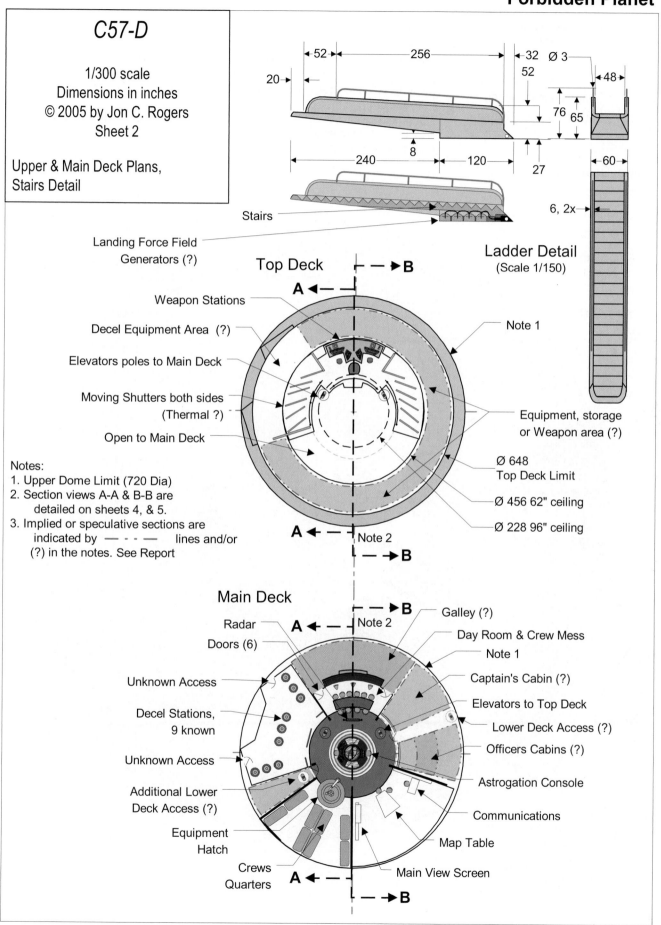

C57-D

1/300 scale
Dimensions in inches
© 2005 by Jon C. Rogers
Sheet 2

Upper & Main Deck Plans,
Stairs Detail

Stairs

Landing Force Field
Generators (?)

Ladder Detail
(Scale 1/150)

6, 2x

Top Deck

Weapon Stations

Decel Equipment Area (?)

Elevators poles to Main Deck

Moving Shutters both sides
(Thermal ?)

Open to Main Deck

Note 1

Equipment, storage
or Weapon area (?)

Ø 648
Top Deck Limit

Ø 456 62" ceiling

Ø 228 96" ceiling

Notes:
1. Upper Dome Limit (720 Dia)
2. Section views A-A & B-B are
   detailed on sheets 4, & 5.
3. Implied or speculative sections are
   indicated by — — · — lines and/or
   (?) in the notes. See Report

Note 2

Main Deck

Radar

Doors (6)

Unknown Access

Decel Stations,
9 known

Unknown Access

Additional Lower
Deck Access (?)

Equipment
Hatch

Crews
Quarters

Galley (?)

Day Room & Crew Mess

Note 1

Captain's Cabin (?)

Elevators to Top Deck

Lower Deck Access (?)

Officers Cabins (?)

Astrogation Console

Communications

Map Table

Main View Screen

Note 2

The Saucer Fleet

C57-D

1/300 scale
Dimensions in inches
© 2005 by Jon C. Rogers
Sheet 3

Lower Deck Plan

Notes:
1. Extreme limit of lower deck.
2. Section views A-A & B-B are detailed on sheets 4 & 5.
3. This is the Exterior Front viewpoint of the ship when the ship is landed.
4. Implied or speculative sections are indicated by ⎯ ⋅ ⎯ ⎯ lines and/or (?) in the notes.
5. The three access Stairs serve as ship stabilizers when landed. Logically, they would enter into compartments with Air Locks and retract into sealed enclosures.

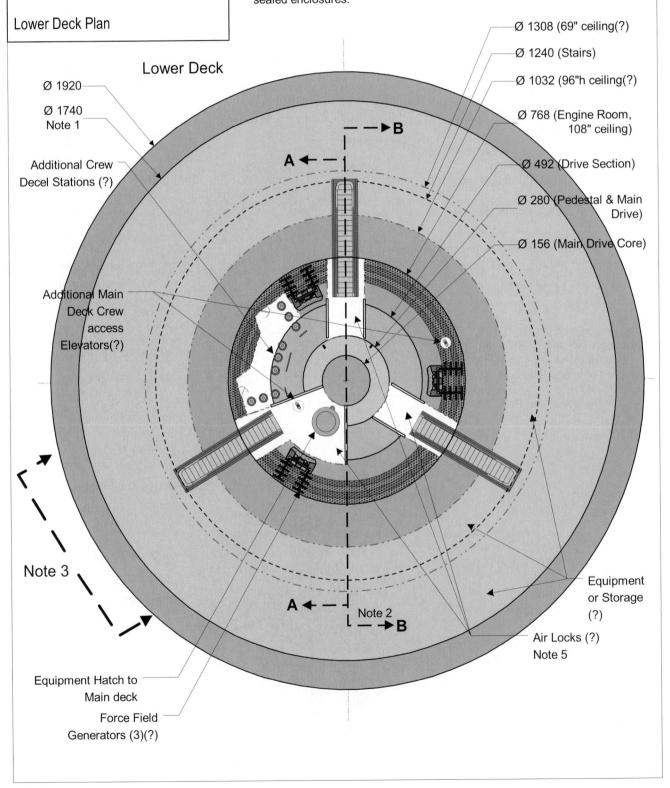

Lower Deck

Ø 1920

Ø 1740
Note 1

Additional Crew
Decel Stations (?)

Additional Main
Deck Crew
access
Elevators(?)

Ø 1308 (69" ceiling(?)

Ø 1240 (Stairs)

Ø 1032 (96"h ceiling(?)

Ø 768 (Engine Room,
108" ceiling)

Ø 492 (Drive Section)

Ø 280 (Pedestal & Main
Drive)

Ø 156 (Main Drive Core)

B

A

A

Note 2

B

Note 3

Equipment
or Storage
(?)

Air Locks (?)
Note 5

Equipment Hatch to
Main deck

Force Field
Generators (3)(?)

C57-D

1/300 scale
Dimensions in inches
© 2005 by Jon C. Rogers
Sheet 4

Dimensioned Views

Notes:

1. Drive is shown here in Dark Red for contrast.
2. Implied or speculative sections are indicated by — — · · — lines and (?) in the notes.
3. This stair is shown in stowed position. It is shown extended on Sheet 5.
4. Astrogator console not shown in this view for clarity. See Section B-B.
5. Pedestal shown here in Flight Position for Reference. When landed it is always extended as it is the main support.

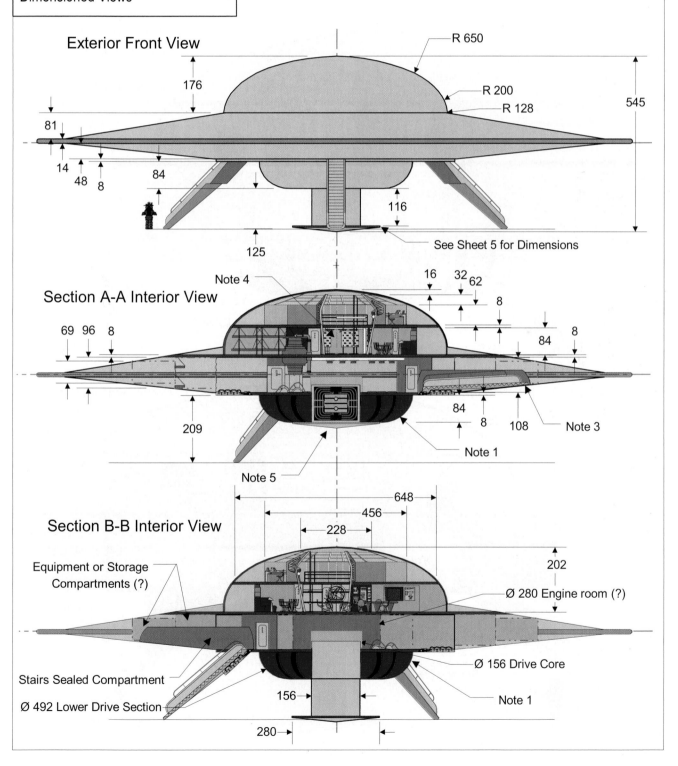

Exterior Front View

R 650
R 200
R 128
176
545
81
14
48 8
84
116
125
See Sheet 5 for Dimensions

Section A-A Interior View

Note 4
16 32 62
8
69 96 8
84
8
84
8 108 Note 3
209
84
Note 1
Note 5

Section B-B Interior View

648
456
228
202
Equipment or Storage Compartments (?)
Ø 280 Engine room (?)
Ø 156 Drive Core
Stairs Sealed Compartment
Note 1
Ø 492 Lower Drive Section
156
280

The Saucer Fleet

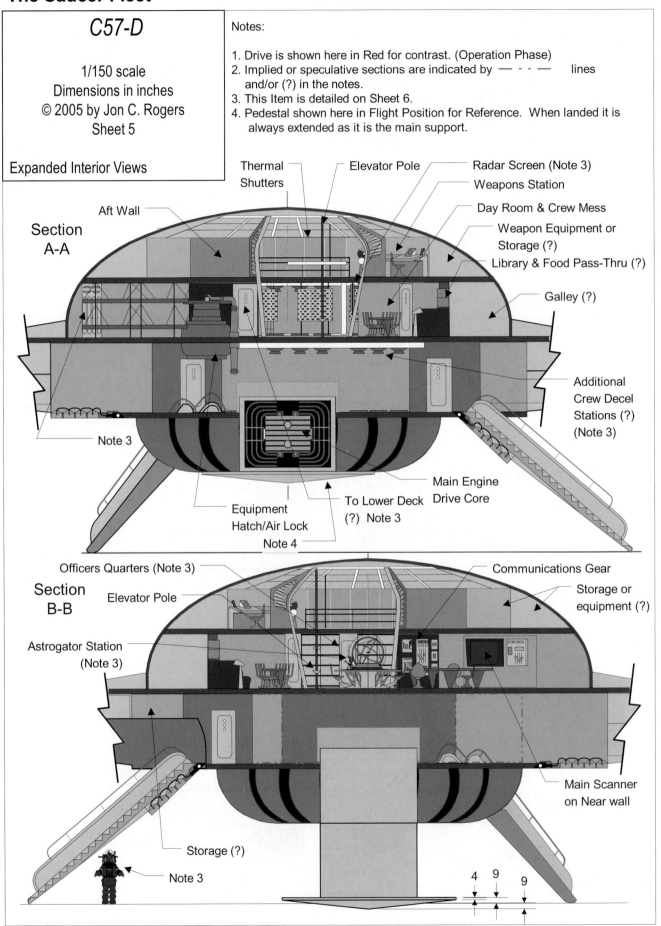

C57-D

1/150 scale
Dimensions in inches
© 2005 by Jon C. Rogers
Sheet 5

Expanded Interior Views

Notes:

1. Drive is shown here in Red for contrast. (Operation Phase)
2. Implied or speculative sections are indicated by — · — · — lines and/or (?) in the notes.
3. This Item is detailed on Sheet 6.
4. Pedestal shown here in Flight Position for Reference. When landed it is always extended as it is the main support.

Thermal Shutters

Elevator Pole

Radar Screen (Note 3)

Weapons Station

Day Room & Crew Mess

Aft Wall

Section A-A

Weapon Equipment or Storage (?)

Library & Food Pass-Thru (?)

Galley (?)

Additional Crew Decel Stations (?) (Note 3)

Note 3

Main Engine Drive Core

Equipment Hatch/Air Lock

To Lower Deck (?) Note 3

Note 4

Officers Quarters (Note 3)

Communications Gear

Section B-B

Elevator Pole

Storage or equipment (?)

Astrogator Station (Note 3)

Main Scanner on Near wall

Storage (?)

Note 3

4   9   9

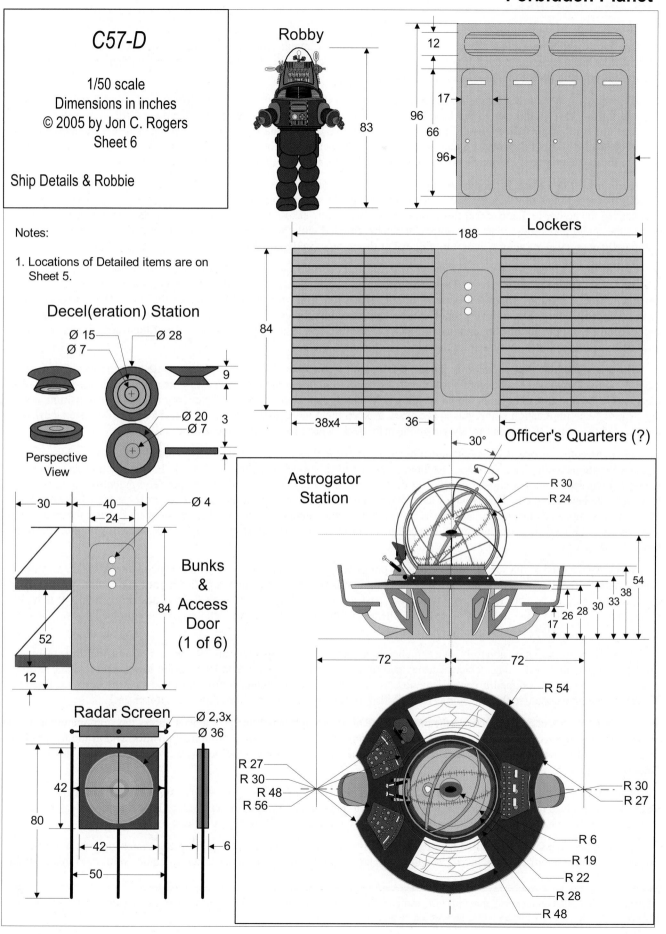

C57-D

1/50 scale
Dimensions in inches
© 2005 by Jon C. Rogers
Sheet 6

Ship Details & Robbie

Notes:

1. Locations of Detailed items are on Sheet 5.

Robby

Lockers

Decel(eration) Station

Ø 15
Ø 7
Ø 28
9
Ø 20
Ø 7
3

Perspective View

Officer's Quarters (?)

Astrogator Station

Bunks & Access Door (1 of 6)

Radar Screen

The Saucer Fleet

One fact we do know comes from a conversation between Captain Adams and Altaira. There Captain Adams clearly states that he has a crew of 18 men. If we add one man to the Captain's head count (say a passenger) there would have to be sleeping accommodations for approximately twenty men.

An analysis of the crew quarters on the main deck proves that there are only bunks for sixteen enlisted men. We know they are for the lower ranks due to the lack of privacy in that space. That leaves four officers, one for each watch and the Captain. The ship's four officers would require different, more private quarters. The large room with the luminescent blue wall is close to the ship's duty stations and fits this description perfectly. Thus it may well belong to the ship's officers and be used for their own quarters. Again we are using logic and known facts to fill in the things we don't know.

Also on the Main Deck is another door leading to a triangular area just to the left of the crew's quarters (Data Drawing sheet 2) that is also undefined in the plans. Although this could be a storage area, the need for an entrance to the main deck area has led the author to theorize that this should be the main crew access to the lower deck. The lack of any access or stairway in the studio plans (other than the equipment hatch), coupled with the fact that you would never design the main deck of a ship without any way of getting in or out of it, makes this a very strong possibility.

One of the other puzzling things in the *C-57D*, are the doors at the rear of the deceleration station (DC Station) area (Figure 1, One Door in the DC area). These two doors are clearly shown both in the film and on the plans for the set (Data Drawing sheet 2, marked "Unknown Access" doors). However, from evaluating the size of the main deck, it appears that these two access doors open into very shallow areas that have insufficient room for a person to enter. Is it possible they are supposed to be an opening into an equipment closet? Other than that, it is only known that one of them was used as a camera location to shoot the deceleration sequence scene.

There is something else mysterious about the DC area. Remember that Captain Adams stated that he has a crew of 18 men. Also recall that when decelerating to enter the Altair system, they announce that everyone aboard ship must get to a DC (deceleration) station. Each man is shown standing on a pad that surrounds him with an energy field for the transition to sub-light speed. The implication is that this is necessary to keep them alive during the severe deceleration. If the ship carries 20 men, and if only nine DC stations are shown on the main deck, where are the other eleven stations?

Since the entire second deck is not shown in the movie or included in the plans, I have concluded that the additional deceleration stations would have to be located there. And the most logical place for them would be directly underneath the existing, known deceleration stations on the main deck. That way they could share any support equipment required with the least distance between them. That is why I have placed them in that position on the lower deck in sheet 3 of the Data Drawing.[4] These stations are theoretical as there is no official documentation to support their existence. But the requirement is there due to the need for a minimum of 20 stations to support the crew (and more for any passengers) for the transition through light speed.

When we descend to the second, or lower deck, things really get challenging. This deck is necessitated by the size of the saucer, and the fact that when the men are seen entering and leaving the ship, they are shown entering into an area not indicated on the main deck. In fact the only thing seen past the stairwell, is a small corridor and possibly a door. It is highly likely that this would be the inside of an airlock. The rest of the lower deck is not detailed and was probably never made up as a set.

What, then, do we know about it? We can already identify some things that must be there. For one, the three staircases must retract into that deck. The center pedestal also retracts into it because it is taller than the lower engine dome it drops down from. To have designed a central engine core, and not have an engine room connected to it, is likewise illogical. Therefore, we may conclude that the engine room is also located in the center of the ship on the lower deck.

The main deck's equipment hatch, (the one with the hoist through which the ID monster enters the ship one night) must also extend into the second deck. Furthermore, the equipment hatch must have a prominent position within the lower deck. In order to be effective, it needs to be located near one of the entrance stairways. In looking at the plans, we find that it is (see Data Drawing sheet 2).

Another important consideration is that each stairway leading out of the ship must have some type of airlock. Since the staircases themselves may not be airtight, it is reasonable to believe that there is either a sealed enclosure or airlock compartment directly above each stairwell. This would be necessary for landing on any planet without a breathable atmosphere.

A necessary characteristic of the lower deck is that it's height decreases as you go from the center toward the outside of the saucer. I have identified two interior heights that place a limitation on the areas crewmen could comfortably use. In the

---

[4] I have only indicated an identical group of 9 DC stations and placed them under the existing stations. I realize there would need to be at least two more, but it is not known where they would be placed.

data drawing I indicate a standard eight-foot ceiling, and a lower 5 foot nine inch ceiling limit. I believe the designers of the *C57-D* would have limited regular crew access to within these areas. There is more space outside these areas but they would be used largely for equipment or storage.

One last thing that must be there by implication is access ways to the main deck. Some way must be provided – besides the main equipment hatchway hoist – for personnel to get back and forth between the two decks. For safety, these main access points would have to open into an airlock area, and/or have the capability of sealing themselves.

In addition I have included two of the previously shown elevator poles for emergency access. One of the poles would enter the main engine room from the officers' quarters and the other would access an airlock area on the main deck to the airlock on the lower deck. Of course, only being implied by the design, these features are speculative.

Quickspec: *C57-D*	
Vehicle Morphology	Saucer
Year	1956
Medium	Theatrical Film
Designer	Arnold "Buddy" Gillespie
Height (gear up)	35.0 ft (10.7 m)
Height (gear down)	45.5 ft (13.8 m)
Diameter	160 ft (49 m)

As the ship is seen coming in for landing, its final approach and final landing are done by a blue force field which appears beneath the ship at a distance of about 20 feet (6 meters) from the center of the craft. The ship settles on this ring of blue light and the three staircases descend after the center pedestal has extended and the drive shuts off (See Figures 3 and 4).

©Warner Bros.

Figure 3

©Warner Bros.

Figure 4

One problem with this blue landing force field is that the location of its generators must cut directly through an area used by the staircases themselves. Therefore, in order to have installed generators in that position they must be situated underneath the very stairs that the men disembark on. I have illustrated what this might look like on Data Drawing sheet 2. As the force field generators must, of necessity, be rather small, the source of the energy is probably located somewhere else and the energy sent to them.

At one point, it was thought that perhaps, just the areas around the landing stairwell would have the force field generators, but a review of the landing sequence does not show any break in the circular pattern of the force field. Therefore, we must assume that the staircases themselves do have landing force field generators in them.

So with the evidence given in the movie, the studio diagrams, and by logic, there are many items that have been discovered to be on the lower deck. This still leaves a great deal of space available for equipment, storage area, or other uses.

Furthermore, this illustrates how some known facts plus logic can predict features that were or should have been present when there is no direct evidence to substantiate them. And, as with all theories, the implied items are clearly identified so that the reader may know what has been actually seen versus what is speculative.

I hope this has helped your understanding of the *C57-D*, one of the most detailed, influential, and important flying saucers of the 1950's.

Modelers' Note

In addition to several small resin kits, there is the gargantuan Polar Lights *C-57D* (sic) styrene kit which, when assembled, spans 28" (71 cm) across the disc. It is a very ambitious kit with a fully detailed interior, and a top dome that is molded in clear styrene to allow the interior to be viewed (or it can be left loose for direct viewing if the modeler wishes to paint the dome for accuracy). While certainly impressive in scope, many

fans have issues with the accuracy of the kit, in particular the edge-on profile of the disc (the portion above the center-line should be taller than, not symmetrical with the lower half) and the layout

The author's personal Polar Lights kit that he swears he'll have time to build once this book is out.

of the interior which differs in many details (Arthur Lonergan's set de-sign was fairly well known to fans and relatively available).

Epilog

Nobody denies *Forbidden Planet*'s landmark status in the genre of science fiction film, but not everyone is convinced of its greatness. Danny Peary acknowledged its significance by including it in the first volume (out of an eventual three) of his *Cult Movies* series. In the section's opening he opines:

> I have long been baffled as to why there is such a strong, loyal cult following for *Forbidden Planet*, just as I have been unable to figure out why film historians and critics almost uni-formly regard it as being not only the best of the many science fiction interstellar films of the fifties, but also one of the few "great" films the SF genre has ever produced. While there are marvelous things in *Forbidden Planet*, they are all of a cosmetic nature, when what the picture needs is a total overhaul.[64]

When searching for the cause of this conundrum he suggests:

> [Writer] Cyril Hume was told by producer Nicholas Nayfack to follow Block and Adler's story outline; and not to insert expensive visuals, but to add dialog, flesh out the characters, and punch it with humor. We must question if he did anything but add dialogue, ad nauseum. It is a very juvenile, trite script. The combined efforts of the cast, director, and writer may be adequate for a low-budget science fiction thriller (which indeed is what they were hired to be part of), but in this million-dollar production their work is overwhelmed by the special visual elements provided by the better-financed MGM units.[65]

And, of course, he knows who to blame:

> It is often overlooked that originally MGM planned to spend only around $500,000 on the project. While [Dore] Schary intended *Forbidden Planet* to be a cut above the average SF film, he didn't have enough faith in it to give it great financing. When Schary eventually became excited about the project and substantially increased its budget, he granted extra money only to the special-effects unit and scenic and art departments while all the other facets of production remained geared for the mak-ing of a relatively cheap film. Schary would have done the picture more justice if he had spread the money out a bit.[66]

Okay, so it wasn't *Gone With the Wind*, but there is no argument that it was the most influential Sci-Fi movie for more than a decade. Sometimes this influence was direct. Designer Bob Kinoshita went on to be Irwin Allen's art director for *Lost in Space*, thus the *Jupiter 2* (and the Robot) could legitimately be called "Forbidden Planet, The Next Generation" (for a de-tailed comparison of the *C57-D* and the *Jupiter 2*, please see the *Lost in Space* section, page 248).

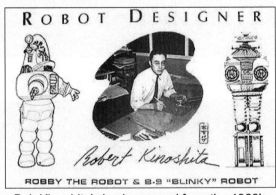

Bob Kinoshita's business card from the 1960's

Beyond that, the movie was so influential that you did-n't even have to see it for it to have the desired effect. In John Carpenter's seminal *Halloween* (1978) he wanted to establish that the children were watching a scary movie on TV. He never shows the screen, but just as the boogey man in his story starts creeping around, we hear the "Beta Beat" of the Id monster coming from the TV. Later, it's the full-blown battle at the ship that underscores Carpenter's own action scene. This was using a film that was more than 20 years old at the time.[67]

---

[64] *Cult Movies*, Dell Publishing, 1981, pg. 94.

[65] Ibid., pg. 95.

[66] Ibid.

[67] Speaking of Halloween, the title character of the recent *Monster House* (2006) is a modern recreation of the Id Monster, at least in the climactic scenes.

One person who was profoundly influenced by this film was *Star Trek* creator Gene Roddenberry: Consider the following similarities between *Forbidden Planet* and his original Star Trek concept:

- Set in 23rd Century.

- United [Federation of] Planets.

- Naval command structure.

- A virile young ship's commander who gets the hots for the young babe on the planet.

- "Trinity" structure of main characters comprising the captain, first officer and ship's doctor.

- "Miracle Worker" chief engineer.

- Beamer/DC Stations with both foot pads and overhead units that surround the occupant with an energetic glow and weird sounds.

- Seeing out through a round-cornered, square main viewplate/viewer (which was made wider with completely square corners later in the series).

- Belt-mounted miniature communicators.

- Circular Command Bridge with stations surrounding the captain.

- Comm station with wireless earpiece.

- Blasters/Phasers that emit blue beams and come in pistol, rifle and artillery models.

In addition to those general conceptual similarities, the original Star Trek pilot, *The Cage*, also owes a considerable debt to *Forbidden Planet*:

- The concept that the power of the mind is paramount in all of human activity, overriding all other considerations.

- A saucer-shaped star cruiser using "hyperdrive" (changed to "warp drive" in the regular series) to travel between stars, not just planets.

- Altair IV = Talos IV

- Beautiful young blonde on the planet who addresses the military men in an awkward, inappropriate way. Alta: "You're lovely, doctor!" = Vena: "You appear to be healthy and intelligent, the perfect specimen"

- Pre-emptive comment on the previous remark. Morbius: "You must make allowances for my daughter, gentlemen, she's never known any human being except her father," (a self-described old scientist) = Illusionary Talos IV scientist: "You must forgive the girl's choice of words. She's lived her whole life with a collection of aging scientists."

- Invisible, destructive monster from Morbius' subconscious = Unbearable pain from Pike's subconscious.

©Warner Bros. ©CBS Paramount Network Television

Ineffective high-power weapons

- All of the ship's power, concentrated through the main blaster batteries, can't stop the Id Monster, since it was being generated by Morbius' subconscious = All of the ship's power, concentrated through a laser cannon, can't blast through the door into the Talosian's underground city because of the illusion planted into the crew's subconscious by the Keeper.

And probably the best one (which is almost certainly a coincidence),

- When Adams asks for a "DC Fix" in the very first dialog in the film, Farman replies, "DC at 1701" = the Enterprise's registration number, NCC-1701.

That's enough to make the point. There are plenty of other examples if someone really wanted to put their mind to it...

©Warner Bros. ©CBS Paramount Network Television

Did the Krell build the Talosian's underground city?

©Warner Bros. ©CBS Paramount Network Television

The young blondes and their "aging scientists"

193

The Saucer Fleet

Props from *Forbidden Planet* started showing up in productions all over the place. Most common was in *The Twilight Zone*, not surprisingly, since that TV series was shot at MGM, and they used the same prop department. The *C57-D*, or at least parts of it, was featured many times. It started right away with the first season episode "Third from the Sun" " (first aired 8 January 1960) where it is an experimental ship that two sci-

entists use to evacuate themselves and their families from an impending nuclear war. We see the ship in long shot at night as they approach it, but it almost looks more like a painting or photo matted against the set. The full-size exterior set is used, although it's starting to show some modifications. The bottom nacelle is gone, replace by a conical skirt between the bottom of the disc and the pylon foot. Also, the disc overhead seems smaller (modified to fit the nar-

©CBS Paramount Network Television

row TV frame instead of the expansive Cinemascope®). We don't see the takeoff, but once underway they use a clip from the movie of the ship banking through space (with the ultrawide Cinemascope® frame squashed down to the width used for TV, rendering the saucer oval!). Inside the ship they have the Astrogator, radar screen and main viewplate from the movie, plus a dozen or so of the Krell power gauges lining the wall.

A couple of months later, the exterior set was used again, very briefly, at the end of "The Monsters Are Due on Maple Street" (7 March 1960). The final scene of the episode shows the *C57-D* flying away...upside down!

©CBS Paramount Network Television

In the second season episode "The Invaders" (27 January 1961), Agnes Moorhead plays a rural woman who spends the entire episode fighting off tiny invaders from outer space.

©CBS Paramount Network Television

©CBS Paramount Network Television

She discovers their flying saucer, which is the 88-inch prop. At the end of the episode she smashes it with an axe. A plywood model was created to show the smashed and smoking wreckage. Not only are there some good close-ups of the underside, but we also get to see one of the landing legs in operation.

The third season episode, the darkly humorous "To Serve Man" (2 March 1962) uses the exterior set again as the Kanamit ship's boarding ramp. It is even further modified with the overhead disc now continuing off screen vertically like a giant cylinder. This episode was an equal-

©CBS Paramount Network Television

opportunity plagiarizer of saucer films as the sequence of the ship arriving at the beginning is lifted from *The Day the Earth Stood Still* (showing Klaatu's "glowing oval" over the Smithsonian and Capitol), and when it leaves at the end, it's Ray Harryhausen's design from *Earth vs. the Flying Saucers* using a short clip from that film showing it lifting off and climbing out.

One month later the ship was back in "Hocus-Pocus and Frisby" (13 April 1962). Andy Devine plays a boorish braggart and spinner of tall tales. Some aliens on a specimen collecting expedition hear him and, because they have no concept of "lying," they take his stories to mean he is the most accomplished human on Earth. They whisk him away to their flying saucer where he has to spin the truth for a change to save himself. The external set seems about the same as in previous episodes, although they shot it from the "back" side of the board-

©CBS Paramount Network Television

194

ing ramp. After he climbs the stairs, we get a never-before-seen shot of the stairs closing from the inside. The interior set owes more to *Earth vs. the Flying Saucers* with the only *Forbidden Planet* piece being the Astrogator (outer sphere only) with a furiously spinning inner frame and wildly gyrating model saucer.

In the hour-long fourth season *Twilight Zone* episode, "Death Ship" (7 February 1963), three astronauts on a survey mission land on a planet only to see a ship identical to theirs crashed nearby, and with their bodies in it! The saucer is called the *E-89* this time and the elaborate landing and liftoff sequences (by the MGM theatrical special effects department, not the TV unit) were done with the 44-inch model that does not have the landing legs. It's a bit disconcerting to watch the ship bump down directly onto the lower drive nacelle! The "crashed" version of the ship was actually a photorealistic painting. When taking off there's a shocking non sequitur of having a bright rocket exhaust streaming from the bottom. The external set has no obvious changes from previous episodes except for the addition of some heavy-duty hydraulic struts on the boarding ramp. When inside the ship, the crew wears the *Forbidden Planet* lightweight "day" uniform, complete with the United Planets insignia on the pocket, but

©CBS Paramount Network Television

modified with spacey, rounded epaulettes and the boots painted silver. The set for the interior of the ship was completely different and owed nothing to the movie.

Its final *Twilight Zone* appearance was in the last episode of the fourth season, "On Thursday We Leave For Home" (2

May 1963). James Whitmore is the leader of a failed human colony on a hellish planet in another solar system. After 30 years, a rescue ship is sent to bring them home. The ship is now the *Galaxy 6,* and the crewmembers wear the same "day" uniforms, complete with the silver boots, from "Death Ship." Also reused from "Death Ship" is the shot of the ship descending near the beginning, and ascending at the

©CBS Paramount Network Television

end (no fire this time, they just ran the "landing" footage backwards). The exterior set has been modified even more from its earlier appearances. Narrow tubular landing legs have been added so the stairs are now just a boarding ramp (incidentally, compare these landing legs with those on the *Terran Space Liner*, page 69). The big hydraulic struts from "Death Ship" are also still there. The *Forbidden Planet* cyclorama (backdrop painting) was even re-used to depict the planet's surface.

This episode is significant for a couple of reasons. First, it is generally considered the best of the fourth season by both critics and by series creator, Rod Serling. Second, it generally follows the premise of *Forbidden Planet* in that the big saucer comes to rescue a human colony on a planet in another solar system, although this time the story is told from the other side.

Robby appears in two *Twilight Zone* episodes. In "Uncle Simon" (15 November 1963), the complex mechanisms under his head dome were replaced with an embarrassingly silly "oatmeal box" robot head which was supposed to look like Cedric Hardwicke (who played the title character), but doesn't. In "The Brain Center at Whipple's" (15 May 1964), his head is restored to its original configuration in the denouement as he replaces Richard Deacon as the head of the factory. (Deacon, incidentally, was the Metalunan helmsman in *This Island Earth*.)

©CBS Paramount Network Television

The Saucer Fleet

©WB Home Video

Adams? Farman? Is that you?

The United Planets formal flight uniforms made it into the *Zone* episode, "The Little People" (30 March 1962), but only for a few seconds at the very end. Prior to their *Twilight Zone* appearances, the uniforms were used in the Zsa Zsa Gabor vehicle *Queen of Outer Space* (1958), an odd movie that is not

©WB Home Video

as bad as you would think by the title. In fact, they went all out to get maximum mileage out of the *Forbidden Planet* connection. In addition to the UP day uniforms, they reused the dress Anne Francis wore in the infamous "kissing scene" and cast actors that bore more than a passing resemblance to Francis, Nielson and Jack Kelly (Farman). Plus, there was a Walter Pidgeon sorta-lookalike playing the main scientist who dressed in a loose fitting brown tunic like Morbius.

George Pal re-used a UP uniform in *The Time Machine* (1960) where it was worn by the Civil Defense official urging everyone into the shelters in the "1966" sequence. Pal used them again, dyed different colors, in his production *Atlantis, the Lost Continent* (1961), and they've since been used in dozens of other productions. Pal used some saucer props in *The Time Machine* as well. Both the Astrogator dome and the main radar screen can be seen as museum exhibits in the "talking ring" scene.

©Warner Bros.

©Sony Pictures

Sic Transit Gloria Mundi

Probably the most offbeat use of *Forbidden Planet* props was in *The 3 Stooges in Orbit* (1962)[68] where the fence posts, radar screen and some Krell power gauges are seen scattered about the Martian headquarters. In one tracking shot a very battered looking Astrogator is placed in the foreground for the camera to move past. The inner dome is missing and the remaining outer dome has had some large cracks (circled in red) crudely patched with tape! Finally, a blaster pistol is used prominently in several scenes, and a blaster rifle (modified) has become a Martian field weapon.

Robby went on to have quite a career all by himself. The year after *Forbidden Planet*, he played the villain in *The Invisible Boy* (1957) starring Richard Eyer (better know to genre fans as Borrani the genie of the lamp in Harryhausen's *7th Voyage of Sinbad*). This was another Nicholas

©Warner Bros.

[68] As goofy as it is, this film, whose central character is an eccentric inventor, followed the best traditions of Sci-Fi in making some interesting predictions. It not only showed the effectiveness of remotely piloted aircraft, but it also quite specifically anticipated the invention of motion-capture CGI animation.

196

Nayfack/Cyril Hume collaboration, and Robby's last appearance in an MGM theatrical release. He did do one more show for his parent company, an episode of the MGM TV series *The Thin Man* titled "Robot Client." In fact, most of his appearances have been on television, starting with various variety shows in 1956 to promote *Forbidden Planet*. He made two appearances on *Lost in Space*; the first season's "War of the Robots" and the third season opener "Condemned of Space" (both of these are discussed in the *Lost in Space* section that starts on page 235). There have been dozens of other appearances, too numerous to list here (such as *The Addams Family*, *Columbo*, etc.), but after the early '70s, it wasn't the original Robby we saw, but one of the very high fidelity replicas created by Fred Barton, one of the film's biggest fans. These appearances continue to this day, some of the most recent being a Robby clone's appearance in a *Tonight Show* skit with Jay Leno and on the short lived Fox network comedy *Stacked* on 11 May 2005.

The original Robby was sold at the big MGM auction in 1970 as the studio was being liquidated. The new owner was the "Movie-World" museum in southern California where he was placed on display next to the "jeep" he drove in *Forbidden Planet*. Unfortunately souvenir hunters started to vandalize the priceless prop; and he began to deteriorate rapidly.[69]

Movie-World commissioned Barton to restore Robby after a 1974 Star Trek Convention where a staff member

©Jack Hagerty
Fred Barton and some of his "children" at the World Science Fiction Convention (WorldCon) in 2006

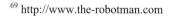

[69] http://www.the-robotman.com

©Jack Hagerty
Bill Malone handing Robby a blaster and saying "do you see that *althaea frutex…*"

saw the first replica that Barton had built. Robby remained on display for several more years until the museum closed its doors, and Robby was sold to film producer Bill Malone, an even bigger fan of the film than Barton (if that's possible). Barton's company, Fred Barton Productions, is today in the business of building and selling (or renting) licensed replicas of not only Robby, but robots from other famous movies and TV shows.

The original Robby got a good home with Malone, where he joined a large number of other props in Malone's collection, including the radar screen, main viewplate, Astrogator control panel, a few blasters (both hand and rifle models) and some of the force field fence posts. He had rescued these from an old MGM storage building literally hours before it was bulldozed. Sadly, he was unable to rescue many of the other props, including the Astrogator itself, which had been destroyed the day before.

Photo by Bill Malone
The author (looking like he really needs to hit the gym) gets to hold one of the *original* blaster props for a few seconds.

All three saucer models escaped destruction and were sold at the MGM auction. The 44-inch model was sold for a pittance ($80) because it had been considerably altered for its *Twilight Zone* appearances and no one recognized it for what it was. After two or three owners it eventually found its way into Malone's collection as well. The 20-inch model was purchased by collector Wes Shank along with one of the blaster rifles. He refurbished the saucer to restore the rotating "main drive" lights and, as of early 1993, still had it on display in his home.[70]

Shank went on to later acquire other props from the film including four of the Krell power gauges, one of the force field fence posts (from the Malone collection),

Wes Shank with the restored 20-inch prop in 1993

and several costumes (including the ones used in *Atlantis*).

The buyer of the 88-inch saucer is unrecorded, but it wound up, of all places, in an automotive museum called "Automobile-o-Rama" in Harrisburg, Pennsylvania. It remained on display there until the mid '90s when the museum closed. After that, the fate of the big prop is unknown.

Finally, let's see if we can resolve this nagging issue regarding *The Tempest*. The author, not being a Shakespearean scholar, feels that this discussion should be left to the experts. In *Filmfax* #89 (February/March 2002), Michael P. Jensen did something that had never been done before. He actually put the claim to the test with a thorough comparison of the two stories:

> Everybody knows that the film *Forbidden Planet* is based on Shakespeare's play *The Tempest*, right? The problem is that while many make this claim, few actually bother to compare the two. Let's clear away the rumor and hyperbole to see the little *Forbidden Planet* does have in common with Shakespeare's play and the vast amount it does not.
>
> It is not entirely clear who is responsible for the bit of Shakespeare that is in the film. *The Tempest* was Irving block's favorite play, and he suggested they use it for inspiration. The problem comes when we learn the story was expanded and given dialog by Cyril Hume... Alas, none of the writers are still around to ask, but it looks like Block is not solely responsible for adding a touch of *The Tempest* to *Forbidden Planet*. It is only a touch.
>
> How about those character "parallels"? All the science fiction books[71] will tell you that Dr. Morbius corresponds with Prospero, his daughter Altaira with Prospero's daughter Miranda. Commander Adams is said to stand in for Ferdinand, since both court the daughter character. Robby is said to be the spirit Ariel, and the Id monster, Caliban. Shakespeareans correctly point out that Robby better corresponds with Caliban, since both are servants. Ariel doesn't really have a counterpart in *Forbidden Planet* [and] *The Tempest* really has no parallel for the Id monster. Prospero and Morbius have this much in common: Both are scholars who are in an isolated place with their daughters. Both have special knowledge which gives them

power. The similarities pretty much end there. Prospero wants to leave his island. Morbius is happy on his planet. Prospero is a usurped Duke who wants his Dukedom back. Morbius wants to be left on Altair IV. Prospero wants his daughter to marry Ferdinand, the son of a king, to consolidate his power. Morbius does not want the astronauts to know about his daughter, then disapproves of the budding romance once it begins.

The Miranda/Altaira correspondence is even more of a stretch. Both live with their fathers in an isolated place, and fall in love as the story unfolds. That's pretty much it. Miranda is very much given to learning, where Altaira wants to play.

Even the correspondence between Caliban and Robby is tenuous. True, neither is human, and both are servants who make their respective households work, but Robby does so willingly and Caliban is full of complaints.

The plots have very little resemblance. Here is everything they have in common: A ship arrives in an isolated place, where there is a father and daughter. Both have comic scenes between a cook and a non-human servant. One member of the ship falls in love with the daughter. Both Miranda and Altaira have nightmares about what happens to those on the ships. Morbius and Prospero are both cold to human feeling, but change and choose not to harm those who have come to their planet/island.

That's it. Five plot points in common, three of them minor. There are literally dozens of differences, but let's not list them all. The list is long and makes tedious reading. Let's just conclude with [the most] telling differences.

Prospero has Ariel create a tempest to bring the ship to his island. Contrast this with Commander Adams and his crew. They come to Altair IV looking for survivors from a lost ship. There is no tempest.

Ferdinand is a minor character, more than a plot convenience, but not much more. Adams is one of the two leading characters, and the character with the most screen time. Ferdinand and Miranda fall in love at first sight. Adams and Altaira, in typical Hollywood style, don't like each other at first, and fall in love later.

While it is true that *Forbidden Planet* is based on *The Tempest*, simply by saying it, an impression is given that really isn't true. What they do not have in common is far more significant than what they do. Block, Adler, Hume, Nayfack and director Wilcox created what many consider to be the best science fiction film up to that time, and for several years after. It would be nice if they received a little more of the credit.

The connection with Shakespeare is further diluted, as Jensen says elsewhere in his piece, with the insertion of a significant amount of Freudian pop psychology, which was all the rage in the mid '50s. The Id monster had no equivalent in *The Tempest* because it, the main villain of the piece, comes from the contemporary, mid century fad of mind probing. Further, Block's addition of classic mythological references (the virgin and the unicorn, and naming Morbius' ship *Bellerophon*) are significant non-Shakespearian plot elements.

There are other items that Jensen missed, for example the offhanded references to IQ's as absolute numbers, all firmly root the story in the time it was written, not Shakespeare's. Additionally, Adler's Sci-Fi concepts, like having Robby obey Asimov's three laws, all add to the texture of this tapestry, but further remove it from the Bard.

There you have *Forbidden Planet*. Is it the greatest science fiction film ever made, or just an overstuffed "B-movie?" You can decide for yourself. History has already made its judgment.

---

[70] *Hobby FX*, Issue No. 1, Summer 1993, pg. 45.

[71] Including this one! See page 158.

Earth vs. the Flying Saucers

Some may consider this film to be the weak sister of the group. It did not have an "A movie" budget like *The Day the Earth Stood Still* and *Forbidden Planet*. It was not made in color like *Forbidden Planet* and *This Island Earth*. It didn't have pretensions towards a serious social message like *This Island Earth* and *The Day the Earth Stood Still*. What it had was the magic touch of the remarkable Ray Harryhausen.

Produced on a budget that would hardly cover the catering bill for a film like *Forbidden Planet*, *Earth vs. the Flying Saucers* need offer no apologies for its "B-Movie" status. It was a film that was true to its heart and knew exactly what it was, namely a highly entertaining (and highly fictionalized) examination of the flying saucer "craze" going on in the mid 1950's.

For most readers, Ray Harryhausen needs no introduction, but for the benefit of those of tender years, who think that all movie special effects animation comes from computers, let's delve a little into his background.

Born in 1920, Harryhausen had a life-defining event happen to him at age 13. He saw *King Kong* at the famous Grauman's Chinese Theater in Hollywood. He was totally captivated by the production and the "stop motion" animation technique[1] used to bring Kong to life. He knew from that point what he wanted his life's work to be. "All of this excitement over a film that some might call 'trivial entertainment' could suggest that I may be a fanatic about the subject," he wrote in his career autobiography, *Film Fantasy Scrapbook*,[2] "But I have found over the years that 'extravagant enthusiasm' can be half of the battle of turning mere desire into actuality."

Remarkably talented as both a sketch artist and model builder, he immediately began making his own films, trying to duplicate what he had seen on the screen using compara-

©Sony Pictures

tively crude wooden armatures (the internal "skeleton" of a stop-motion figure) covered by latex rubber skins.

While still in high school, he met another "Ray" also destined to become a genre giant, author Ray Bradbury. They were both 17 at the time and instantly became fast friends. Of his life-long soul mate, Bradbury says:

> Harryhausen and I, at 17, were like most teenagers. But unlike many, we had large dreams that we intended to fulfill. We used to telephone each other nights and tell the dreams back and forth by the hour: adding, subtracting, shaping and reshaping. His dream was to become the greatest new stop-motion animator in the world, by God. Mine, by the time I was 19, was to work with Orson Welles.
>
> Somewhere along through the years, Ray was best man at my wedding.
>
> Somewhere along through the years, we realized our dreams.[3]
>
> We made a pact that while we might grow old, we would never grow up, and that we would love dinosaurs forever![4]

Harryhausen's dream was temporarily put on hold while he served a hitch in the army during WW II. After graduating from high school, he got a job working for George Pal on the "Puppetoons" that Pal was producing at Paramount.[5] Harryhausen worked on the first thirteen shorts, but towards the end, with World War II starting he knew that he would be going into the military, so he started taking night classes at Columbia on how to be a combat cameraman.[6]

Fortunately (for us), it wasn't necessary for him to go into combat. He knew that stop motion animation could be an effective way of making training films for the military, so while taking the class, he produced, on his own time and at his own expense, a four-minute sample film, *How to Bridge a Gorge* and showed it to his teacher. The teacher passed it on to Frank Capra, who was just organizing his special army film production unit. Capra had Harryhausen

---

[1] This is an incredibly laborious method of animating a subject by photographing a static model of it one frame at a time, moving pieces of the model almost imperceptibly in between frames.

[2] *Film Fantasy Scrapbook*, Second Edition, Revised, A.S. Barnes & Co., 1974, ISBN 0-498-01632-3.

[3] *Film Fantasy Scrapbook*, pg. 11.

[4] Video special *The Harryhausen Chronicles*, 1997.

[5] *Ray Harryhausen; The Early Years Collection*, 2005.

[6] *An Evening With Ray Harryhausen*, 28 April 2004.

The Saucer Fleet

A very young Ray Harryhausen at work on *How to Bridge a Gorge*, and a frame from the resulting film.

assigned to the unit where he worked not only on training films, but also on Capra's *Why We Fight* propaganda series.[7] Another notable in the unit was Ted "Dr. Seuss" Geisel with whom Harryhausen worked on the *Private Snafu* educational cartoons.[8]

After being discharged in 1945, he returned home and started working once more on stop motion animation films. He did a series of short films based on Mother Goose stories and other fairy tales using the techniques that he had learned while working on the *Puppetoons*. It was a family project with his mother sewing the costumes and his father, a mechanical engineer, helping with the props and armatures.[9] These were distributed commercially by Bailey-Film Associates and are still in use today in many grade schools.

Ray and Joe

Even before going into the Army, Harryhausen had met his mentor and hero, Willis O'Brien; the legendary effects wizard who had animated *King Kong*, by the simple expedient of calling him at his office and being invited over to the studio. It wasn't until after he had some commercial production experience, though, that he mustered up the courage to show O'Brien some samples of his work. O'Brien was impressed enough

to hire Ray as his assistant on a new production called *Mr. Joseph Young of Africa*, later re-titled *Mighty Joe Young*. Even though O'Brien is the one that got the screen credit, it was Harryhausen who designed all of the armatures and did about 90% of the animation on the film,[10] which won the 1949 Academy Award for Special Effects.

It may or may not be significant that the first film on which he was completely in charge of the special effects was based on a Ray Bradbury short story, *The Beast From 20,000 Fathoms* (1953). Of course, they had to add quite a bit to the plot as the entire original story is summed up in the one scene of the creature attacking the lighthouse. Made for only $200,000, it was a surprise hit, and one of the largest grossing films that year. Animator Phil Tippett, whose career was inspired by Harryhausen, put an homage to this film into *Jurassic Park* (the famous "lawyer" scene with the T-Rex).[11]

His next film, *It Came from Beneath the Sea* (1955) continued in the same vein. Rather than a lizard-like creature from the ocean depths, it was a gigantic octopus that attacked San Francisco. Actually, it wasn't really an octopus, but rather a "sixtopus." As Harryhausen explains:

> The octopus only had six tentacles. Two tentacles less to build and animate did save quite a bit of time. And in Hollywood, time is money. I sometimes wonder if the budget had been cut anymore if we might not have ended up with an undulating tripod for the star villain of the picture.[12]

---

[7] *Fantastic Films*, Vol. 1, No. 4, October 1978, pg. 8.

[8] *An Evening With Ray Harryhausen*, 29 January 2005.

[9] Ray thought it was unprofessional to have the credit screen filled up with all "Harryhausens" so his father used the name "Fred Blasauf" and his mother "Martha Reske."

©Ray Harryhausen

NARRATION WRITTEN *by* **CHARLOTTE KNIGHT**
Narrator · · · · **DEL MOORE**
Costumes · · **MARTHA RESKE**
Associate · **FRED BLASALIF**
Music Editor · **WALTER SOUL**
Photographed in three Dimensional Animation

[10] *Fantastic Films,* updated by Harryhausen phone interview, August 2003.

[11] *An Evening With Ray Harryhausen*, 28 April 2004.

[12] *Film Fantasy Scrapbook*, pg. 37.

©Ray Harryhausen

The "Sixtopus" model attacks writing collaborator Charlotte Knight (see credit screen previous page)

This film was significant, too, in that it was the first one where Ray collaborated with Charles Schneer, whose company would produce most of Harryhausen's films from that point on, although Ray would occasionally "freelance" on other projects for different producers. It was Schneer who wanted to move away from the "monster on a rampage" type film, and do something to exploit the UFO phenomenon sweeping the country at the time. This lead directly to our subject film, which we'll get back to in a moment, but first let's finish up Harryhausen's career.

While working on *Earth vs. the Flying Saucers* he also did one of those side jobs, animating the dinosaur battles on Irwin Allen's *Animal World*, whose special effects were being supervised by Willis O'Brien. This gave Ray one final opportunity to work with the man who inspired his career. In 1957 he did *20 Million Miles to Earth*, an homage to King

©Ray Harryhausen

Early rough concept sketch for the Ymir. For the face he used his design for the Martian in his *War of the Worlds* (see page 80).

Kong in that a strange creature is brought from its natural habitat into a major city where it wreaks havoc and has to be shot down off of one of the city's architectural landmarks. But rather than being a giant ape from a tropical island being shot off of New York's Empire State building, it was a reptilian, but still humanoid, alien from Venus being killed on Rome's ancient Coliseum.

His next film, *The Seventh Voyage of Sinbad* (1958) had two significant "firsts." It was his first feature film done in color[13] and it was his first collaboration with composer Bernard Herrmann. The music score would set the standard for fantasy films for over a decade, and Herrmann apparently enjoyed it enough to do the music for Harryhausen's next three films as well. Such was his attention to detail that Harryhausen took a sword fighting class before animating the classic dual between Sinbad and the Skeleton. Ray liked this genre of story and eventually did two more "Sinbad" stories; *The Golden Voyage of Sinbad* (1974) and *Sinbad and the Eye of the Tiger* (1977). *Seventh Voyage* was also the first film to fly under the flag of "Dynamation," the trade name that he and Schneer coined to describe his integrated effects techniques.

©Ray Harryhausen

Animating the dragon that guards the entrance to the "Valley of the Cyclops" in *7th Voyage of Sinbad*

The Three Worlds of Gulliver (1960) let Harryhausen stretch his effects expertise in a different direction. Using all live-action (except for two brief animated scenes, one a battle with an alligator and the other a giant squirrel) he makes the worlds of Lilliput (tiny people) and Brobdingnag (giant people) completely believable with seamless mixtures of giant/miniature props, forced perspective and matched camera angles all tied together with flawless traveling mattes.

Mysterious Island (1961) was a re-telling of Verne's sequel to *20,000 Leagues Under the Sea*, combining stop-motion creatures (giant crab, bee, bird,

©Ray Harryhausen

Pre-production art of the giant crab fight

[13] Although *Animal World* was in color, it was an Irwin Allen production to which Harryhausen only contributed one ten-minute sequence. Also, all of his "Mother Goose" and fairy tale shorts were done in color thanks to his finding a large batch of expired Kodachrome® that he picked up for next to nothing.

The Saucer Fleet

octopus) and live-action miniatures (balloon, Nautilus, volcano, pirate ship) with regular humans in seamless process shots.

Jason and the Argonauts (1963) is considered by most to be his masterwork in this genre. Harryhausen himself said, "Of [all] the fantasy features I have been connected with, I think *Jason and the Argonauts* pleases me the most."[14] In a loose telling of the Greek legend of Jason and the Golden Fleece (some of the adventures depicted actually come from Homer's *Odyssey*) it is a tour de force using all of his techniques learned up to that point: forced perspective live-action miniatures (the gods in Olympus, Triton and the clashing rocks), full scale sets (the Argo itself) and a whole pantheon of mythological creatures done in stop-motion (the Harpies, Talos, the seven-headed Hydra, and his signature work, the battle with seven skeletons, the "Children of the Hydra's teeth").

First Men in the Moon (1964) was a somewhat whimsical adaptation of the H.G. Wells novel. It was his only film using the wide screen Panavision process, and it required his abandoning the rear-projection effects technique, which is not suited to the process.[15] The animated individual Selenites and "moon cow" were, as always, first rate, but the budget didn't allow him to use stop motion animation to give the impression of thousands of Selenite workers, so he had to resort to using child actors in rubber suits. Still, the present-day framing story showing the first moon landing was so well done that footage from it was used in early NASA films to depict what the Apollo program would look like.[16] Laurie Johnson, better known for his scoring of TV's cult favorite *The Avengers* (and friend of Bernard Herrmann), provided music for the film.

Both ©Ray Harryhausen
Above, Harryhausen joins King Triton amid the "Clashing Rocks."

Below, the armature design for the fighting skeletons, and one of the finished products. This particular skeleton is the one built previously for *The 7th Voyage of Sinbad*.

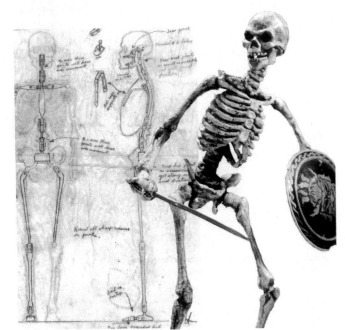

©Ray Harryhausen
Harryhausen and producer Charles Schneer (center) are joined by Frank Wells (H.G.'s son) on the set of *First Men In The Moon*. The rather stoic individual between Schneer and Wells is a Selenite, who has already been cocooned for his hibernation period.

[14] *Film Fantasy Scrapbook*, pg. 85.

[15] *Fantastic Films*, pg. 17.

[16] Jeff Rovin, *Ray Harryhausen: Special Effects* event flier, pg. 3.

One Million Years B.C. (1966) was another non-Schneer production that Ray lent his skills to realizing. Yes, it's a little late in the game for the producers to be making the unforgivable mistake of showing humans and dinosaurs living at the same time, but with Raquel Welch in a fur bikini, who cares?

The Valley of Gwangi (1969) was not a big success, but it wasn't the fault of the effects work. The premise seemed strained at best. It's almost as if they said, "Kids like dinosaur films and cowboy pictures. Let's put the two together!" They set the film in Mexico (but it was actually shot in Spain) for yet another retread of the *King Kong* tale. Cowboys searching in the "Forbidden Valley" find Gwangi, a surviving Allosaurus, and capture him for a circus. He eventually escapes and starts demolishing the town before being destroyed in a burning church (sound familiar?).

The film has origins that are actually more interesting than the movie itself. It was started by RKO in 1942 with Willis O'Brien doing the effects. It was cancelled after almost a year of preproduction where nearly all of the effects setups were constructed and ready to go.[17] Harryhausen had no connection with that production since he was in the Army at the time, and he doesn't reveal why he decided to make it (with all new effects) 27 years later.

Effects legend and Harryhausen's mentor, Willis O'Brien, works on a matte painting for the never-completed RKO version of Gwangi.

We've previously mentioned his next two films, *The Golden Voyage of Sinbad* and *Sinbad and the Eye of the Tiger*. *Golden Voyage* is most noted for having a score by the remarkable Miklos Rozsa,[18] and yet another signature sword fight against an enchanted opponent. Rather than multiple skeletons, this time six swords were being wielded at once by the six-armed Hindu goddess Kali. *Eye of the Tiger*, though, showed that the franchise was

running out of inspiration. With a trek to the North Pole at the film's climax, the tale had long since abandoned any connection with its Arabian Nights origins. Maltin's film guide only gives it 1½ stars. It also had the surprisingly wooden Patrick Wayne (John Wayne's son) in the title role along with Taryn Power (Tyrone Power's daughter).

Harryhausen was back in form for his last film, *Clash of the Titans* (1981) where he returned to classic mythology, trying to recapture some of the magic of *Jason and the Argonauts*.

Harryhausen and Jim Danforth having a good time learning to animate Pegasus

With an all-star cast, lead by Lawrence Olivier in one of his last roles, it is a fitting cap to a brilliant career. He mixes up the mythologies again (he has the protagonist, Perseus, taming the flying horse-god Pegasus rather than Bellerophon as in the original Greek legend), and there's more than a little of *Star Wars'* R2D2 in the character of "Bubo," a short rounded mechanical owl who speaks in clicks and whistles. Bubo was sometimes stop-animated and sometimes a real mechanical prop. Non-genre film fans will be interested to know that he based the lighting of the climactic fight between Perseus and the Medusa on the Joan Crawford film *Mildred Pierce*.[19] The Kraken, seen at the end of the film, was again based on his Martian design for his never-made *War of the Worlds* project (see page 80).

The terrifying medusa

All ©Ray Harryhausen
The Kraken

---

[17] *Film Fantasy Scrapbook*, pg. 111.

[18] Rozsa is probably best known for his massive, Oscar-winning score to *Ben Hur*.

[19] *An Evening With Ray Harryhausen*, 28 April 2004.

The Saucer Fleet

The most astounding aspect of all Harryhausen's effects work is that, with the exception of *Clash*, he always worked alone when doing the animation. Naturally, he worked with a regular crew when shooting the live action portions, but except for the occasional visit from a cinematographer, he was by himself, working sometimes around the clock (he always had a cot in the studio for naps) on the effects. Compare this to the long list of names on the effects teams for the other films in this book to get some idea how remarkable this is. Ray Bradbury comments:

> Looking through these scores of photographs [in *Film Fantasy Scrapbook*] reminds us once again of the creative powers of single individuals in the world. Not groups, but lonely creative spirits, working long after midnight, change the cinematic and aesthetic machineries of civilization.[20]

With this well-known penchant for doing the effects work all alone, many people got the impression that he was doing everything on the film. Harryhausen, like Alfred Hitchcock, was that rare behind-the-scenes worker whose name had marquee value. You'll note that in the synopses above they are referred to as "Harryhausen films" even though he never directed the actors (except for those sequences that had to match up with his animation) and was, at best, co-producer along with Charles Schneer.

Returning to our main subject, Harryhausen recalls the origins of the movie:

> The flying saucer reports were predominant in the news and Charles Schneer wanted to develop a film about an invasion of our planet by saucers from outer space. Flying saucers have always been a fascination to me, which made the film doubly interesting to work on. A great deal of research was necessary to keep the whole idea as credible as possible and within the scope of reported sightings. Of course, imagination had to play its part in the script as well. This examination of facts brought me into contact with various groups and individuals who claimed to have had actual contact with these beings, supposedly from outer space. I still have an open mind about the subject and am eagerly waiting for more concrete evidence of their existence.[21] I got an opportunity to talk with George Adamsky who at that time claimed he spoke to alien beings from another world.[22]

For story material, they used events described in the nonfiction book *Flying Saucers*

George Adamsky

Donald Keyhoe

from Outer Space (1954) by retired Marine Corps Major Donald Keyhoe,[23] most notably a sighting of UFO's over Washington DC in the summer of 1952 which inspired the film's climactic scenes. When forming the story with writer Curt Siodmak (probably best known for his screenplay of the original *Wolf Man*), Harryhausen realized both the limitations and opportunities of having such a limited budget:

> I think some of the picture's latent dramatic and visual values would have been brought out more had there been time to refine the script as well as present a more imaginative musical score. A prime fascination to me was the challenge of seeing just how interesting one could make an inanimate object such as a rounded metal spaceship. Although the variations were limited for stop-motion they did have great potentials for doing something a little different than the other "saucer pictures" of the time.[24]

This was also, as it turned out, Harryhausen's only pure "hardware" picture. There were no animated creatures at all, not even the aliens, which were played by actors in suits.[25] All of the stop motion effects were of the title characters and the destruction of the buildings at the end.

> We had to work with an inanimate object like a flying saucer and still give it the impression of having some sort of character with alien beings inside it. We had to show the Treasury Building being hit by a ray gun, and we didn't have the money to go into high-speed photography because you have to build everything on a very large scale and have a big crew to photograph it. So I ended up animating it brick by brick and it took me about five days, but we got the result.[26]

Incidentally, in order to stop-motion-animate pieces of a building falling to the ground, each piece had to be suspended by wires to hold it up in the air between frames. The reason we don't see the wires is that Ray painted them to match the background![27] They sometimes required touch-up painting between frames as the pieces were moved imperceptibly down the wires. This is yet another example of the brilliantly simple (but laborious) effects techniques he invented.

---

[23] Keyhoe was a notorious conspiracy theorist who wrote many books during the '50s. In addition to *Flying Saucers from Outer Space*, he wrote *Flying Saucers Are Real* (1950) and *The Flying Saucer Conspiracy* (1955). With his continual accusations of government cover-ups, he probably did more to create the saucer "cult" (e.g. "Area 51," Roswell, New Mexico) than any other individual.

[24] *Film Fantasy Scrapbook*, pg. 41.

[25] He probably would have been happier if he *had* animated the aliens. Even though he designed the alien costume, they were built by the studio's prop department outside of his control. In an interview (March 2005) he told the author that he was very disappointed in the way the costumes turned out. "They could not bend their elbows or knees," he said, "and walked like stick figures. If I could have animated them, they wouldn't have looked so artificial."

[26] *Fantastic Films*, pg. 10.

[27] Interview by Joe Dante included on the DVD version of the movie.

---

[20] *Film Fantasy Scrapbook*, pg. 11.

[21] *Film Fantasy Scrapbook*, pp. 41, 42.

[22] *Fantastic Films*, pg. 10.

The executive producer at Columbia (who was bankrolling the picture) was the notoriously cheap Sam Katzman. With his approval, they hired Fred F. Sears as director due to his ability to shoot a picture on a very short schedule (in addition to *Earth vs. the Flying Saucers*, he directed eight other movies in 1956!). Principal photography was done in an astoundingly short ten days, and that included locations on both coasts.[28] Legendary voice artist Paul Frees provided the voice of the aliens, but strangely, he didn't get a credit.

Paul Frees

©Sony Pictures

Hugh Marlowe and Joan Taylor

Hugh Marlowe was cast as the protagonist, Dr. Russell Marvin. Marlowe had played Tom Stevens in *The Day the Earth Stood Still*, the bad guy who betrays Klaatu's whereabouts to the army. Here he plays the good guy, but a somewhat frumpy one. Television and "B-movie" stalwart Joan Taylor plays his secretary, and new wife, Carol (the film opens, after a prologue, with the two of them returning from town where they have just been married). Prolific character actor Morris Ankrum played Carol's father, General Hanley. Ankrum seemed to be in just about every movie about Mars or alien invasion in the '50s. In addition to General Hanley, he was the Secretary of Defense in *Red Planet Mars* (1952), Colonel Fielding in *Invaders from Mars* (1953), Dr. Flemming, head of the R-XM project, in *Rocketship XM* (1950), and, switching sides, he played the Martian leader Ikron in *Flight to Mars* (1951).[29]

©Sony Pictures

Morris Ankrum

After the screenplay was written, Harryhausen went out and photographed desert and beachside locations. To save time when storyboarding the film, he drew many of the storyboard illustrations directly on top of the photos. Art Director Paul Palmentola used many of these composite photo-drawings to design the sets, costumes and makeup.[30]

Another way of saving money on the production was the liberal use of stock footage. Harryhausen was able to create some pretty amazing illusions by incorporating real scenes of rockets exploding, buildings exploding, and footage of airplanes disintegrating by animating in some saucer rays.[31] Director Sears also used quite a bit of footage from *The Day the Earth Stood Still* starting by reusing the very first opening shot of nebulae in space as his opening, and continuing by sprinkling many of the "Earth standing still" montage shots throughout the picture.

When designing the saucers themselves, Harryhausen already had a head start. When the whole flying saucer phenomenon broke in the late '40s he was fascinated by it. He produced this concept drawing of explorers coming across a wrecked saucer in a remote jungle area. This was not for any particular film project at the time, just one of those "file for future reference" ideas.

©Ray Harryhausen

A couple years later, in 1949, he started his *War of the Worlds* proposal (see page 79). As part of that produced this pre-production drawing of a War Machine attacking a village shortly after emerging from the cylinder. The flying saucer motif is just as strong here as it was in Al Nozaki's

©Ray Harryhausen

concept drawing (see page 83) and it's clear that many of his saucer design concepts were already starting to gel. The basic proportions, the domed top and spinning section with the radial strakes are all present in this artwork.

---

[30] Interview by Joe Dante.

[31] The use of this particular scene is somewhat questionable since it used footage of a real air show crash in which two jets collided and both pilots were killed.

---

[28] Filmfax #60, pg. 56.

[29] Outré #32, pg. 53.

The Saucer Fleet

When *Earth vs. the Flying Saucers* became a real project, he dusted off his saucer design and updated it with features from a popular photo of Adamski's flying saucer taken in 1954. The photo showed a circular saucer with three bumps on the 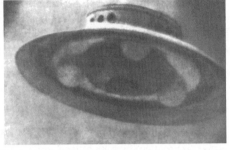 underside. These bumps would provide a touchstone to reality for the audience, since most of them had seen the photo.

Other than being circular and having three bumps on the underside, the finished saucer design bore little resemblance to the rather clunky looking craft in the photo.[32] Harryhausen's design was far sleeker and followed the standard for movie saucers of the time with a central dome leading to a more flattened outer rim. Since the entire raison d'être for his career was making inanimate things move, he couldn't just have static discs zipping around. He added dynamic elements in the form of two rings inset into the outer rim, one on the top and one on the bottom, which spun relative to the stationary part of the rim and dome. With these he could create a sense of effort and purpose as the rings spun faster or slower as the saucers maneuvered.

There were four different size saucer models used for different purposes. Models of 8-inch, 10-inch and 12-inch in diameter (20 cm, 25 cm and 30 cm, respectively) were made out of aluminum, and were built by Harryhausen's father on his old Craftsman lathe (in the same home shop where he built the armatures and props for the Fairy Tale films).[33] There was also an 18-inch (45 cm) model made from wood. That one was used for the close-ups, such as the "hanging miniature" on the beach; the other models were used for medium and long shots. For the scene of the whole fleet arriving over Washington DC all three of the smaller versions were used to create a forced perspective of the fleet covering a large area. Only the large wooden model had the operating pylon in the base.[34]

For location filming of the rocket base where Dr. Marvin works, they first visited Point Mugu Naval Station, a real rocket and weapons testing facility on the coast north of Los Angeles. Ironically, they thought it didn't look enough like a "real" rocket base, or at least the popular conception of one.[35] After some more scouting, they wound up at the Playa Del Rey treatment plant that processed most of the sewage from Los Angeles at the time. Its labyrinthine network of pipes was just what they wanted to represent the base interiors, visually, plus it had another side benefit:

> The Playa Del Rey Sewage Disposal Plant was used for one location. Its structure acted as the rocket base in the film. The many weird noises of the underground motors of the plant became the basis of the saucer sounds.[36]

So remember, when you hear the sounds of the saucers swooping through the skies in this picture, it's really the sound of high-speed sewer sludge coursing though pipes!

When the film was released, it was a huge hit. It earned nearly five times its $250,000 production costs, grossing $1,250,000 in initial release.[37] It was one of the biggest Sci-Fi movies of the year, second only to *Forbidden Planet* in revenue, although that landmark film's production costs were so high that it actually lost money (see page 165).

The film has remained popular with fans for over fifty years, although its impact has faded in the face of the effects-laden extravaganzas of the latter part of the century. It pretty much defined "alien flying saucer" in the public mind, probably because the design was deliberately based on a photo of a "real" saucer that had been fully exploited. Orson Welles, himself no stranger to "alien invasion" productions, even used footage from this movie in his exposé film *F for Fake*.[38]

Harryhausen shows Joe Dante one of the 8-inch filming models during the 2002 video interview.

The Story

The movie opens with a black screen with a narrator giving a voiceover prologue. The screen only stays blank for a couple of seconds before we see the image of a familiar looking nebula; the same nebula shot that's used to open *The Day the Earth Stood Still*. This is the first hint of the "recycled" nature of many of the film's visuals. The pro-

---

[32] Later revealed to be a fuzzy photo of a chicken incubator lamp!

[33] *An Evening With Ray Harryhausen*, 28 April 2004.

[34] *Ray Harryhausen, An Animated Life*, Billboard Books 2004, pg. 80.

[35] Interview by Joe Dante.

[36] *Film Fantasy Scrapbook*, pg. 44.

[37] *Cinefantastique* double issue, V8, N2/3, Spring 1979, pg. 65.

[38] Interview by Joe Dante.

logue opens, "Since Biblical times, man has witnessed and recorded strange manifestations in the sky and speculated on the possibilities of…visitors from another world!" As it continues, we see shots of a P-80 *Shooting Star* being paced by saucers at close to the speed of sound, farmers in both the US and Asia seeing "things" in the sky, and a DC-6 airliner being "buzzed" by the same type of craft.

We cut to the "Air Intelligence Command," where wildly gesticulating people report to officials sitting calmly behind desks. The narrator informs us that 97% of all sightings are explained as natural phenomena, but the remaining 3% remain "unidentified flying objects." Moving on to the "Hemispheric Defense Command" in Colorado Springs (an obvious stand-in for NORAD), we are told that they have issued a "shoot first" standing order against "all objects not identifiable." Still, the military wonders out loud how they would fare "in any battle of The Earth vs. The Flying Saucers!"

After the opening credits finish, we fade in to an aerial shot of a desert highway. A lone car speeds along it. Moving directly inside, we see a man dictating into a portable tape recorder.[39] He is Dr. Russell A. Marvin starting a report on "Project Skyhook." The woman driving gleefully identifies herself to the microphone as "Mrs. Dr. Russell A. Marvin." She's his secretary and they are returning from town where

©Sony Pictures

they have just been married. After some corny newlywed dialog ("you don't have to sneak up on me, we're married," she says as he makes a pass at her while she's driving), he continues dictating. It's the introduction to a report on a satellite-launching project. While he speaks, the film moves to stock footage of a V2, then an MX-774 taking off, followed by animation of satellites in orbit. Marvin explains that they will be studying the upper atmosphere, meteoric dust, and the sun. As we dissolve back to him dictating, he's just explaining that they've already launched 10 of the eventual 12 satellites "or 'birds' as we call them," when he is interrupted by a weird sound. The tape reels keep spinning as the car is buzzed from behind, and then again from the front by a flying saucer about 60 feet (20 meters) across.

Carol pulls to the side of the road and Marvin gets out.

---

[39] An incredibly advanced technology for 1956.

©Sony Pictures

He looks up to see the saucer ascending directly vertically then shoot off to the side at high speed. He gets back into the car on the driver's side as Carol obediently slides over to "let the man drive" in this now abnormal situation. She notices that the tape recorder is still running, and shuts it off. "It…it *was* a saucer; a **flying** saucer?" Carol stammers. "We saw what appeared to be a flying saucer, that's all we can say," declares Marvin, trying to maintain some scientific detachment. "But we both saw it!" implores Carol. Marvin replies, "We have to have time to think and evaluate this," already trying to talk himself out of what his senses have shown him.

They continue on to Project Skyhook and down into the bowels of one of the buildings amid a maze of complicated piping. He calmly asks her to transcribe his notes from the tape recording as if nothing unusual had happened on the way that afternoon. As she is doing so, they are startled to hear the saucer noises on the tape (remember, it had been running the whole time the saucer flew over them). Their reverie is broken by a control voice over the PA announcing, "Rocket Number 11 will be launched in 20 minutes!" They leave the tape and head to the control bunker.

©Sony Pictures

Meanwhile at the west gate to the project, we see an Army staff car pull up to the guard station. In the back is General Hanley who tells the guard that he must speak with Dr. Marvin. "No one is allowed in, sir," says the guard, "there's a rocket taking off." "That's just what I want to stop!" snaps the general as he bolts from the car. He commandeers the MP's phone and calls the bunker. Marvin answers, and after some pleasantries (where we learn that the general is Carol's father) Hanley tells him that he should abort the launch until after he hears what he has to say. Waiving off this serious message with a perfunctory "we're tied to a definite schedule of launchings," he hands the phone to Carol so that she can tell him about their being married. They almost seem more excited about that than the launch activity going on around them. A warning buzzer and a glare from a technician

©Sony Pictures

brings them back to reality. They get back to work, but not before inviting Hanley to dinner.

Marvin gets back to his console barely in time to do the last 20 seconds of the countdown. Looking at a monitor, we see the Skyhook launch vehicle (in reality a late-series Viking sounding rocket) on the pad. Marvin pushes the button for a textbook launch and ascent.

We dissolve directly to the Marvin household that evening. Dr. Marvin, puffing on his pipe and looking very scientific, is discussing the launch with his new father-in-law. "How many 'birds' have you sent up so far?" the general asks. "Eleven, counting today's," Marvin replies. "And how many are you in contact with?" continues Hanley. "Just one, today's," says Marvin dejectedly. But then, upbeat, he continues, "but we'll correct that. We can certainly tune them in if they're up there!" "What I'm trying to tell you," says the general, "is that they're no longer 'up there'." Hanley goes on to list the results of an Army investigation. "The burned remains of Number 7 fell on Panama, Number 1 and Number 3 fell over Africa, Number 5 around the North Pole, and Numbers 9 and 10 along the Andes. The rest are presumed lost at sea." The general explains that they apparently explode. "But there's nothing explosive in them," protests Marvin, somewhat defensively. "When a rocket's blasted off it should circle the Earth for a long time…unless something shoots them down as fast as we send them up."

Their serious discussion is interrupted by Carol calling from the back yard that the BBQ is ready. As they sit down to eat, Marvin tells the general about their close encounter that afternoon. Carol mentions the tape recording with the saucer sounds. Just then, a jeep pulls up with a dispatch from Skyhook HQ saying that they've lost contact with Number 11. As they mull over the report, glowing lights appear over the house. The general is concerned, but Marvin gives some scientific hand-waiving explanation about electric particles in the air. "There have been so many around the project the past couple of days," Carol adds, "that we all just take them for granted."[40] She

©Sony Pictures

continues, "Number 11 ought to be visible in the sky right now on its second lap around the world." They all look up just in time to see a bright meteor flash across the sky, right on cue. "Was that it?" asks Hanley. "Yes," replies Marvin, gloomily.[41] Surely you're not going to send up Number 12 tomorrow!" implores the general. "I have to!" says Marvin. "With television and sound pickups, I'm going to know what's happening up there, or know the reason why!"

The next day they're shown checking out the cameras on Number 12. Marvin brags to Hanley how soundproof his underground bunker-office is. The general is unimpressed and leaves to watch, as he prefers, from ground level. On the way out, he passes Carol coming down with a bag of groceries. She and Marvin are going to take turns monitoring this one personally, "for days, if necessary." As the control voice intones a simplistic "Prepare rocket for launching," we get a great montage of stock scenes of the latest high-tech equipment of the day (patchboards and paper tape readers).

©Sony Pictures
High-Tech c.1956

As nervous technicians work their consoles, a radar operator reports a UFO due west and approaching fast. The sentry at the west gate calls in to report a "flying saucer" hovering over the gate. The tower officer, Sergeant Nash, chastises him for fooling around only two minutes from launch. Radar reports again that the UFO has stopped over the west gate. "Does it look like a flying saucer?" Nash asks sarcastically. "Yes, sir, it does!" comes the unexpected reply.

We cut to the gate where a large saucer, identical to the one that "buzzed" the car earlier, is hovering overhead. It starts moving over the base towards the control bunker. As it approaches the ground, a large cylindrical pylon emerges from the bottom for it to set down on. Hanley (who is also in the tower) and Nash race from the tower to get a closer look. A force field drapes the lower part of the saucer from

---

[40] It's interesting that a phenomenon that they originally thought was a once-in-a-lifetime event (St. Elmo's fire) starts showing up every day and they dismiss it. Where is the scientific curiosity?

[41] How does he know? Does he keep the vehicle ephemeris in his head? Besides, it was moving much too fast. Since this film was made before any actual man-made satellites had been launched, it can be forgiven for not knowing that satellites move **much** slower than meteors, taking at least three or four minutes to cross from horizon to horizon.

©Sony Pictures

the rim to the ground as three aliens emerge from a thin rectangular door in the side of the pylon. As per the standing order mentioned at the beginning, a truck-mounted anti-aircraft "Bofor gun" arrives on scene and sets up.

As the first alien leaves the force field, the battery opens fire. One of the first shots hits the exposed alien and he goes down, but the gun is useless against the force field. A jeep full of infantry arrives and they start firing at the saucer with

©Sony Pictures

their 30-06 Springfields. The aliens retaliate with energy rays emitted from their arms. Their first target is the Bofor gun followed by all the foot soldiers (another plot event "borrowed" from *The Day the Earth Stood Still,* although these aliens target the soldiers as well, not just the weapons). The two remaining aliens retrieve their fallen comrade and pull him back into the saucer.

Meanwhile, in the heavily insulated firing room and Marvin's underground bunker, they don't hear any of this, and are puzzled by all the communications going dead. Carol looks up at the TV monitor (which is still working) and exclaims, "Look!" just as the rocket falls over and explodes. Cutting back outside we see that it's the aliens who are zapping everything with their ray, causing massive destruction. Marvin and Carol race through the labyrinthine pipe runs trying to escape their bunker, but find the exit blocked by debris. Back outside, an alien lumbers over and picks up General Hanley, who had been knocked unconscious in a blast, to carry him to the ship.

The aliens take off and speed skyward, waiting until they're quite high before retracting the pylon. We then move to the inside of the saucer where the general is just waking up on the floor. The room he's in is enormous, almost as big across as the entire saucer looks from the out-

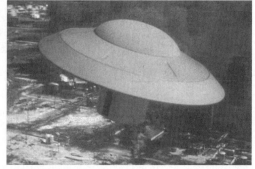

©Sony Pictures

©Sony Pictures

side.[42] The ribbed dome ceiling looks about 20 feet (7 meters) high. The floor is bare except for two large rectangular screens. A large glowing crystal, shaped something like a rose, descends from the center of the ceiling, rotating slowly on its shaft. It is a translating device. "Perhaps you can explain why, after contacting Dr. Marvin, we were met with violence," demand the two aliens who have appeared in small booths set into the wall. "You contacted Dr. Marvin?" asks the general, still not fully conscious. "We spoke to him," reply the aliens. "All he heard was meaningless sounds," returns Hanley. The aliens explain that they made "adjustments for the time differential between us," but apparently it still needed refinement.

The general says nothing more than his name and rank. The glowing crystal emits a beam that strikes Hanley on the head, causing his brain to be exposed! (This is an interesting foreshadowing of "real" alien abduction reports, the first of which didn't happen until 1961.)

©Sony Pictures

Before we know what they're doing to him, we cut from this grisly scene to Marvin in the bunker, gabbing away into his tape recorder. He's just finishing his report and notes that the batteries are failing, slowing the recorder. For no apparent reason (except that it's necessary to the plot), rather than shutting the tape off, he switches it from "record" to "play" and keeps it running. Suddenly, the whining saucer noises can be heard as a highly distorted voice saying, "Dr. Russell Marvin, it is very urgent that we meet. We will appear tomorrow at 'Operation Skyhook' when your sun is exactly over your local meridian" (i.e. noon). This is the "contact" they supposedly made the day before. With power failing and prospects looking bleak, he says to Carol, "If I'd only figured it out before, we wouldn't be trapped down here!"

---

[42] As we've seen in other chapters, both the *C57-D* from *Forbidden Planet* and the *Jupiter 2* from *Lost in Space* have the same problem, but here it was deliberate. In those other two vehicles it was a case of the interior set designers trying to stuff too much equipment into the size established by the exterior sets. Here, the interior set is almost completely empty and the size is intended to show that the aliens operate in a different space-time continuum, meaning that vehicle size can be a variable.

The Saucer Fleet

We never see their rescue, but rather cut straight to newspaper headlines that blare: "Skyhook Wiped Out!" with the sub-heading "Two Survivors Rescued." Since Marvin and Carol were in the bunker (thus didn't see the saucer attack) and are the only survivors, no one knows what happened. The narrator advances the plot significantly by telling us "An aroused public demanded an answer, and the federal government dedicated all its branches to finding one." The Marvins fly to Washington, bringing the reel of tape for the investigation at the Pentagon.

©Sony Pictures

Marvin presents his evidence to General Edmunds (left foreground) and Admiral Enright at the Pentagon

In a military conference room, we see the tail end of the message being played. Marvin explains, "A landing at the project was proposed on the day of the disaster. If I couldn't keep the appointment, a message was to be sent by ordinary short wave transmission." The assembled military officers, including General Edmunds, the Chief of Staff, start debating his interpretation, and even the validity of the tape. Marvin explodes in frustration that he and his wife have been dragged all over Washington and no one wants to believe him and take action. Mr. Cassidy, the under Secretary of State sitting to Marvin's right, says that he is inclined to believe Marvin, after which the military men agree. However, Marvin won't be allowed to contact the aliens because they are "a hostile and unknown power," and that "any decision to meet with them must be made at the Cabinet level." Marvin pleads with them to allow a private meeting while the Secretaries of State and Defense are returning from overseas. The best that the generals will do is recommend to the Cabinet that he be allowed to contact the aliens.

As the meeting breaks up, Admiral Enright asks Marvin that he not leave his hotel since, "you may be needed at any time." They assign him a liaison, Major Huglin, who defuses Marvin's objections by saying that he's "just doing my job." He is actually quite personable. They arrive at the hotel, and Marvin has to slip quietly into the room so as to not wake Carol (the military apparently trusts Marvin enough to put Huglin in a separate room). He sits down at a portable short wave radio (which he had ordered earlier in the day for just this eventuality) and starts to contact the aliens as instructed on the tape: "Dr. Russell Marvin calling on 225.6 megacycles…" After a couple of tries, he makes contact. The alien voice instructs him to meet in 90 minutes at the north end of Chesapeake Bay. Marvin protests that he

won't be allowed to meet with them for a couple more days yet, but the aliens are gone.[43]

©Sony Pictures

Carol wakes up and hears the tail end of the conversation. She, of course, doesn't want him to go (she's already lost her father). He rushes off, against orders, exclaiming, "I've got to go through with it!" After he's gone, she calls Major Huglin (who is sitting up in bed reading, fully dressed, at 4:30 AM) who tells her to call the garage to prevent Marvin from getting his car. The garage attendant tries to stop him, but Marvin decks him and speeds off. Carol and Huglin give chase. Naturally, their pursuit attracts a motorcycle cop who just happens to be manning a speed trap on a country road at 5 AM. Marvin gets to the designated spot on Chesapeake Bay followed closely by the other car and motorcycle.

While the others are catching up, Marvin walks across the beach to the saucer quietly sitting there. When Carol, Huglin and the cop come up, the alien voice invites them all in. The saucer spools up enough to rise up on its pylon, exposing the door. They walk in and take an elevator-like platform up (or maybe it's just the pylon retracting up) into the

©Sony Pictures

---

[43] There is a confusing continuity error in this scene. The Pentagon meeting is obviously conducted during the day (you can see a daylight scene out the windows), but it's night when the car pulls up to the hotel. When Marvin contacts the aliens they give the time in "Greenwich meridian time," which translates to 4:30 AM in Washington. Why was Marvin just getting back to his hotel room at that hour?

saucer proper. One of the large screens is showing clouds flying by. The translator device descends from the ceiling. The screen images are now the opening "nebulae shots" from *The Day the Earth Stood Still*, only played in reverse to give the impression of moving away. "We must be thousands of miles away from the Earth!" Huglin gasps. "And in a matter of seconds!" Carol adds. The alien confirms that they are speeding away from the Earth, but not in just seconds. The voice explains again that they operate in a "different time reference" and that all of this was happening "between the ticks of your watch." (Do young people today understand this analogy?)

When discussing the attack on Skyhook, Marvin suddenly realizes "You shot down our 11 rockets, why? "We had no way of knowing that they were only primitive observation posts. We thought they might be weapons directed against us," explains the alien (this is classic Cold War thinking and, you'll notice, exactly the same posture that our military took at the beginning of the film). "Who are you? Where are you from?" continues Marvin, now the interrogator. "We are the survivors of a disintegrated Solar System," says the alien, not really answering the question. He continues, "At this moment the remainder of our fleet is circling your globe." The screen now shows saucers flying over New York City, Paris and London (another reference to *War of the Worlds* which had similar scenes during an alien invasion).

©Sony Pictures

"What do you want with me?" Marvin asks, finally getting to the point. They say they want him to arrange a meeting of all world leaders in Washington DC. Just like Mr. Harley in *The Day the Earth Stood Still*, Marvin protests that they won't listen to him. To give him some proof, they move in an instant from beyond the moon to the Earth (the screen now shows the cloudy Earth-moon shot from late in the *Day the Earth Stood Still* title sequence). Zooming down on a Navy destroyer in the Atlantic, they blow it up with a singe burst from a ray projector that emerges from the underside of the saucer. Huglin is aghast, "There were 300 men on that ship!" he protests. The aliens are quite smug in their demonstration of superiority. "With weapons like that, why don't you just land and take over?" Marvin asks. They reply that it's a simple matter of logistics. There

aren't enough saucers to control all of us at once, and the aliens realize that by the time their small force subdues us the planet would be wrecked. They'd rather we just hand it over now without a struggle, thankyouverymuch, since it's inevitable that we'd eventually lose.

Marvin fires off one last question, "How do you know so much about us?" At least the answer is different than the old "we have been monitoring your radio and TV broadcasts." This time it's more interesting: They basically suck all the knowledge out of the brains of abductees (another "real" point from George Adamsky's and other books) and place it into an "infinitely indexed memory bank."[44] When testing the aliens' knowledge with a question-and-answer session, the voice answering is different. Carol recognizes it as her father's. Turning around, they see General Hanley stagger out, zombie-like. The policeman draws his gun and fires at the two aliens in the "booth" along the wall. The same beam from the translator that we saw earlier with the

general, now strikes the policeman, exposing his brain. "This is the beginning of the process," the alien explains calmly. Continuing, as if the horrific scene they're watching is some sort of threat, "Will you arrange a conference for us? You have two lunar days (months) or 56 days Earth time."

©Sony Pictures

We are not shown how Carol, Marvin and Huglin are returned to Earth. Rather, we fade directly to the Pentagon conference room where Admiral Enright is writing "56 days" on his note pad. Marvin is trying to convince the military men that they (the abductees) weren't all hallucinating. Even giving them the example of the sunken destroyer only elicits an admission that "the Navy had lost contact with one of its vessels." Admiral Enright continues to rake Marvin over the coals for contacting the aliens in the first place against their direct orders. They discuss using nuclear weapons against the aliens, but rule it out since they would destroy the cities and surrounding countryside as well (some pretty rational thinking under the circumstances).

Marvin pipes up that he got an idea of how the alien technology works from his time aboard the craft, and he thinks a type of "ultrasonic gun" would work against them. Enright is incredulous that a new weapon could be built in less than 56 days, but Marvin thinks he can do it, with enough help. While he's talking, an aide brings in a dispatch confirming the sinking of the destroyer *Franklin Edison* at

---

[44] That's not a bad description of the World Wide Web coupled with a good search engine.

The Saucer Fleet

exactly the coordinates the aliens had specified. The generals give Marvin all the help he needs.

We move directly to Marvin in the lab with Carol and a few assistants. He is adjusting a crude prototype of his invention, a parabolic dish with a central emitter mounted on a tripod. Huglin arrives for the demo, just as they're about to test it. They fire it at a large concrete block, which bursts

Kanter (with cane) and the Marvins test the prototype

apart. The device burns up its generator in the process, but the test is a success. Marvin, though, grumps that the device, based on sound waves, can't be scaled up with current technology, and they're down to 27 days. One of his associates, Professor Kanter, remembers a report from a scientist in New Delhi who suggested an approach using electric fields to interrupt the alien's magnetic field. "Of course!" bursts out Marvin, interrupting Kanter. "We cut the ultrasonic wavelength into the circuit and knock them down like clay pigeons!" (Marvin is one quick study!)

Next comes another montage scene of stock footage showing heavy industry mobilizing while the narrator intones, "From all parts of the globe came every facility and scientific help the governments of the world could furnish.

Dr. Marvin and his staff assembled these materials in a concealed laboratory." This lets us jump straight to the test of the new prototype at the "Belmont Lab." They suspend a steel ball between two magnets, then aim the device at it. This is a larger unit than the previous one with a bigger reflector and what looks like a small Newtonian telescope in the center. They turn it on and the steel ball thuds down.

As they are busy congratulating themselves, Huglin notices a glowing sphere near the ceiling. It "buzzes" them a few times before Huglin shoots it down with his service revolver. Carol recognizes it as one of the "glowing lights" that were hovering over Skyhook previously. They figure

that it's a reconnaissance device and realize that the aliens now know about the new weapon.

They load the prototype onto a truck to drive it to Washington. As they pull away, a saucer appears overhead and lands outside the lab. One alien gets out and goes into the building, apparently looking for them. Marvin realizes that this is the perfect opportunity to test his prototype, so they stop their little caravan and fire up the generator while Huglin radios the Air Force for support. Two more aliens emerge and start walking around. As they aim Marvin's invention at the saucer, it starts wobbling around on its central pylon. The two aliens look up in alarm then rush (sort of) back in and take off. While the device keeps the saucer wobbling around in the sky, it doesn't have enough power to actually bring it down.

The lone alien left behind emerges from the building and sees the saucer in distress. Professor Kanter inexplicably gets out of the car and starts walking towards the alien. It raises its ray-arms and vaporizes the good professor followed by the truck with Marvin's prototype.[45] Huglin fells the alien with three or four quick rifle shots (without the saucer's force field, their suits apparently don't provide much protection). They rush over to the alien body, which they find "light as a feather," and remove the helmet.

A shriveled and desiccated-looking humanoid head is revealed that simply disappears as they watch.

---

[45] We want to give credit for some careful scripting here. Since the alien destroyed Marvin's prototype, how can they build more? Well, one scene earlier when instructing the lab personnel to load the device on the truck, Marvin says, "I'll take the diagrams (blueprints) [in the car]." This is also why they had to get Kanter out of the car before zapping him, as illogical as that is for a man who walks with a cane. Had the alien vaporized the whole car, then the plans for the device would be gone as well.

A B-29 bomber appears overhead, responding to Huglin's call. The saucer, freed from the beam of Marvin's device, flies up and destroys it with the ray projector. It turns the ray onto several buildings (presumably the lab) then, for good measure, sets the forest on fire. As Marvin, Carol and Huglin run into a drainage culvert for protection, the saucer comes into the fierc-

©Sony Pictures

est center of the flames, lowers its pylon while still hovering, and pushes out two bodies. Carol rushes over only to discover that it's her father and the policeman. The aliens had kept their word about returning them. They never said alive…[46]

We dissolve to Marvin and Carol walking into yet another laboratory. Huglin is already there with the Joint Chiefs along with Dr. Alberts, a civilian scientist. They've been analyzing the alien helmet and other artifacts. One of the devices in the helmet is a miniature version of the translating "crystal." Using it, they have built up the entire alien language by reading the dictionary into it and analyzing the resulting output. The translator only works one-way, though (from human to alien), so they needed to use the "electronic translator," which Professor Alberts developed, to translate the communications intercepted from the alien fleet. Some of what they translated was a plan of attack!

Alberts gives a demonstration of his "electronic" translator, which, in reality, is the mechanical UCLA computing machine used by George Pal as the "DA" in both *Destination Moon* and *When Worlds Collide.* They crank up the machine, and although we don't hear any alien input, the

©Sony Pictures

The "Electronic" Translator

output is written (in cursive script!) by a pen plotter. The message is not shown in its entirety, but it involves the Sun and uses Mercury's perihelion as a time reference.[47] General Edmunds goes on to talk about an improved version of Marvin's device being developed at Aberdeen Proving Grounds, but even that one would have an effective range of only 1,500 yards (1,500 meters).

"Has anyone tried that helmet on?" asks Marvin, unable to contain his curiosity any longer. "Yes, try it," urges Alberts. In what is a direct conceptual lift from *War of the Worlds,* where they hook up the Martian "eye" to the epidia-

©Sony Pictures

scope, Marvin puts the thin shell on his head and suddenly he has super senses in vision and hearing (this is probably the least effective scene in the movie; he looks ridiculous). Alberts says that it's made of "solidified electricity" and that it resists everything they've used on it (it's a good thing Huglin wasn't aiming at the head when he shot the alien!). They speculate that the aliens must be physically quite frail if they have to build such capabilities into their outer suits.

Just then, the alien leader's voice comes booming out of the loudspeakers in the lab and, as we see, all over the world. "People of Earth, attention!" it demands several times. "Following eruptions on your Sun," it continues, "there will be eight days and nights of meteorological convulsions." So rather than the Earth "standing still" for half an hour, they are going to cause a week of massive storms, after which they will land in Washington so that we can turn over the keys to the planet. As the message repeats, there is another montage sequence showing people all over the world hearing the message in their own language. Many of the shots are, indeed, from the "Earth standing still" sequence from *The Day the Earth Stood Still.* The message continues for twelve hours after which there is a huge prominence on the sun, which the Bureau of Meteorology confirms will cause a week's worth of hurricanes and tidal waves. General Edmunds reminds everyone that there are only nine days left before the aliens land, and that we will be preparing for their attack under the worst possible weather conditions.

Back at the hotel room, Marvin says goodbye to his wife. He is leaving for Aberdeen to help with the construction of the more powerful versions of his weapon, which will at least get him away from the coast some. Carol is supposed to head for Palm Springs where she will hopefully

---

[46] Of course, this means that had Marvin's device had enough power to bring the saucer down, he would have unknowingly killed his father-in-law, presuming he was still alive before being pushed out.

[47] Mercury's perihelion, or closest approach to the Sun, happens once during every 88-day orbit. This is mistakenly described by Professor Alberts as happening "twice every three months."

The Saucer Fleet

be completely safe from the approaching storms. This leads directly into another montage of stock footage showing huge storms affecting transportation; shipping lanes, airways, rail and highway, all stopped. Telephone and telegraph were knocked out when the lines went down, and radio is useless due to the sunspots. The narrator breathlessly adds, "The world, crippled by these events, waited for the first sign of an invasion…from outer space!"

Once the storms lift, the army and civil authorities start evacuating Washington DC, which oddly shows no signs at all of storm damage. Carol shows up at General Edmunds office, having never gone to Palm Springs. She asks if there's been any news of her husband, but all communications are still out. Just then the phone rings (?) and sirens start blaring. A Red Alert!

More stock footage of radar units turning, air crews scrambling, a pair of P-80's taking off and a battery of Nike -Ajax coming up to the ready on their elevators. Batteries of 155's (artillery) raise their barrels ominously skyward. A saucer sweeps in over the Atlantic, just like

©Sony Pictures

Klaatu's ship, but this one's first act is to take out two of the jets with its ray projector. Several rounds of Nike and other missiles are launched and strike their target to no effect. The anti-aircraft artillery is likewise useless.

Just as Edmunds is sending Carol to the shelter, the phone rings again with the news that the convoy from Aberdeen has entered the city with Marvin and a fleet of his disruptor devices. The saucer fleet has also entered the city, scattering what's left of the population in panic. We see Marvin's convoy, which aims a disruptor at the first saucer it sees, sending it into the Potomac. Carol runs out of the Pentagon amid thunderous explosions and over to Marvin's part of the convoy, which is conveniently right out front. Word comes over the radio that saucers are hovering over the White House and Capitol. They hop into a jeep and take off along with a couple of disruptor trucks.

©Sony Pictures

While they travel, other saucers start laying waste to gas storage tanks and other buildings. One of the buildings is the old Post Office administration building whose square tower turns into the Los Angeles City Hall the instant it explodes, thus allowing the use of a snippet from *War of the Worlds*.

©Sony Pictures
Aliens on the lawn of the White House

The saucer in front of the White House hovers for a landing, but before extending its pylon, it lowers its heat-ray projector and takes out the disruptor truck that had just arrived. Only then does it extend its pylon, land, engage its force field and have two aliens emerge. A couple of army sharpshooters arrive and start plinking away at them. Since the aliens are still within the force field, the bullets are useless, but it does convince them to turn around and get back in the ship (but not before vaporizing the soldiers).

The Marvins pull up with two more disruptor trucks close behind. They hop out of the jeep and stand by the fence as the saucer takes off. It lowers its ray-projector and destroys the jeep the Marvins had just left. The two disrup- tors go to work and cause the saucer to crash into Union Station. Lots of battle scenes with the saucers flying around monuments, and the disruptor trucks causing some of them to crash; most famously into the Washington Monument, which topples over crushing visitors under the falling debris. The saucers also score hits, taking out several of the disruptor trucks.

©Sony Pictures

214

Russ and Carol, now driving in one of the trucks, pull up in front of the Supreme Court building. Rather than hopping on the back to fire up the disruptor (why else would they be there?) they inexplicably run away from the truck and up the stairs of the building. Naturally, a saucer takes out the truck then turns its attention to the building

©Sony Pictures

where it demolishes the facade. The Marvins barely escape the falling debris. They rush back to another truck in time to hear a radio report that a saucer has landed in front of the Capitol. They head over with all the remaining trucks. When they arrive, they see one saucer idling on the ground and another hovering overhead. Two aliens are walking around at the top of the stairs.

A whole platoon of sharpshooters shows up and opens fire. Since they are away from the saucer's protective force field, the two are quickly

©Sony Pictures

dispatched. Marvin picks up the radio mike and issues the completely redundant command "Keep firing at saucers!" (like they weren't going to otherwise?)The disruptors find their mark and one saucer crashes into the portico of the House of Representatives, and the other into the Capitol dome, which explodes![48] That, apparently, was the last one as an

©Sony Pictures

---

[48] It's interesting that none of the other saucer crashes produce an explosion.

authoritative voice on the radio repeats "The present danger is ended," over and over. In one of the iconic images from the film, the Marvins stand heroically in front of the

©Sony Pictures

wrecked Capitol with a saucer sticking out of the side.

We dissolve to the beach where the Marvins are enjoying their long-delayed honeymoon. Carol is reading the paper and, reading aloud, announces that the president had ordered Skyhook rebuilt "under the direction of Dr. Russell A. Marvin," and that the United Nations has awarded him a gold medal. They walk to the edge of the water and gaze out as Carol says, "I'm glad it's still here," to which Marvin replies, "and still ours!" They skip off to frolic in the surf.

The Vehicle

The saucers in this film are kept deliberately vague so that they would seem that much more threatening. While dramatically it's good not to reveal too much about a mysterious invader, it makes documenting them a little difficult.

©Sony Pictures

Externally, they are quite a bit more interesting than most movie saucers from the era. Whereas some filmmakers strove for featureless perfection (Klaatu's ship comes to mind) to depict advanced alien technology, Harryhausen created ships with some interesting surface detail. Most prominent are the rotating rings on the top and bottom surfaces of the disc. Watching them spin, you eventually realize that they are counter-rotating (i.e. they spin in opposite directions), which is not only more interesting, but creates a sense of balance and control. Both rings have radial strakes on them to make sure you see them turning.

Centered on top, inside of the rotating ring, is the ubiquitous dome that every saucer seems to need. Harryhausen's dome is very similar to the one Buddy Gillespie's crew put onto the *C57-D* for *Forbidden Planet*. It wasn't a case of copying as much as convergent evolution. Both pictures were in production at the same time, and by the time *For-*

The Saucer Fleet

bidden Planet was released, Harryhausen was well into filming the saucer models, if not already finished. Outside of the rotating rings, top and bottom, is a non-rotating area that forms the saucer's rim.

The bumps on the underside of the disc, in the slightly concave, outer stationary part of the disc, add a "reality check" to the design, having been adapted from similar features on Adamski's famous UFO photos. Inside of the lower ring is a non-rotating circular area that projects down slightly and is fixed. It contains the retractable landing pylon

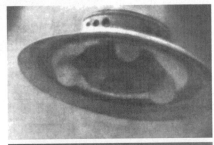
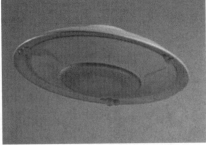

©Sony Pictures

that emerges from the center. This again duplicates similar thinking by the crew at MGM who designed the same feature into the *C57-D* at the same time. The pylon has two other functions: it provides the doorway into the ship, and is the storage area for the ray projector. This means, of course, that when Dr. Marvin and friends walked into the ship, they should have had to walk around the projector mechanism, which has to be retracted for landing. Jon has a suggested design solution to this in the "Archeologist's Report" coming up next.

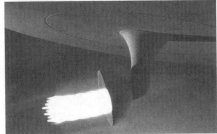

©Sony Pictures

Both the pylon (top, shown partially retracted) and the Ray Projector emerge from the center of the saucer bottom.

The saucers are highly dynamic when they are flying. Not only are the rotating rings zipping around, the whole vehicle banks and pitches as it changes direction, just like an airplane. This is part of Harryhausen's effort to give the saucers a personality and create the sense of an intelligence behind their operation[49] by giving a real-life reference point. Unlike airplanes, though, the saucers in other scenes accelerate, decelerate and change direction instantaneously, as described by many real life UFO witnesses.

Just as dynamic is the sounds they emit. We've already mentioned previously in the "production" section that the basic sound came from the sewage pumps at the Playa Del Rey treatment plant, but that was further processed by filtering and adding other sounds. The final sound effect could be varied in pitch to give a very solid sense of effort when taking off and landing.

Determining how the saucers actually operate is a bit more problematic. There is almost nothing to go by in the movie itself. This was partially to keep them mysterious by not giving away too much, and partially because it really wasn't necessary for the story. We get basic operating principle from the scene where Dr. Marvin is on board talking to the aliens. The interrogator says, "We generate a magnetic field stronger than the gravitation field on your Earth. This is the principle by which we move though space. We have adjusted the magnetic field to compensate for the normal loss of gravitational effect and atmospheric pressure." If they are using magnetism to counteract gravity, it would seem that the aliens had cracked the Grand Unified Field Theory problem, or at least the part that links gravity and electromagnetism. From what little we know today of this relationship, the amount of energy needed to create a negative gravity field for propulsion, or a positive gravity field to counteract inertia and weightlessness, would be unimaginably stupendous. It would require a power source beyond nuclear fission, fusion or even antimatter. It obviously uses some completely unknown (to us) principle.

A couple more thoughts before we leave this section. We know the aliens distort space and live in alternate dimensions by the scenes inside the saucer. Not only is the room they are in larger than the outside of the saucer, but the elevator they ride up in is distinctly odd. We assume that they are going up because the wall next to

©Sony Pictures

them is going down. The first thought is that they are riding the bottom of the pylon up as it retracts into the saucer. If that were true, though, the walls should be moving up with them. So where is the wall of the pylon stored when retracted? Like the disappearing ramp on Klaatu's ship, this adds to the eerie, alien nature of the design.

[49] There was…his!

Regarding the ray projector, Dr. Marvin had determined that it is an ultrasound weapon (after the first test of his prototype based on this principle, he says, "Now we know what hit 'Skyhook'"). Again we are looking at some pretty advanced development here. Ultrasound (sound waves above the range of human hearing) is not an especially energy dense medium, not at all suitable for projecting destructive energy. For example, if all the sound energy emitted by a large symphony orchestra in a two-hour concert could be concentrated under the feet of the conductor, it would raise him only a few inches off the po-dium! It is difficult, therefore, to imagine what sort of technology the aliens have that cause sound waves to be emitted from the ray projector and the arms of their body suits in a glowing, collimated beam.

After Jon's Archeological Report, we'll be back with some final thoughts on this very interesting and enigmatic spaceship design. As we shall see in the next section, it produced one of the most recognizable flying saucer designs of the decade, and influenced a couple of generations of filmmakers.

Archeological Report on the Saucers in *Earth vs. the Flying Saucers*

By Jon Rogers

Ray Harryhausen's flying saucer, first featured in *Earth vs the Flying Saucers,* succeeded so well that it became somewhat of an icon in its own right. It has been mimicked in other movies and featured in several documentaries since its introduction. This is because it was so believable, somewhat unique, and very recognizable as to what the public in 1956 thought a flying saucer *should* look like.

At first glance, his saucer design seems to be a rather simple, straightforward one. This is what we already know about it.

- It was central to a 1950's "B" movie created quickly on a tight budget.
- It had features similar to George Adamski's famous saucer that had been widely publicized by the media in 1953 & 1954. Specifically, three half domes on its bottom side.[1] On Harryhausen's saucer, these may have been force field projectors–otherwise, their function is unknown.
- The other major design features were an upper dome and a central pylon that extended down for landing. Its interior was one large, mostly empty room that seemed to be inconsistent with its outer shape. This room was similar to those in other saucers and spaceships seen in movies during this era.
- It had circular sections on the top and bottom of its disc that rotated when it flew or was actively powered up. This was different from other previous saucer designs.
- It flew by using some mysterious application of magnetic fields and by changing the space-time continuum. Its main weapon was a large ultrasound ray projector that emerged below the ship when needed.

There were several different sized filming models used to create this alien ship. However, there were no formal drawings made, being so short on time. A few of these models survive and have been seen on subsequent documentaries.

But what was the ship they were trying to create? That is what I want to document–the alien saucer that appeared in the movie. In doing so, I will also identify the proportions, but not the exact dimensions, of the models used to create it.

In recreating the saucers in *Earth vs. the Flying Saucers*, the first job is to establish the actual size of the ship. During the movie, the alien saucers are seen in many shots with different backgrounds. This is the first problem, how to identify which scene(s) to use to derive the actual size of the saucer.

Figure 1

---

[1] These are plainly visible on the cover of Adamski's 1953 book, "*Flying Saucers have Landed.*" (See Figure 22, page 294)

The Saucer Fleet

There is one scene where Dr. Marvin, his wife, his military guard, and a policeman are all approaching a saucer sitting on the beach. (Figure 1) They are all directly in line with the Saucer's centerline and can therefore be scaled to the ship with little to no error. In reality, this is a superimposed picture of a saucer model and live actors on the beach, but it will be taken at face value as representing what Harryhausen was trying to create in the audience's mind.

By comparing the lower ring and pylon diameter to the heights of the men and Carol Marvin accurately, some of the craft's dimensions can be identified. The height of the ship is not available because the viewpoint of the camera distorts it. However, once the ring and pylon dimensions are known, the rest of the saucer's dimensions can be derived from a later scene that shows the whole ship with the camera viewpoint very close to its beltline.

Right after the scene on the beach shown in Figure 1, there is another scene where the four are seen entering the ship together. (Figure 2)

This scene allows us to accurately determine the size of the opening in the saucer's pylon. It is interesting that the pylon wall is fairly thick and that the opening on the inner wall is a little smaller than the outer wall.

Once inside, they ride an elevator up into the saucer's interior "Dome" room.[2] But, as Jack has already said,

> …but the elevator they ride up in is distinctly odd. We assume that they are going up because the wall next to them is going down. The first thought is that they are riding the bottom of the pylon up as it retracts into the saucer. If that were true, though, the walls should be moving up with them. So where is the wall of the pylon stored when retracted?

This intriguing observation led me to examine the elevator and the main room very carefully.

This room was already seen when General Hanley was captured. When that scene opened, the General was lying stretched out on the elevator platform. (Figure 3) This allows us to scale the size of the elevator itself.

Not counting his outstretched arms or feet, when I carefully measured General Hanley's body I discovered that he takes up almost exactly half the elevator. That makes the elevator floor about 12 feet (4 meters) in diameter.

In a later scene, when our four humans are facing their alien captors, a clearer understanding of the size and make up of the room is obtained. (Figure 4)

Figure 2

Figure 3

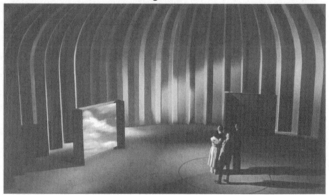

All ©Sony Pictures

Figure 4

Like so many of Hollywood spaceship interiors, the Dome room is too tall. The camera is about 15 feet (5 meters) above the people. If the ribbed lines in the wall are extrapolated to where they come together, the Dome room becomes almost 25 feet (8 meters) high. The real saucer's Dome room cannot be this high because the whole saucer is only 28 feet (8.5 meters) high and there must be a 14 foot (4 meter) tall ray projector (shown in Figure 5) stored somewhere below the floor.[3]

---

[2] I have named this area the Dome room because it was constructed to give the appearance of being right inside the saucer's upper dome. It is the only interior room we are shown.

[3] In the elevator we pass 7 ribs in the back wall. Each is about 16-18" apart which implies there are 10 (plus 6" at the top) in all. This indicates that the shaft is about 14 to 15.5 feet deep which agrees with my other measurements.

Knowing that the elevator was 12 feet in diameter permits this interior room to be scaled. This shows that it was at least 16'd x 30'w x 25'h (5 x 9 x 8 m). It may have been larger, but this is the way it was shown in the scenes in Figures 3 and 4. Also, the camera location was in *exactly* the same position for almost all wide shots in the scenes inside the ship. The other interior shots show little detail of the Dome room. This suggests maximum use of a limited set.

©Sony Pictures

Figure 5

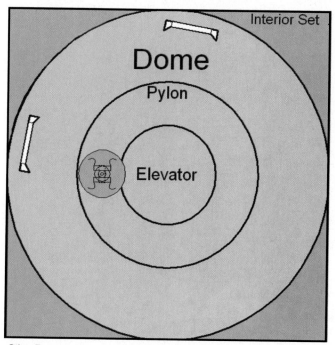

©Jon Rogers

Figure 6

More evidence that this is just a movie set is that the back wall and left wall are actually straight walls, 90° to one another and joined by a curved section to make it look as though the room is circular. Therefore, I concluded that this is a soundstage modified to look like the real saucer interior, which would only be 14 ft high in reality.

So now that we know the designers, in making the set, were constrained by money and time into modifying a standard sound stage to approximate the interior of the upper dome, what more can we find out about the elevator?

One problem with the elevator is the big ray projector location. In Figure 5, it is obvious that it emerges from the center of the bottom of the saucer pylon. Note that there are no other markings in the bottom of the pylon--no indication of where an elevator might be.

Based on the scene in Figure 1, I had determined that the upper dome and bottom ring protrusion were 40 feet (12 meters) in diameter. The pylon was 22.5 feet (7 meters). The next problem was to find the location of the elevator the humans rode in.

The first assumption was that the elevator was in the center of the pylon. However, that put the elevator out of position between the two walls and screens in the Dome room as seen in Figures 3 and 4. It also interfered with the ray projector when it extended down through the center of the saucer. Even if the projector were moved aside, there would not be enough room for it within the pylon. (Figure 6)

Since the Projector is 6 feet (2 meters) in diameter and the humans did not see it on their way up, it might be behind the unmoving ribbed wall they did see. Also, moving it out of the way for the elevator platform allows the elevator to be repositioned closer to the side of the pylon. This allows the elevator to be correctly positioned inside the dome room. Now the proportions and the dimensions all begin to fit (Figure 7).

In this plan, the elevator is merely a platform in a tunnel, one side of which is the door to the ray projector.

One result of offsetting the elevator pad is that a small passageway (or the thick wall we see in Figure 2) is needed to connect to the outside of the pylon. If the elevator were

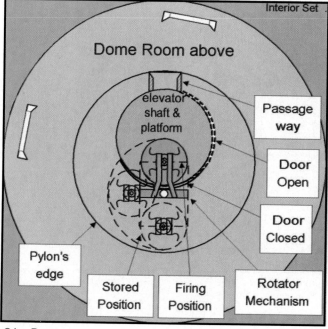

©Jon Rogers

Figure 7

The Saucer Fleet

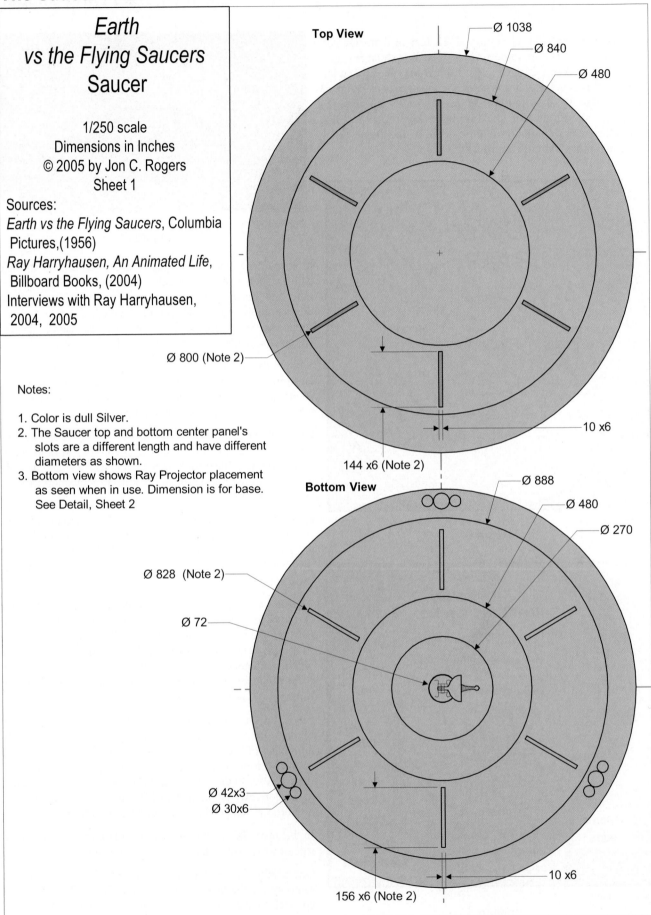

Earth vs the Flying Saucers Saucer

1/250 scale
Dimensions in Inches
© 2005 by Jon C. Rogers
Sheet 1

Sources:
Earth vs the Flying Saucers, Columbia
 Pictures,(1956)
Ray Harryhausen, An Animated Life,
 Billboard Books, (2004)
Interviews with Ray Harryhausen,
 2004, 2005

Notes:

1. Color is dull Silver.
2. The Saucer top and bottom center panel's
   slots are a different length and have different
   diameters as shown.
3. Bottom view shows Ray Projector placement
   as seen when in use. Dimension is for base.
   See Detail, Sheet 2

Top View

Ø 1038
Ø 840
Ø 480
Ø 800 (Note 2)
10 x6
144 x6 (Note 2)

Bottom View

Ø 888
Ø 480
Ø 270
Ø 828 (Note 2)
Ø 72
Ø 42x3
Ø 30x6
10 x6
156 x6 (Note 2)

Earth vs the Flying Saucers Saucer

1/250 scale
Dimensions in inches
© 2005 by Jon C. Rogers
Sheet 2

Side views, Pedestal, & Details

Notes:

1. The Saucer top and bottom center panels counter rotate. Analysis shows the top panel rotating Clockwise with the bottom panel rotating Counter Clockwise as seen from the side, most of the time.
2. Force field is approximately 900" (75 ft) in dia.
3. The doorway appears to have two dimensions, an inner and outer one going through the thick pedestal wall.

Hemisphere Projections detail (3ea)

12.0°

R 21
R 15 x 2

Ray Projector detail
(extended)

102
82
Ø 72
Ø 69
Ø 23
Ø 12
25
Ø 23
62
Ø 12

168
180°

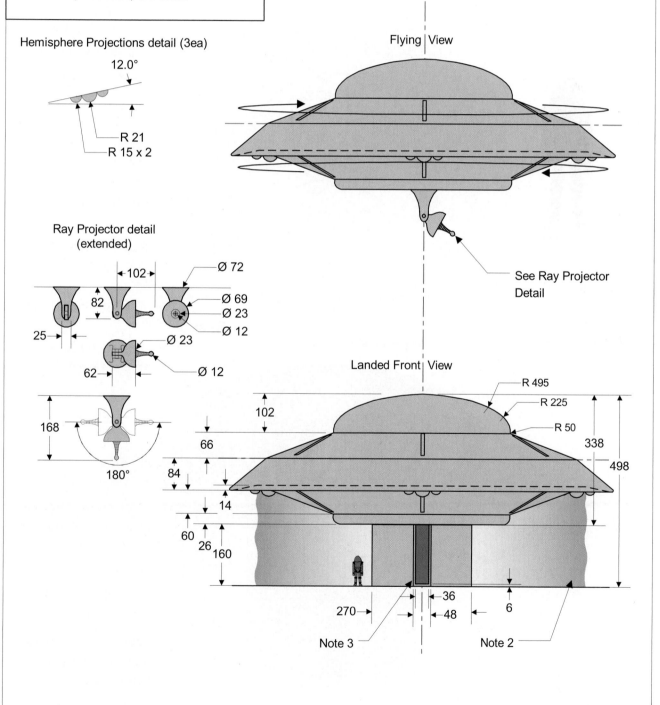

Flying View

See Ray Projector Detail

Landed Front View

R 495
R 225
R 50

102
66
84
14
60
26
160

338
498

36
48
270
6

Note 3
Note 2

The Saucer Fleet

Earth vs the Flying Saucers Saucer

1/150 scale
Dimensions in inches
© 2005 by Jon C. Rogers
Sheet 3

Interior Details

Notes:

1. Interior is shown correctly scaled.
2. Theoretically, the ray projector must be offset when stored. (See report).
3. The Interior Rear view shows the landing pedestal retracted and extended. It also shows the elevator platform extreme up and down positions. The short passage to the entrance is implied.
4. The rest of the saucer interior is undetermined.

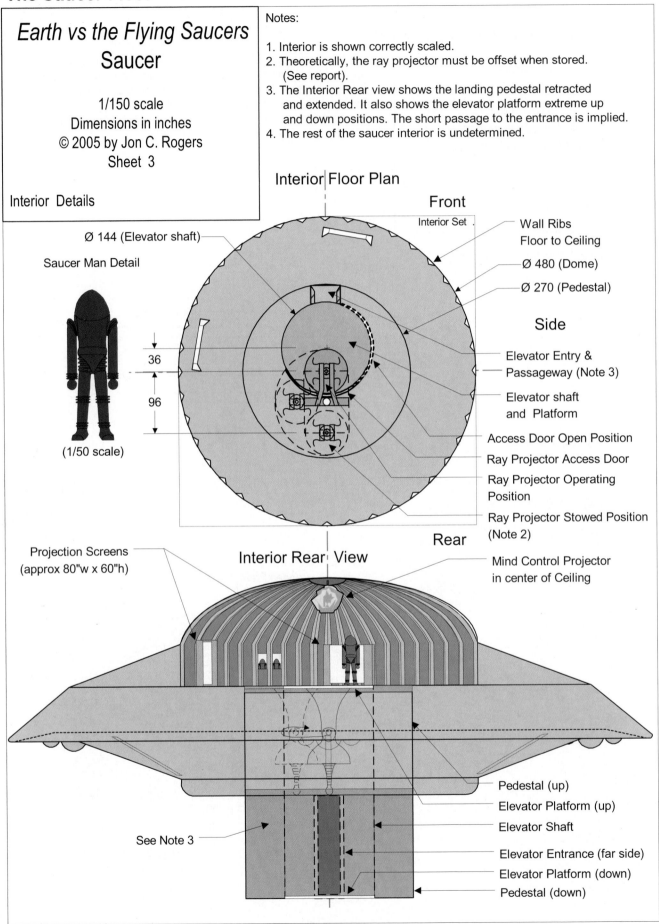

Ø 144 (Elevator shaft)

Saucer Man Detail

36

96

(1/50 scale)

Interior Floor Plan

Front

Interior Set

Wall Ribs
Floor to Ceiling

Ø 480 (Dome)

Ø 270 (Pedestal)

Side

Elevator Entry &
Passageway (Note 3)

Elevator shaft
and Platform

Access Door Open Position

Ray Projector Access Door

Ray Projector Operating
Position

Ray Projector Stowed Position
(Note 2)

Rear

Projection Screens
(approx 80"w x 60"h)

Interior Rear View

Mind Control Projector
in center of Ceiling

See Note 3

Pedestal (up)

Elevator Platform (up)

Elevator Shaft

Elevator Entrance (far side)

Elevator Platform (down)

Pedestal (down)

placed up against the outside of the pylon, the elevator's edge would conflict with the center line of the saucer and not allow space for a mechanism to rotate and lower the ray projector into place. With the small connecting passageway, the projector can be placed so it can use the same space in the center of the saucer.

The fact that the projector must rotate out of the way for the elevator to work can be solved. As long as the elevator platform is in the "up" position, ample room exists for the ray projector to be rotated into the center of the pylon, lowered into position, and fired. The elevator wall that the humans see as they are being lifted up into the main domed chamber is most likely a rotating door in its closed position. That door moves out of the way around the elevator shaft to allow the projector to use the elevator shaft and emerge out the bottom of the saucer (Figure 7).

Quickspec: Alien Flying Saucer

Vehicle Morphology	Saucer
Year	1956
Medium	Theatrical Film
Designer	Ray Harryhausen
Height (pylon up)	28.2 ft (8.6 m)
Height (pylon down)	41.5 ft (12.6 m)
Diameter	86.5 ft (26.4 m)

With this arrangement, if one ignores the excess ceiling height and other Hollywood stage economies, the dome room does fit within the forty-foot saucer dome's dimension and can combine a 12 ft elevator platform and a ray projector within the pylon space. Considering the shortage of both money and time they were working under, Harryhausen came up with a pretty ingenious design.

Another implication is that, although the upper dome room can fit within the saucer's outside dimensions, there is no room for the aliens to talk to the humans as we see them doing in Figure 5. That is, there is no space left for a small passageway for them to stand in without being outside the saucer dome's wall. Therefore, they are not really there. Most likely, the images of the aliens that we see on the wall are holographic projections--possibly from the thought control device that descends from the ceiling.

(Note: These theories are all logical derivations of the saucer's physical design. This is the way it *could* have worked. No evidence is known to show how it *did* work.)

A final discovery was made during the review of this saucer. While watching the counter-rotating rings to ascertain the direction of rotation, it was observed that they do rotate as indicated in the notes on the data drawing most of the time, i.e., clockwise on the top, counter clockwise on the bottom when seen from a side view. In several scenes however, they were seen to rotate the other way, for example, during the final approach and landing at Skyhook. There were also a couple of isolated seconds where they actually stopped rotating completely.

I mention these discrepancies not to criticize Harryhausen's work but to emphasize the tremendous difficulty he faced in creating and animating all of these sequences *single-handedly*, over the film's entire production schedule. No one was there (but him) to check his continuity and consistency. Furthermore, none of these minor flaws are seen when one views the finished film in normal space-time. The fact remains that this is handiwork in the tradition of the great American artists. It only serves to increase one's appreciation of the creativity, dedication, and skill of one of the greatest early animators, Ray Harryhausen, someone who could also design a great little flying Saucer.

Modelers' Note

Many models of the ship and aliens from the film are available, most of them resin "garage kits." Monsters in Motion is a good source for many of these kits, including one of just the alien figure (far right).

Skyhook models (a very appropriate name in this case!) makes several versions. One is of just the saucer at approximately 1/96 scale, shown here by itself (left) and on a scratch-built base representing the White House lawn. They also have a mini diorama of the saucer slicing into the Washington Monument. This is the whole kit, all three pieces of it! The larger circle is actually the base of the monument, the saucer is the small circle.

Epilog

Earth vs. the Flying Saucers was a surprise hit. As is the fate of most popular movies, it was imitated and plagiarized for years. While the *C57-D* from *Forbidden Planet* may have held the technological edge for movie saucers, Harryhausen's creations from this film went much further in crystallizing the public's perception of what "real" saucers looked like. We've already mentioned that Orson Welles used footage from it in his 1974 exposé film *F for Fake*. Snippets of the saucers flying have shown up in TV shows almost continuously since the '50s, from the Twilight Zone's black comedy episode *To Serve Man* (where it is used as the Kanamit ship taking off), to the recent Discovery Channel program *Unsolved History* (November 2003), an examination of the whole UFO phenomenon.

The 1962 opus *3 Stooges in Orbit* used props from *Forbidden Planet* to illustrate the Martian's hand-held technology, but when they wanted to show their major weapons systems, they used a clip of Harryhausen's saucers over the Potomac.

For a film that was itself exploitative and derivative of earlier alien-invasion films, it's only fitting that *Earth vs. the Flying Saucers* was remade and parodied by other filmmakers, although it took them a few decades to get around to it. *Independence Day* (1996) was a synthesis of both this film and *War of the Worlds*. With shots of the alien saucers hovering over different world capitals, and the destruction of Washington DC landmark structures, there's no doubt where the source inspiration was from.

In Tim Burton's genre homage/spoof, *Mars Attacks!* later that same year, the saucers were deliberate copies; right down to the counter-rotating discs set into the rim, although now they took up the whole rim with no stationary outer portion. There were some other detail changes, though, made possible by the large budget. For one thing, the saucers have a real folding/retracting landing gear instead of the simple pylon. The boarding ramp that emerges from the underside seems to be halfway between the *C57-D*'s stair ramp and the magically emerging ramp on Klaatu's ship.

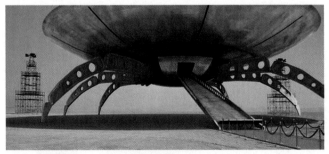
©Warner Bros.

Thanks to the miracle of computer generated imagery (CGI), the Martian saucer fleet consists of hundreds of ships rising off of Mars and heading towards Earth (visible as a globe in the Martian sky) during the opening credits. While the film starts out in a *Day the Earth Stood Still* vein with a lone ship landing, this time in the Nevada desert, and the aliens emerging claiming to "come in peace," it quickly turns into a bloodbath (both red and green). And, you guessed it, by the end of the film they are laying waste to Washington DC, right down to toppling the Washington Monument on top of tourists (Cub Scouts!) trying to run away. Harryhausen, though, didn't seem especially impressed by the homage. "I knocked down the Washington Monument long before *Mars Attacks!*" he said.[50]

©Warner Bros.

Earth vs. the Flying Saucers is a significant and influential film, made with almost no budget; and the things that made it memorable and significant were almost completely the work of one talented man. A pretty good performance by a "weak sister," wouldn't you say?

[Author's Postscript: After completing this chapter, I sent review copies to Ray Harryhausen and his agent, Arnold Kunert. Arnold had a few minor corrections plus the following note. I'm very happy that I was able to include it - JH]

Something I hope you can include in your chapter is a "tip of the hat" to screenwriter Bernard Gordon (listed in the original credits under his blacklisted pseudonym Raymond T. Marcus), who did the final draft of the screenplay and is most responsible for the majority of memorable scenes and lines of dialogue in the film. It was Bernie who came up with "Infinitely Indexed Memory Bank," a neat term for what we would eventually call a computer or, as you point out, the World Wide Web.

Because Bernie was blacklisted in the 1950s, he never received credit for writing the script during his lifetime.

Shortly after his death last year, I was able to get permission from the Writers Guild of America and the Legal Department at Sony Pictures to put Bernard Gordon's name into the front credits of *Earth vs. the Flying Saucers* for the first time on the DVD release of the colorized/black-and-white version of the film. We replaced the original screenwriting credit of "Raymond T. Marcus" with "Bernard Gordon."

Sadly, not a single reviewer of the new DVD has mentioned Bernard Gordon or his contribution to the *Earth vs. the Flying Saucers* screenplay.

---

[50] *An Evening With Ray Harryhausen*, 28 April 2004.

Disneyland's Flying Saucers

Here's a question for you: What do all of the other saucers in this book have in common? Some of them were built by aliens and come to Earth. Some were built by humans and travel to other planets. Some bring messages of peace while others bring conquest and destruction. But the one thing they all have in common is that they are all fictional. *You can't really ride in any of them.* In 1960, Walt Disney and his design team of "Imagineers" set out to rectify that situation.

In the early '50s, Bob Gurr was working as a designer for Ford and Kaiser-Willys. In 1954, he took on a part-time job with WED Enterprises, the company Disney had set up to build Disneyland, to design the miniature cars for the "Autopia" ride. He immediately found himself in great demand designing all sorts of vehicles for the nascent theme park, including all of the "antique" trucks and buses used on Main Street, U.S.A (which had to look old, but run with modern reliability and safety). Within a year he had joined Disney full time with the title of "Director of Special Vehicle Development."

By 1960, Gurr was working on one of the many upgrades already changing the face of the park, which Walt had said, "will never be finished." In particular, he was given the task of replacing the "Phantom Boats" attraction. These were small, powered, self-directed boats that were conceptually much like Gurr's own Autopia cars, except on water. The Phantom Boats were bothersome to park operators since the ride's throughput was low and the waterway

©Jack Janzen, Used with permission

they used took up a lot of space. The ride had actually already been removed after only a year of operation and, as part of the 1959 expansion, the Submarine Lagoon was put in the same spot. Gurr's job was to create a new, much smaller, water ride that individual guests could still operate. It was a variation on the bumper car idea, only on water. Called the "duck bumps," the vehicles were basically large inner tubes with a small pivoting motor in the center.[1] The rider would steer the "bump" around a small pond using the motor and, well, bump into other riders. This idea wasn't new to Gurr; duck bumps had previously been installed in many amusement parks around the country, and they are still popular today. Gurr's job was to create a new version in the Disney idiom for Tomorrowland.

It's not recorded how far Gurr got on the duck bumps,[2] but it came to a sudden halt one day in early 1960 when a German inventor stopped by the Studio to show off his creation, a one-man hovercraft powered by a small gasoline engine. "Walt asked me to have a look at it," recalls Gurr, "and I tried it out, but I could see nothing but danger in it."[3] As Jack Janzen relates:

> Gurr inspected the German saucer, started it up and was off. He fought to control the craft as it careened down the narrow alley between sound stages. He soon got the feel of the controls, as he blew dust on everyone that came out to see what all the noise was about. Though it handled well, and was truly a thrill to ride, the hovercraft was rejected. Individually powered [hovercraft] could fail, blocking the other riders and the high speed, razor sharp fan blades were unsuitable for the public.[4]

---

[1] *The "E" Ticket* #8, Winter 1990, pg. 5.

[2] An updated variation of the duck bump ride was eventually installed 40 years later in the Tokyo "DisneySeas" park.

[3] *The "E" Ticket* #39, Spring 2003, pg. 36.

[4] *The "E" Ticket* #8, pg. 6.

©Disney Enterprises, Inc., Used with permission
The Phantom boats

The Saucer Fleet

Even though that particular design was rejected, Disney loved the idea of a hovercraft ride. Not only could it be developed into a type of bumper car attraction, it fit thematically into Tomorrowland much better than the "duck bumps," especially after they started thinking of them as "flying saucers" rather than earth-bound hovercraft. Walt instructed Gurr to find some way to make it work, so he (Gurr) started talking with Arrow Development in Mountain View, California (significantly, in the context of this story, adjacent to NASA's Ames Research Center). Arrow was a manufacturer of vehicular amusement park rides, specializing in roller coasters and other tracked vehicles. They had already built many of Disneyland's attractions, most notably the Matterhorn Bobsleds (the world's first steel-tube-tracked, urethane-wheeled coaster).[5]

Joe Fowler, director of all new construction at Disneyland, gave Arrow the go-ahead to develop a prototype for a "safe" hovercraft ride, and sent the prototype that Gurr had ridden to Arrow for evaluation as a starting point. Ed Morgan, the chief designer at Arrow, had an even more dire assessment of it than Gurr did:

> There was not a question that it would work. We got to thinking that there was gasoline in each of these vehicles, and a tremendous air movement. What would happen [if] somehow, no matter how remote, you'd get some gasoline on the floor and it gets touched off by some kind of spark. You'd have a blowtorch. We would have cremated everybody![6]

The design that Ed and his partner Karl Bacon came up with to solve these problems and make a safe ride was truly inspired. The major safety concern, outside of the admittedly remote possibility of fire, was the spinning fan blades, and the most worrisome failure mode was an individual craft failing and becoming a maneuvering hazard to the others. Morgan solved both of these problems by moving the motive power off of the individual vehicles, and placing it under the floor. He took his inspiration from air hockey, a game where a blower under the table forces air through pinholes in the tabletop allowing a puck to glide almost completely frictionless on a thin cushion of air.

A laughing Walt Disney examines the prototype saucer floor at Arrow with Joe Fowler (in hat), Dick Irvine and Dick Nunis in 1961. Beyond Nunis is one of the huge scroll case blowers that supplied air for the ride. Stacked up behind them are some fiberglass Autopia car bodies under construction.

Of course, a fully loaded hover-Saucer with one or two guests would weigh upwards to 500 lbs (230 Kg) so it would need a lot more than "pinholes" to supply the air to lift them. No blowers in the world could pressurize an entire arena with dozens of Saucers floating around. Arrow's solution was the "Morgan Valve" (named after Ed) which keeps the air hole in the floor closed at all times, until the Saucer gets right on top of it, then it

Uncle Walt tries out a Saucer Prototype

almost magically opens, only to re-close once the Saucer has passed. We'll be discussing this brilliantly simple design further down in the "Vehicle" section.

The Flying Saucer attraction opened on 6 August 1961 and was an immediate hit with the public. It was an "E-Ticket" ride, Disneyland's highest rating, which put it on a par with the Matterhorn Bobsleds and the Submarine Adventure. Not only was it a visceral thrill floating and skittering about the floor, but it was, and remains today, the only ride in the history of the park where you got to completely direct your own craft in any way you wanted (the closest other rides came to letting you steer were the Autopia, and the short-lived Phantom Boats,[7] but

---

[5] In addition to the Flying Saucers and Matterhorn Bobsleds, Arrow developed and built the vehicles and track systems for the following Disneyland attractions (a partial list):

- Mad Tea Party (spinning Teacup ride)
- Casey Jr. Circus Train
- Dumbo Flying Elephants
- The "Dark Rides" Snow White's Adventures, Mr. Toad's Wild Ride and Alice in Wonderland
- "it's a small world"
- Pirates of the Caribbean
- Adventure Through Inner Space
- Haunted Mansion

[6] *Roller Coasters, Flumes and Flying Saucers*, Northern Lights Publishing, 1999, pp. 108-109.

[7] The Phantom Boats have the dubious distinction of being the very first ride ever retired from Disneyland, and are known today only among hard-core Disneyland fans.

even they were restricted to allow only a small amount of left and right deviation from your predetermined path). To the operational and maintenance staffs, though, they weren't so great.

The analogy with an air hockey game only goes so far. For one thing, there's only one puck on the game table, not dozens, and that puck slides quickly and smoothly across the table. It doesn't start, stop and bounce around in the middle. With all of those Saucers out there moving and stopping at random, the airflow from the huge blowers could get unbalanced, sometimes in destructive ways. We will go into more detail about this later, but for the moment let's just say that it wasn't at all unusual for the air underneath the floor to get so unbalanced that all of the Morgan valves would open up at once, venting the whole ride with a tremendous "BOOM" that rattled the windows of the administration building which, at the time, was located behind the adjacent "Rocket to the Moon" attraction. All the Saucers would then drop down to the floor and all the riders had to be lead off so that they could begin the laborious process of re-starting the whole ride from scratch.

Then there was the noise. Even if it wasn't rattling the nearby windows, the ride was still really loud. Even well muffled, 600 HP worth of air blowers are going to make a lot of noise, and the thousands of Morgan valves made "poppity-pop" clacking noises as the Saucers glided over them.

Even though the ride remained very popular with guests, between the noise, high maintenance and unacceptable system crashes (to use a modern term), it was a major headache for Park operators. When drawing up the plans for the 1967 overhaul of Tomorrowland, there was a short list of attractions that would be staying. Unlike the Monorail, Submarines and Autopia, the Saucers weren't on it. On 5 August 1966, five years almost to the day that it opened, the ride was shut down for the last time. It was removed, along with the *Moonliner* pylon[8] during the big "tear-out" that

©Disney Enterprises, Inc., Used with permission

An aerial view from behind the Saucer ride in 1964. The large blower inlets are indicated by the red arrows. The green-roofed structure at the bottom right is the administration building that took the brunt of the ride "crashes." The two domed structures to the right are the theaters for the "Rocket to the Moon" attraction. Any large "booms" getting through their walls would only enhance that ride!

followed (although the "Rocket to the Moon" attraction itself remained after being updated and renamed "Trip to the Moon"). The large excavations for the plenums (air chambers) below the ride floor were reused to make the "Tomorrowland Stage," an entertainment center where the entire stage was stored below ground when not in use, and raised to appear "out of nowhere" at the start of a show. That lasted until the late '80s when the stage was itself replaced with the Tomorrowland Theater, a state-of-the-art special effects movie theater that originally showed the 3-D Michael Jackson film *Captain EO*, and currently features the *Honey, I Shrunk the Audience* attraction that's part movie, part Audio Animatronics and part audience involvement.

The Ride

As with most Disneyland rides, the Flying Saucer attraction was divided into two equal parts so that one could be loading/unloading while the other side was riding. There were 64 Saucer vehicles, 32 on each side. This ride was unique in that, just by its nature, each half was further subdivided.

The ride took place in a roughly circular arena 100 feet (30 meters) in diameter divided by a wall across the center that split it into two sub-arenas. Sweeping across each half of the arena was a boom that separated its 32 Saucers into two groups. While 16 of the craft were out floating and bumping into each other, the boom trapped the other 16 in two rows of eight up against the central wall separating the two arenas. The boom, plus the central wall, formed walkways that allowed access to the confined Saucers for getting guests on and off. When the Saucers were loaded with riders, the boom would sweep across the arena, simultaneously releasing the new riders from one side while gathering up

[8] See *Spaceship Handbook* for the history of this spaceship.

The Saucer Fleet

©Jack Hagerty. Used by permission from Disney Enterprises, Inc.

The author's father pilots a Saucer in 1965

the ones whose ride was over on the other. Cast Members (ride attendants), riding on the boom, used poles to help push the Saucers into two neat rows during the last few feet. There were at least 32 Saucers active on the ride at any one time, but that increased to 48 when a boom was sweeping. Since the two booms rarely moved at the same time, they almost never had all 64 Saucers floating at once.

Piloting a Saucer was quite intuitive, although there were subtleties that required practice to maximize your performance, and be the best Saucer jockey on the ride. If you sat with your weight directly over the center of the Saucer, it would sit level and float motionless since the air from underneath would be escaping equally around the whole rim. When you leaned to one side, the Saucer would tilt, thus narrowing the gap between the rim and the floor on that side, while simultaneously widening the gap on the other side. The wider gap meant more air would escape from that side while less escaped from the narrow side. This created more thrust on the wide side, which Sir Isaac says will push you in the other direction, i.e. in the direction you were leaning. The result is that the Saucer moved in whichever direction you leaned. Very simple.

If you were to lean too far or too quickly, however, you could run the "down side" of the rim into the floor, stopping you cold (this happened mostly with adults since the lighter kids floated higher). Even if you avoided that, too much gap on the "up side" could exhaust the air too quickly, resulting in a loss of support, and the Saucer would drop suddenly. It wouldn't drop all the way to the floor since there was still a lot of air coming up from below, but it would over-

compress that air, resulting in the Saucer then being pushed up briefly higher than its normal ride height. It would come right back down, though, resulting in an oscillation called "hopping" or "bouncing." Many riders would get their craft hopping on purpose, just for the fun of it. This, however, had some severe effects on the ride, which we've already hinted at, and will discuss in detail in the "Vehicle" section next.

Experienced riders soon learned that the Saucers to pick were the ones near the tip of the boom, since those were released and could start maneuvering as soon as the boom started to move, while those down by the pivot had to wait until the boom was halfway across the arena before they were released. Conversely, at the end of the ride, you wanted to be near the outer wall, since those were the last Saucers to be swept up and gathered to the loading area.

The Vehicle

When discussing these Saucer vehicles, we have to also include the arena itself, since all the real action took place under the floor. The Saucers themselves were completely inert. They had only two moving parts: the lap belt that held the rider in (the same leather-and-brass-grommet belt as used on the Autopia cars), and the rider himself. The body was a fiberglass shell that was white on top leading, after an edge break, to a rounded red/orange rim. A grey rubber bumper surrounded the rim down near the lower edge. At the very bottom, mostly invisible, an inverted-bowl shaped skirt trapped the air to provide lift.

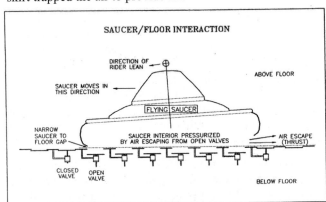

©Jack Janzen, Used with permission

The original prototype Saucers were pretty bland looking, so Gurr added some styling features to give them visual interest. A vane motif was used on the transition section leading from the flat "deck" of the Saucer up to the seat. This suggested the turbine blades and air guides used in jet engines and the inlets to real (i.e. self powered) hovercraft. Textured square panels were used around the deck itself. None of this had any functional purpose, although the textured panels helped offer some slip resistance if the surface was wet. Two small handles with round knobs were added on either side of the seat. These, too, didn't do anything directly, other than giving you something to hold onto when leaning backwards, although some Saucer jocks swore that you could finesse the craft's performance by twisting them one way or the other.

On the back of the seat was a logo showing a stylized inner solar system: a yellow Sun in the center and three white circles in blue orbits around it with a horizontal line holding them all together. The third circle out, the Earth, also had a vertical line through it forming a

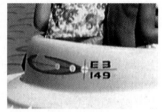

©Disney Enterprises, Inc., Used with permission

crosshair indicating where your Saucer happened to be at the moment. Continuing out from the Earth circle, and above the horizontal line was "E3" (meaning, presumably "Earth, third planet") in a very clean "space age" font. Below the line in the same font was the Saucer's three-digit serial number.

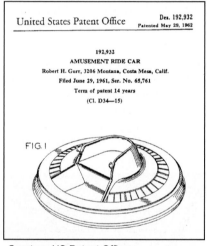

Courtesy US Patent Office

For his design work on the Saucer, Gurr was awarded US Patent D192,932, but the story really gets interesting when we go down below the floor.

As we've already mentioned, there were no blowers in the world powerful enough to pressurize the entire 8,000 ft^2 (730 m^2) arena if the 7" (18 cm) diameter vents were just open holes like on an air hockey game. Even if there were, no one would want to ride on something with that sort of vertical hurricane going on all the time!

To control the flow of air, each of the 18,000 holes had its own valve to control the airflow. The valve's job was to open up only when a Saucer was directly above it, and remain closed the rest of the time, thus conserving air until it's directly needed. This seems almost magical at first. What kind of sensor was used to detect the presence of a Saucer, and how was the valve actuated so quickly? The design shows the brilliance of simplicity.

As we can see in this drawing, the valve consists of a disc that fits into the underside of the hole it has to cover. The disc is attached to a spring that is trying to pull it open. Pressure from the air beneath the floor, though, keeps the valve shut when there is nothing above it in the arena (i.e. just regular atmospheric pressure). But when a Saucer passes over, it has that high-pressure bubble of air underneath supporting it. When it gets over a valve, the pressure

is roughly equal on both sides, thus the only force on the valve comes from the spring's downward pull. That pulls the valve open, adding its air to the underside of the Saucer. When the Saucer leaves, the pressure on top goes back down to atmospheric, so the greater pressure below in the plenum (air chamber) forces the valve closed again.

You'll notice that the valves don't seal tightly, but allow a small amount of air to bleed around the edge, even when closed. This was done for two reasons. First, it improved valve speed by giving each of them a "head start" towards opening by adding a small amount of air to the Saucer's underside even with the valve closed. Second, it helped prevent the valve sticking closed when the valve rim got dirty or rusty by keeping deposits (especially water) from settling on the rim and getting squashed into the floor by a tightly closing valve.

One consequence of this design is that the valves are all open when the blowers are off. You couldn't just turn the blowers on to close the valves since that would be trying to pressurize all 18,000 holes at once, which we already know couldn't be done practically. To make the ride workable, the floor was divided into a series of narrow plenums arranged side-by-side, all feeding from a central corridor. The corridor was supplied by four 150 horsepower blowers, two at each end blowing towards the center. Each plenum was only six valves wide, and air entering it was controlled by a door that separated it from the central feed passage.

©Disney Enterprises, Inc., Used with permission

Central pressurized corridor with plenum doors leading to the sides

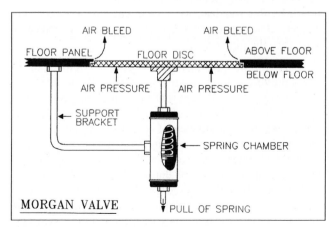

©Jack Janzen, Used with permission

The Saucer Fleet

The view down an individual Plenum with a hundred or so of the thousands of Morgan Valves on the ride.

To start the ride, the blowers were cranked up with all the control doors closed. The doors would be opened one at a time, pressurizing that plenum and closing all of its Morgan valves. Once the valves were closed and the plenum pressurized, they could move on to the next one. Eventually, the whole floor would be up to pressure with all of the valves closed (except the strip directly under the Saucers in the loading area, which were kept unpressurized so that the Saucers would sit still on the floor while loading and unloading). Once the riders were seated, that door would be opened as well, lifting the Saucers and closing the valves not directly under a Saucer.

This all worked well as long as the Saucers behaved themselves and just glided across the floor. But remember the "hopping" and "bouncing" mentioned earlier? To understand why this was bad, we have to examine a musical instrument invented by the Romans a couple of thousand years ago: the pipe organ.

Each of the plenums was a narrow, rigid channel with a moving column of air in it, essentially no different than an organ pipe, except that the "note" played by this pipe would have been well below the range of human hearing, having a frequency of only a few cycles per second. As it turns out, a hopping Saucer bounces up and down a few times a second as well, the actual frequency depending on the weight of the rider(s). Each time a Saucer bounced, it sent a pressure pulse back down through the floor into the plenum. If the Saucer was "hopping" at close to the natural frequency of the plenum, a standing wave could be generated (i.e. causing the plenum to "play" its "note").

Usually these waves just damped themselves out and steady airflow resumed. If several Saucers were hopping at the right speed though, the standing wave could become divergent and the amplitude would continue to increase rather than damp out (the "note" would get louder rather than softer). Remember that sound waves are just like ripples on a pond. They have alternating crests (higher pressure) and troughs (lower pressure). If a wave was divergent, the crests kept getting higher and the troughs lower. If the pressure in the trough got low enough, there wouldn't be enough to hold the Morgan valves closed, and a whole section of them would open up. That would vent the plenum enough to open all the valves, which, in turn, would vent the central feed corridor connecting all of the plenums. The end result is that in a matter of a few seconds, every valve in the entire arena floor would open up, completely venting the ride with the huge "BOOM" mentioned earlier. At that point there was nothing to do but lead everybody off the arena, give them their "E-Ticket" back, and begin the tedious process of re-starting the ride.

This makes it sound like a lot of coincidental things have to happen in the right order before the ride would crash, and, actually, that's correct. But with three to four dozen Saucers out there bouncing around at random, that particular set of circumstances happened once or twice per day. The irony is that it wouldn't have happened at

The ride plenums were large enough for maintenance men to walk through

Disneyland Flying Saucer

1/25 scale
Dimensions in Inches
© 2005 by Jon C. Rogers

Sources:
The "E" Ticket, Winter 1990, Spring
2003
*Roller Coasters, Flumes and Flying
Saucers,* Northern Light Pubs, 1999
Robert Gurr, Inventor, US Patent D192,932

Notes:
1. The seat is part of the molded upper body shell. At times
   3M gray non-skid tape was added.
2. 3M gray non-skid tape was also used as Floor mat material.
3. Control knobs color varied. Different colors, such as Black,
   White, and Orange have been identified.
4. A Seat belt across the lap was included (Not shown).
5. The Logo was only present for the first few seasons. The top
   number was constant; the bottom number varied.

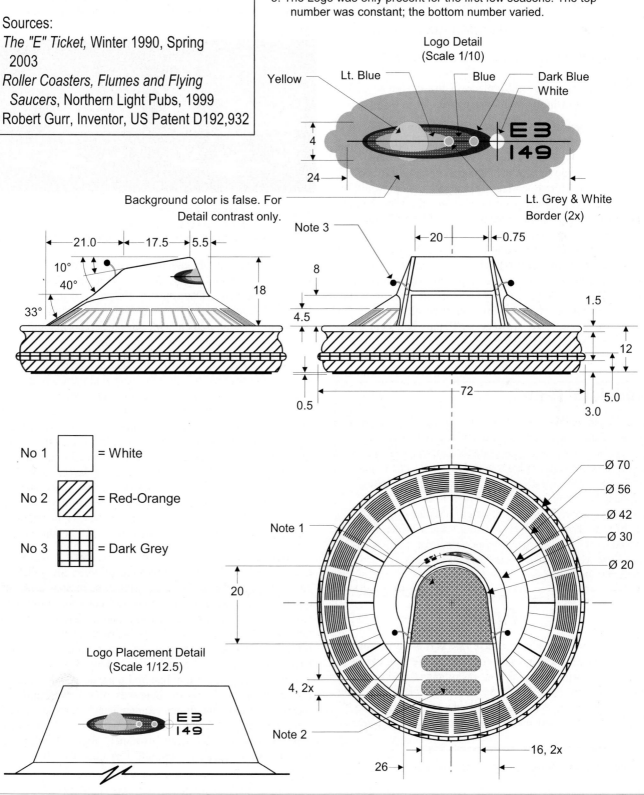

Logo Detail
(Scale 1/10)

Yellow   Lt. Blue   Blue   Dark Blue
White

Lt. Grey & White
Border (2x)

Background color is false. For
Detail contrast only.

No 1 = White
No 2 = Red-Orange
No 3 = Dark Grey

Logo Placement Detail
(Scale 1/12.5)

The Saucer Fleet

Quickspec: Disneyland Flying Saucers

Vehicle Morphology.............Saucer

Year ....1961 to 1966

Medium...............Theme Park Ride

Designer..........Gurr/Morgan/Bacon

Diameter...........................6 ft (2 m)

all if not for a change they made in the design to make things easier for the maintenance people.

In the original prototype, the plenums were five feet (1.5 meters) wide, just like in the final ride, but they were only about a foot (30 cm) deep.[9] This was sufficiently large to conduct all the air necessary to run the ride, but as Ed Morgan correctly anticipated, there would be a lot of maintenance involved in servicing all of those thousands of valves. To give better access, and keep the maintenance crews from having to move around on mechanic's creepers, he suggested making the plenums in the final ride over seven feet (two meters) deep so the crews could walk around underneath the arena unimpeded. The natural frequency of the prototype plenum was in the tens of cycles per second range, an order of magnitude higher than the "bounce frequency" of the saucers. If they had kept them the original size (or at least much smaller than they finally built it) the problem would never have happened.

Modelers' Note

To the best of our knowledge, no one has ever produced a model of the Disneyland Saucer ride, or the individual Saucers themselves. That makes the opportunity here unique! If you start with an air hockey table, it wouldn't be too hard to "dress up" a puck with some vacu-formed topside details to make a working Saucer. In fact, the amount of air in the table could probably support a much larger "Saucer" than the standard 2-inch (5 cm) diameter puck, if you made it a little lighter. It would seem a fully functional Flying Saucer ride model, complete with working booms, is not outside of the reach of the dedicated Disneyphile.

Epilog

The Disneyland Flying Saucer ride was a typical case of not knowing what you have until it's gone. "We didn't know it was a classic when the attraction was running," said Gurr in a recent interview. "The Flying Saucers became legendary long after it was gone, and nobody thought to save anything."[10] Legendary is a good way of describing it, if the reminis-

cences of park guests are any indication. Ask anyone who visited Disneyland between late 1961 and 1966 about the Saucers, and you will see their eyes get that faraway look as they remember. It was bumper cars of the future, that slightly sociopathic activity combined with the deeply significant social icon of the saucer. It was "floating on air" in the most literal sense, while being free to direct your craft wherever you wanted.

L   R

Image courtesy of Elliot Swanson
Used by permission from Disney Enterprises, Inc.
A stereo view of the Saucer ride circa 1962

The influence has endured, even if the ride is not well known to the general public. When Robert Reynolds wrote his history of the Arrow Development company, he titled it *Roller Coasters, Flumes and Flying Saucers*,[11] even though of the dozens of large amusement park rides the company built, there was only one of them that had "flying saucers."

There's even an urban legend that has grown up around the ride. On several web sites you can find the claim that the "Morgan valves" were actually built for the second stage of the Jupiter missile, and that the manufacturer had grossly overestimated the number that the government would buy, thus had thousands and thousands of them sitting in a warehouse. Arrow Development, the story goes, somehow found out about them and bought the lot dirt cheap for use in the Saucer arena floor. When they went to refurbish the ride a few years later, the manufacturer didn't have any left, and to re-start the production line was prohibitively expensive, so that's why the ride was shut down. Well, let's see what we can do to dispel this myth.

Starting from the end, we have seen that the reason the ride was removed was a combination of noise, low reliability and high maintenance. But even with those reasons aside, it's quite clear that the Morgan valves did not come from aerospace. The author, having a background in that industry, noticed several things about the legend that don't ring true. First off, the Jupiter was an intermediate range missile that didn't have a second stage. Second, even a fully deployed missile system requires only a few hundred missiles, so why would a manufacturer build tens of thousands of valves? Third, and most significantly, the design requirements for aerospace hardware are vastly different than for amusement park ride hardware. Missile pieces require extremely high reliability, extremely low weight, but durability isn't important; if it works

---

[9] *Roller Coasters, Flumes and Flying Saucers*, pg. 113.

[10] *The "E" Ticket* #39, pg. 41.

[11] Northern Lights Publishing, 1999, ISBN 0-9657353-5-4

©Jack Janzen, Used with permission
Bob Gurr in 2003

for an hour, that's all that's needed. Ride hardware also needs extremely high reliability, but weight isn't important (especially for something attached to a steel floor), but durability is extremely important. The piece must keep working for years and years. To anyone who has worked in the aerospace industry, it's obvious that something designed for use in a missile wouldn't last a week in a theme park ride.

To verify this, we asked Bob Gurr directly about the valves. He replied: "Ed Morgan and Karl Bacon invented that valve just for Disneyland. Writers make up all kinds of stories...the Jupiter deal is an example."[12]

Will they ever bring the Saucers back? It's possible, but not at Disneyland where space has always been extremely limited. "People have talked about it over the years," Gurr continues, "but when it went out of service, everybody breathed a sigh of relief. The problem [was] an air control issue and we just didn't have the technology at the time."[13]

But what about one of the other Disney parks in where there is more space available? Gurr speculates on what would have to be done to make the ride successful. "What was needed

©Disney Enterprises, Inc., Used with permission
View of the Saucers' rarely seen bottom skirt that holds the air in. Note the increased gap to the left as these riders lean to the right.

were electronic controls and high speed doors so that the flow of air could have been monitored and damped. With a few electronic feedbacks, and a control system which could have worked the doors quickly, the ride could have continued. The whole thing could probably be operated with one PC today."

Ed Morgan concurs: "They keep talking about redoing that ride. We were talking about it this year [1998] at the trade show again. I told them that Karl and I are not getting any younger, and we're the guys who hold the knowledge to bring this thing off. If you want to do it, do it soon."[14]

Author's postscript

One of the reasons that this book took so long to complete is that new material kept popping up after I thought that I was finished with a chapter, requiring constant rewriting and editing. But none of the others came close to matching this 11th hour addition in either timing or impact.

After this chapter had been completely written, formatted and shipped off to Disney Worldwide Publishing for review and to secure permission for the images, I happened to visit Disney's California Adventure (DCA, the new theme park built in Disneyland's former parking lot) while chaperoning my daughter's high school orchestra, which was performing there. Talking to the Cast Members (employees), as I always do, I learned something that was at the same time heartening and annoying with respect to this chapter.

In 2007 Disney started a billion dollar makeover of DCA that won't be completed until 2012. This will upgrade many of the sections to theme areas related to Pixar films and the largest of these will be built around the movie *Cars*. With the working name of "Cars Land," I learned was that one of the planned attractions would be an air cushion ride centered on Luigi, the excitable Fiat 500 who runs the tire store in the film's imaginary town of Radiator Springs.

After returning home, I went back into research mode and tried to contact anyone at Walt Disney Imagineering (WDI) that might be able to shed some light on this. After being passed off between four or five offices, all of whom were extraordinarily polite while claiming not to know anything, I wound up with the office of media relations. This was a little different, in that the response was the classic "I can neither confirm nor deny that rumor" mode. Of course, she didn't say it exactly that way, it was more like, "Where did you hear that? We haven't released any of that yet." She said that she would see what she could do, but I had the feeling this book would be out and into its second printing by the time I heard anything.

Moving outward from the official sources, I checked with several of the Disneyland fan websites and instantly found a description of the new attraction that was quite specific, if a bit light on details. It was soon obvious that these sites were all quoting the same article from the LA Times online edition. In the article, they confirmed that it involves Luigi, and gives the name of the attraction:

---

[12] Interview, 18 Apr 2003.

[13] *The "E" Ticket* #39, pg. 41.

[14] Roller Coasters, Flumes and Flying Saucers, pg. 116.

The Saucer Fleet

Luigi's Flying Tires is an update of Tomorrowland's original Flying Saucers futuristic bumper-car ride. Instead of piloting a flying saucer, riders in the new Luigi ride will steer a big-rig truck tire through tilting towers of retreads.[15]

The article is accompanied by some deliberately low-rez conceptual art, which, despite the assertion by the media relations office that none of this had been released, is credited to "Disney."

Further confirming the name of the attraction is the fact that Disney has already filed a trademark for it:

> On March 18, 2008, Disney Enterprises, Inc. filed to protect the trademark *Luigi's Flying Tires* in relation to "amusement park services." We suspect that *Luigi's Flying Tires* will be what Disney previously announced as *Luigi's Roamin' Tires*.[16]

©Disney Enterprises, Inc., Used with permission

OK, so where are we with this exciting possibility that the saucer ride will be brought back? Here's what's known for certain:

- Disney is doing a massive update of DCA
- The largest theme area of this upgrade will be dedicated to the movie *Cars*.
- One of the proposed attractions in Cars Land is *Luigi's Flying Tires*, which had a previous working name of "Luigi's Roamin' Tires."

Here's what we don't know:

- Whether this attraction has been "green lighted." Just because they've filed a trademark on the name doesn't mean it will be done now.
- What the technology behind the ride will be, if they will use the under-floor air source, or if they've found some other way to levitate the, uh, vehicles.
- If the nature of the ride will be bumper cars like before, or a one-at-a-time navigation of an obstacle course.

We can speculate on that last point based on the conceptual art published in the LA Times article. To the right is a map of DCA showing the planned upgrades. Cars Land is the entire bottom right corner of the park. The tire ride (still called "Luigi's Roamin' Tires" on this map) is on the right edge of the theme area. From the tiny riders' eye view of the attraction (above), it sure looks like the old Flying Saucers with lots of riders bouncing around at random. On the other hand, looking at the ride layout illustration (right center), the riders seem to be moving around enormously tall stacks of precariously tilting tires. While there is a striped barrier in the middle to separate the arena into the two ride areas, it's obvious that you can't have a sweeping boom to end the ride when the floor's a rectangle and there are permanent obstacles on the floor.

Speaking of the vehicle, from this illustration it's quite obviously a retread (pun intended) of Bob Gurr's saucers.

©Disney Enterprises, Inc., Used with permission

To try and get a handle on the technology behind the ride, I contacted Gurr again to see if he was involved in any way, even as a consultant. "They told me about it," Bob said. "They've been itching to get some sort of air cushion ride back into the park for years. All I know is that it's not an independent design. Some Danish company has perfected this type of ride over the years, so they'll probably be doing it."[17]

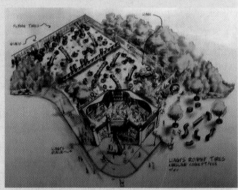

He then gave me the name of a project manager at WDI who would know the specifics. I started the usual contact dance with e-mails, voice mails, etc., but with the publishing deadline upon me I never had a chance to complete this round, so we'll have to leave it with speculation.

©Disney Enterprises, Inc., Used with permission

The next four years will be interesting. You can be sure that if they build this, I'll be first in line on opening day!

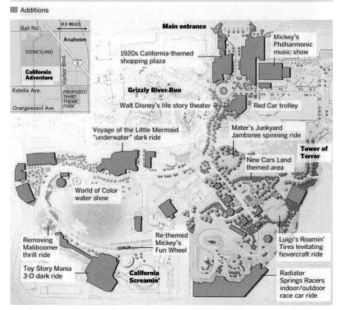

©Disney Enterprises, Inc., Used with permission

---

[15] http://travel.latimes.com/daily-deal-blog, 14 November 2007

[16] Ibid.

[17] Interview, 21 May 2008

Lost in Space

If it weren't for *Star Trek*, The CBS series *Lost in Space* would probably be remembered today as **the** premier outer space Science Fiction show of the '60s. With a one year head start over the Roddenberry creation, this Irwin Allen production was well on its way to cult status before Kirk & Co. were even out of the starting gate. Ironically Roddenberry had pitched *Star Trek* to CBS executives[1] before he tried NBC, the network that eventually bought it. CBS rejected it, though, because *Lost in Space* was already in the works (but not yet on the air), something that Trekkers gleefully point out as if CBS was offered the golden fleece, but turned it down and got "stuck" with a second rate kiddie show. However, even the most ardent Trekkie has to admit that "Danger! Danger, Will

©20th Century Fox Television

Robinson!" has entered the lexicon just as surely as "Beam me up, Scottie!" has; and with the added bonus that the Robot actually said it.

Lost in Space began as a comic book, *Space Family Robinson*, published by Gold Key starting in 1963.[2] The premise was based, of course, on the classic 1813 Johann Wyss novel *Der Schweizerische Robinson* (*The Swiss Family Robinson*).[3] The comic involved the Robinson family; father Craig, mother June, children Tim and Tam, dog Clancy and parrot Yakker, leaving Earth to conduct research on a space station in the

Asteroid Belt. The station is hit by a violent cosmic explosion that throws it off course by 30,000 miles (50,000 Km), a distance hardly worth mentioning in real cosmic terms, but so far off course in a comic book that the family becomes "lost in space." While attempting to get back, they visit and explore many different planets with exotic names like Orious, Kliklag Norica, Kormat and Zytrox.[4] The comic lasted 59 issues, and starting with #15 was re-titled *Space Family Robinson Lost in Space*, then, with #37, it changed again to *Space Family Robinson, Lost in Space, On Space Station One*". That's quite a mouthful, so beginning with issue # 38 the cover was redesigned so that "Lost in Space" were the largest words in the title, but this very long title was used to the end. The comic book series continued right through the three-year run of the TV show and kept on going into the early '70s, but there was never any commonality between the two in terms of characters or plots.

Irwin Allen, fresh off of the success of *Voyage to the Bottom of the Sea* (both the 1961 film and the TV series just going into production), was looking for a property to develop as his next project. Outer space was a natural subject for the early '60s so he combined that with the family oriented action-adventure of the Wyss novel. Like the comic book, his original title for the series was *Space Family Robinson*. It's not clear if he knew about the comic book when he started (there was never a formal licensing agreement with Gold Key), but some sort of informal understanding was reached along the way. Disney, however, thought the title too close to their popular movie, *Swiss Family Robinson* (1960), also, of course, based on the Wyss novel, and threatened legal action if he didn't change it.[5] Allen changed the title to *Lost in Space*, with which Disney had no problem. Note that the comic book changed its title to the same thing at about the same time, indicating that some sort of correspondence was going on between the two productions.

In Allen's version the characters were named John (father), Maureen (mother), Judy (daughter) and Will (son). A second daughter, Penny, was inserted between the two siblings. The Robinson surname was kept, naturally, to

---

[1] Steven Whitfield, *The Making of Star Trek*, Ballantine, 1968, pp. 37, 38.

[2] Jeff Rovin, *Aliens, Robots, and Spaceships*, Facts on File, Inc., 1995, pg. 231.

[3] "Robinson" in this case was more than just a surname. Going back nearly a century earlier, Daniel Defoe's *Robinson Crusoe* (1726) was so popular that any adventure story, especially if it was set in an exotic locale, was called a "robinsonaid." An alternative meaning of the title is, therefore, "the survival adventures of a Swiss family in an exotic place," which applies quite well to the comic book and TV show as well (substituting "Space" for "Swiss").

[4] This story sounds more like the premise of *Space 1999*.

[5] Idak episode guide: http://members.ozemail.com.au/~idak/lis/

maintain some connection with the original title. Allen also kept the story closer to the Wyss novel's premise. Rather than just orbiting the sun in a space station, this family was leaving the Solar System completely to start a colony on another planet. This required yet another addition to the cast, a non-family member, Don West, to pilot the spaceship.

To select actors for the show, Allen cast his net pretty wide over episodic TV. For John Robinson he picked Guy Williams, still very well known from playing the title character in Disney's *Zorro*. For mother Maureen he selected June Lockhart, already known to millions as the mom in the TV version of *Lassie*. Billy Mumy was cast as Will Robinson. Mumy, only 10 at the time, had already logged more than 50 television appearances, mostly in high profile series like *Twilight Zone* and *Alfred Hitchcock*. To play Will's slightly older sister, Penny, he chose Angela Cartwright. Having appeared in only a dozen TV shows, she would appear to be less experienced than Mumy, except that one of her credits was as a regular character on a major series, *Make Room for Daddy*, as Danny Thomas's daughter, Linda. She also had a major big screen credit having just completed *The Sound of Music* where she portrayed Brigitta von Trapp. Cartwright is often confused with her sister, Veronica, who is the one in *Alien*, a completely different form of space monster show.

In the role of older sister, Judy, Allen selected the young actress Marta Kristen. Although Judy was a teenager, Kristen was already 20 when cast. A Finnish war orphan, she had been acting about five years and had only a dozen small screen appearances to her credit plus one theatrical role as the mermaid in *Beach Blanket Bingo* earlier that same year. To play the ship's pilot, Don West, Allen found Mark Goddard, a relatively new actor and, as it turns out, a distant cousin of rocket pioneer Robert Goddard.[6] Like Kristen, Goddard had only been acting about six years and had appeared in a couple of dozen TV shows, nearly all westerns and crime dramas, and one theatrical film, *The Monkey's Uncle*, released that year.

In the novel, the Robinson family, tired of the problems that accompany European civilization, leave to join the emerging British colony of Port Jackson in Australia. Along the way they encounter a fierce storm and the crew abandons ship, leaving them behind. They are shipwrecked on an uncharted Pacific island where, with the provisions salvaged from the wrecked ship, they start a small colony of their own. They not only survive, but actually prosper in this new land due to their ingenuity and hard work. They are eventually rescued, but the parents and two of the sons decide to stay in their new home (the middle son returns to Europe to continue his education).

Allen's pilot, which cost a record-setting $700,000, follows this story outline, allegorically at least, fairly closely. The Robinsons are selected to be the first family in space to start a new colony on Alpha Centuri (well, presumably on a planet orbiting Alpha Centuri), as a response to the growing overpopulation of Earth. They, and *Dr.* West (not yet the pilot, but a doctor of "Planetology" and a colleague of John's) take off from Earth in the *Gemini XII*.[7] Shortly after takeoff, they encounter a meteor storm that knocks them off course and they crash land on an uncharted planet. The *Gemini XII* is wrecked beyond repair in the landing. Being in the upper latitudes of the planet (where the temperature drops to -150°F (-100°C) at night), they salvage what they can, pack it into the tracked surface exploration vehicle called the Chariot, and head south. They eventually reach the more temperate zone near the planet's equator, but not before having many adventures with local aliens and natural obstacles. This is where the pilot ends, and presumably from here they would go on to build their colony using their ingenuity and hard work while awaiting rescue. An interesting footnote for genre fans, the music Allen chose for the score (pilots almost always use existing music to save money) was pieces of various Bernard Hermann scores, most appropriately the main title from *The Day the Earth Stood Still* for the title and liftoff sequences.

The screening for CBS executives didn't go exactly as Allen though it would:

> When the pilot film was completed, Allen proclaimed it as his best work ever. He showed the film to CBS but when the

©20th Century Fox Television

The intrepid cast clockwise from upper right: Guy Williams, June Lockhart, Angela Cartwright, Billy Mumy, The Robot, Jonathan Harris, Marta Kristen and Mark Goddard

---

[6] Internet Movie Database: www.imdb.com/name/nm0323813/bio

[7] This was, of course, almost two years before the real *Gemini XII* flew. We'll be discussing this more later.

high-brow executives began laughing at the film, Allen was horrified. Furious, he bolted from his chair to stop the screening. Story editor Anthony Wilson pulled Allen back down. "Irwin, they love it," Wilson whispered urgently to his irate friend. Laughter or not, the brass was enjoying the show. The show sold.[8]

This pilot sold CBS on the series, but they felt some more "fine tuning" was needed. They thought natural disasters and encounters with aliens did not provide enough dramatic tension and the stories would basically condense down to our stalwart family fighting off the peril of the week.

To provide some internal conflict, Anthony Wilson (story editor) suggested they add another character.[9] Series writer, Shimon Wincelberg, came up with Dr. Zachary Smith, a medical doctor and undercover agent whose mission it is to make sure the Robinson expedition fails. Who Smith works for, or why they want the mission terminated is never explained, but it doesn't really matter since the point is moot. Once they leave Earth, they're not going back.

Smith was played to perfection by veteran character actor Jonathan Harris. There is considerable controversy among *Space* fans as to whether Dr. Smith was to be killed off sometime in the first few episodes after he had served his purpose (i.e. sabotaged the mission and gotten the family "lost").

©20th Century Fox Television

This would have made sense since it would leave the screenwriters unfettered when using the footage from the pilot, which did not contain Smith. But according to Wincelberg, Smith was always supposed to be a continuing character, "valuable as a source of additional storylines and constant skullduggery."[10] Harris, though, remembers it differently:

> The original Smith was a deep, dark, scowling villain and I hated him. I thought they might have to kill me off in five episodes because he was just so damn rotten. I decided to take a chance and sneak in the thing for which I am justly famous, comedic villainy.
>
> I did it hoping that Irwin wouldn't notice, but he noticed! One day he came to my dressing room with his finger waving at me under my nose. He said, "I know what you're doing." I said, "Yes." He said, "Do more," and stormed out. Irwin had no loyalty to anything but the ratings, and the crazier I got, the higher the ratings went up. It worked.[11]

---

[8] Mark Phillips, *The History of TV's Lost in Space*, https://writer.zoho.com/public/ziffen63/Lost-In-Space-story1/fullpage

[9] *Starlog* #57, April 1982, pg. 36.

[10] *Starlog* #96, July 1985, pg. 78.

[11] *Starlog* #248, March 1998, pg. 29.

So, once the series was out of the "pilot" phase, and into the "monster of the week" era, the character of Smith changed rapidly from the cold, efficient enemy agent, to the bumbling, cowardly, return-to-Earth-at-any-cost schemer we loved to hate.

Aiding Smith in his acts of evil, at least at first, was the final new cast member, the Model B-9 Environmental Control Robot who, through the entire series, was referred to only as "Robot." Genre fans have always noticed the strong family resemblance between the Robot and Robby from *Forbidden Planet*. That's because both of them (plus the *Jupiter 2*, as we will see below in the "Vehicle" section) sprang from the fertile mind of art director Bob Kinoshita (see the *Forbidden Planet* chapter, page 157 for more on Robby's background).

A comparison of the two is quite revealing as to how close they really are. Both the Robot and Robby keep their "thinking parts" visible under clear plastic "heads." Robby had big mechanical relay contacts that flipped when he was thinking (usually just before speaking), whereas the Robot had a constellation of twinkling lights in the metal framework inside his bubble, analogous, perhaps, to neurons firing. At the top of his "head," Robby had a platform of rotating gyroscopes while the Robot had a conical spinning mirror disc. Sprouting from the sides of the "neck," Robby had a pair of ring antennae, common at the time in radio direction finders, which were constantly twirling, one vertically and one horizontally (suggesting that they were navigation devices). The Robot had two rotating paddle shaped "feelers" (both spun vertically, but one was "tall" and the other "wide"). Both the mirror disc and the feelers only rotated for the first few episodes, after which they were stationary. The reason why they were stopped varies with the source. Some say that the mechanisms broke down too often and the prop department just got tired of fixing them. Others sources maintain that they were too noisy, and required time consuming re-recording of dialog (called "looping") for any actor standing close to the Robot. They did operate the mirror and feelers one last time in the opening scene of the second season premier episode "Blast Off Into Space."

©20th Century Fox Television   ©Warner Bros.

Below the head, both had a barrel-shaped cylindrical body with their most prominent feature, the ribbed chest light that flashed in cadence with the voice when talking (Robby's was blue and the Robot's red, once the show went to color in the second season). Robby was fully anthropomorphic with legs made up of spherical joint segments so

The Saucer Fleet

©20th Century Fox Television ©Warner Bros.

[Note: Those who don't want to spoil the illusion of the Robot's personality should skip the next two paragraphs]

It took two people to bring the Robot to life. Inside the costume was veteran actor Bob May, wearing black makeup on his face to prevent the camera from seeing him, as he looked out through slots in the ribbed "collar" on top of the body (see page 160 for a similar photo with Robbie's Frankie Darro). When speaking, he operated the chest light by means of a small switch in the hand space of the left arm. The Robot's speech was very halting and mechanical at the beginning (it was May who suggested that it not use contractions, something that Bret Spiner adapted for his android character "Data" on *Star Trek, The Next Generation*), but became more fluid, without losing its mechanical quality, as the show progressed. May had to be in excellent physical condition since the "Bermuda shorts" costume weighed over 200 pounds (100 Kg), all of which he carried on his shoulders. One way to tell if May was wearing the Bermuda shorts in a shot is that he usually held his arms out to help balance the heavy suit instead of drawing them in to the normal "home" position when not in use.

that he could walk in a human, if somewhat tottering, manner. He had gripper hands consisting of three equally spaced digits (two fingers and a shorter thumb) at the end of jointed arms that had only minimal extension capability. The Robot's arm and leg mechanisms were inside pleated rubber coverings, much like the "way covers" on industrial equipment. His "hands" were two fingered pincers patterned after industrial grippers of the period which could extend three feet (~1 meter) or more from the fully retracted "home" position. In a couple of the early first season episodes, the Robot is seen playing chess, and his pincers have bent clip on extenders to allow him to grip the chess pieces. These extenders are never seen again after those episodes.

With the shift to color in the second season, the Robot got some color accents to replace the dull silver and gray of the first season: The pincer hands were painted red, the square lights in the central chest panel went from white to multiple colors (the round lights at the bottom were always colored) and the panel itself was darkened to give the lights some contrast.

While Robby walked with a lumbering gate, the Robot had at least two locomotion modes. Mostly, he drove around on his tractor tread "feet" (pulled on a wire by stage hands) although early on, especially in the very first episode, he could be seen shuffling along by moving his treads alternately. Sharp-eyed viewers will notice that in close-ups, the robot seems to walk with a human-like gate. That's because there was a "Bermuda shorts" version of the Robot costume that ended at the "knees" which allowed much faster camera setups.[12]

The Robot did have a couple of abilities that Robby didn't. First, the bubble-head could rise up about a foot indicating surprise or alarm. Second, the body could rotate 360° relative to the legs, which made it possible for him to turn around and face completely to the rear while traveling forward (Robby could only do this with his "head").

Dick Tufeld (left) and Bob May with friend in 2001

Even though it was Bob May who spoke the Robot's lines during filming and operated the chest light, it was not his voice that you hear on the show. All of the Robot's lines were dubbed in post-production by announcer Dick Tufeld. Tufeld was no stranger to Science and Science Fiction programming having narrated the first two installments of the Disney *Man in Space* series and been the announcer on *Space Patrol!*[13] On *Lost in Space*, in addition to speaking for the Robot, Tufeld also did the voiceover at the beginning of each first and second season show, "Last week as you recall…" For all their hard work, neither May nor Tufeld received a screen credit.

Lost in Space premiered on 15 September 1965 and was an immediate hit. *The New York Times* praised it for the high quality and believability of the special effects, and the serious nature of the story. The series now had its own well-integrated, character driven music by "Johnny" Williams (yes, it's the same John Williams of *Star Wars* and *Harry Potter* fame).

---

[12] *Starlog* #57, pg. 40.

[13] See *Spaceship Handbook* for descriptions of both of these shows.

The melodramatic nature of the show lasted for about five episodes, until the supply of footage from the original pilot was used up. Starting with Episode 6, "Welcome Stranger," the show started moving in a new direction. Episode 6 was the first one produced without any footage from the pilot,[14] and the difference is quite noticeable. Rather than having the *Jupiter 2* sitting atop a sizable mound of dirt (presumably pushed up during the crash landing) with a long, rubble-strewn dirt ramp leading up to the side hatch, the ship is now right down on the flat, and a mini-version of Klaatu's wedge ramp leads the short distance down to the ground.

©20th Century Fox Television

The J2 crash site shortly after landing in "Island in the Sky" (left) and later in "Welcome, Stranger"

The "stranger" in the title was a space cowboy who had been traveling around the stars for 15 years in what looks like an oversized Mercury capsule he calls "Travlin' Man" (for 15 years? What about air? Food? Propellants?). He lands near the Robinson camp after mistaking their distress radio call as a homing beacon. He spends a few days with our heroes before returning to his own ship to leave. He discovers, though, that some "space spores" he'd picked up on his travels had germinated and grown to completely engulf the ship and surrounding area.

This episode introduced two plot devices that would be used by (some would say plague) the series to the end. First, that space is absolutely brimming over with travelers who are mostly human (or at least humanoid) and stop by this particular planet on a weekly basis. But, like the castaways on Gilligan's Island, the Robinson party never quite manages to be able to leave with them, for one reason or another. Second is the "monster of the week" style plot. The aliens who aren't humanoid, in this case the spores, always manage to be something huge and roaring that have to be fended off.

While that cliché may be a bit harsh (if mostly accurate), the first season did have some very thought provoking

episodes. In "My Friend, Mr. Nobody," Penny communes with an incorporeal being, thus questioning the nature of existence and intelligence. The show paid homage to its Sci-Fi roots in its only two-part episode, "The Keeper," starring Michael Rennie (Klaatu from *The Day the Earth Stood Still*[15]) as an intergalactic zookeeper who's more interested in the Robinsons as specimens. "War of the Robots" was a sort of family reunion for the Robot when his cousin, Robby, plays the part of an evil robot intent on (what else?) capturing the Robinsons for its masters. This episode, incidentally, provided yet another link to *The Day the Earth Stood Still* in that it used Bernard Herrmann's "Gort" theme extensively for the scenes with Robby.

Perhaps the eeriest thing to happen during the first season was an astounding example of life imitating art. The episode "His Majesty Smith" aired on 16 March 1966, which is also the day that the *Gemini VIII* flight of Neil Armstrong and Dave Scott was launched. An hour or so before *Lost in Space* aired that evening; the astronauts completed the first-ever docking between two spacecraft when they mated their craft with a Lockheed *Agena* over the Pacific. Unfortunately, a short circuit in one of the Gemini's thrusters caused the capsule to start rolling uncontrollably. To save themselves, Armstrong had to turn off the main control system and switch to the thruster system normally used only for reentry. Once he did that, mission rules forced the cancellation of the mission and they had to be brought down immediately. CBS had to pre-empt *Lost in Space* in order to cover this real life drama of a crew we nearly lost in space, but this, ironically, brought howls of protest from *Space* fans!

The second season brought some major changes thanks to the influence of a couple of other programs. The most obvious, was the shift from black & white to color, which was a major event if you were one of the 20% in the country lucky (or rich) enough to have a color TV.[16] Then there was the competition from *Star Trek*, which debuted on NBC that September. Seeing that Roddenberry was taking the dramatic high ground and going for significantly more realism than *Lost in Space* did even in its first episodes, Allen decided to push the other way and move from just campy to

---

[14] Technically, Episode 2, "The Derelict," owed nothing to the pilot except for some connecting shots of the *Jupiter 2* moving through space.

[15] The connection with *The Day the Earth Stood Still* goes both ways. Jonathan Harris had a bit part in that landmark film as one of the scientists attending the meeting at the end. He can be seen in a couple of "reaction" shots, but most clearly in the "three-people" close-ups with the Russian officer to the right.

©20th Century Fox Film Corporation

[16] 1966 was the first "full color" season where every program on all three networks was broadcast in color.

The Saucer Fleet

high camp. This was guided by the other show to influence the series, *Batman*.[17] The Caped Crusader was a surprise smash hit as a mid-season replacement in the 1965-66 season. The two shows were filmed on the same lot at Fox, so there was plenty of mixing of casts and crews. The techno-pop nature of the "Bat Cave" and tech-savvy villains fit right in to this new direction for *Space*.

One thing that Allen had to do was get them off the planet. It's one thing to do an outer space show where your characters never go back into space, when you're the only game in town; but when faced with competition from a show where the ship is **always** in space and never lands, you're going look really silly. So, on the very first show of the second season, "Blast off into Space," John and Don discover that, gee, the *Jupiter 2* isn't wrecked after all, and

they manage to make it into space just ahead of the planet's total collapse. They also discover that they aren't lost, either, and get back on track for Alpha Centuri. In the second episode, "Wild Adventure," Smith (John and Don no longer call

©20th Century Fox Television

him "Doctor") gets hypnotized by an exotic green girl who lures him outside to get him to do her bidding (hmmm, an exotic green girl who hypnotizes men to get them to do what she wants. Do you think anyone on the show may have slipped into the *Trek* camp for a viewing of *The Cage*?). Rescuing Smith knocks them off course, and they are, again, lost in space.

The third season got even more updates. There was all-new, jazzier theme music to go with the revised art direction which was taken quite "mod." New costumes and themes reflected the "psychedelic" movement that was taking root in 1967. They even ran into "space hippies" (just like the *Star Trek* crew) in one episode. But in between the love beads and giant carrot monsters (from "The Great Vegetable Rebellion," everyone's favorite choice for worst episode[18]), there were some interesting and thought-provoking shows. Allen wanted more episodes actually in space, which was welcome news to fans wanting to give *Trek* a run for its money. In "Visit to a Hostile Planet" they manage to get back to Earth, but in 1947, exactly 50 years before they took

©20th Century Fox Television

off. Naturally they are reported as a flying saucer when they leave, thus being the cause of the whole saucer phenomenon (see Chapter 1). There were a couple of other time travel episodes, including one where they meet their own descendants a couple of hundred years hence, and one where John runs into his evil twin made of (gasp) antimatter. One casualty this season was the demise of the "cliffhanger" ending with its "To Be Continued" title cards.

In the first episode of the third season, "Condemned of Space," Robby the Robot returns as the warden of an automated space prison. The third episode, "Kidnapped in Space,"

introduced yet another new piece of hardware, the Space Pod. This was a small shuttle-like vehicle, remarkably similar to the Apollo Lunar Module (which was in the news quite a bit at the time). It was used for excursions between the *Jupiter 2* and other spacecraft, or down to a planet's surface when they didn't want to land the whole ship. Just where it lived inside the *J2* is ambiguous at best. The exterior shots of it being launched show it emerging from a bay at the rear of the saucer, about midway between the central drive nacelle and the outer edge.

©20th Century Fox Television

Of course, this puts its storage bay right in the middle of the lower deck living quarters. By this time, though, the producers had long since stopped worrying about such details:

The final show, although it was not supposed to be, was "Junkyard of Space" where both the Robot and the *Jupiter 2* are sold as scrap to be melted down. It seems appropriate, somehow, but it wasn't supposed to be the end. The ratings were still quite decent (certainly better than *Star Trek*'s at the end of its second season), and they started signing up the cast for a fourth season. The end came unexpectedly and quite suddenly after a production meeting. As Jonathan Harris recalls:

> We had a fourth year ahead of us at CBS. The ratings were high enough and the fan mail was still coming in from all over

---

[17] Billy Mumy interview, *Starlog* #48, July 1981, pg. 80. Repeated in TV special *Lost in Space Forever*, 1998.

[18] The Carrot Monster was played by Stanley Adams, who also played the part of Cyrano Jones in the *Star Trek* episode "The Trouble with Tribbles," conversely one of that series best liked episodes.

the world. But CBS and Fox had a meeting with Irwin and said, "You're spending too much money for the fourth year. We're going to cut the budget." He said, "Over my dead body. No cutting the budget." Irwin was that way. He loved the set and all the gadgetry and he wanted it in every episode. They wanted a big budget cut, which would have affected that. And Irwin said no. *That's* why we were cancelled.[19]

The show went into syndication where it remained quite popular. In fact, it is more popular today than ever, with a fanatical following of fans, most of whom weren't even born when the show originally aired. We'll touch on that a bit more in the Epilog.

The Story

Space does not allow us to present even a small fraction of the plot summaries for the 83 episodes produced over the three years this series was on the air. As a compromise, we will present the usual detailed account of the premier episode, then somewhat condensed versions of the next four. These were the ones that incorporated the footage from the original pilot, and actually form one long, continuous episode. They were broken at the end of each show by a campy "To Be Continued NEXT WEEK!... Same Time, Same Channel" title card, done in the antique "Playbill" lettering popular with the early movie serials. The announcer who gave the story synopsis at the beginning of each episode was Dick Tufeld (taking a break from being the voice of the Robot). This "cliffhanger" style of ending each episode continued through the second season.

Episode 1, "The Reluctant Stowaway," begins with a date, "October 16, 1997,"[20] superimposed over a shot of a large control room where white-coated technicians and engineers work at complex looking consoles. A voiceover narrator (Tufeld) tells us, "This is the beginning. This is the day. You are watching the unfolding of one of history's great adventures. The colonization of

©20th Century Fox Television

space beyond the stars." We learn further that this is the first of what they hope will be up to 10 million families leaving the Earth every year to relieve massive overpopulation.[21]

The Robinson family emerges into the glare of TV lights. We are told that they have been selected from over two million volunteers for this flight due to their "high level of scientific achievement, emotional stability and pioneering resourcefulness." They will be put into a state of suspended animation just prior to liftoff, and not be thawed out for 5½ years until they reach the atmosphere of the new planet. They are headed for a planet orbiting Alpha Centuri, "the only planet within the range of our technology."

The camera moves on board the *Jupiter 2*, for a narrated tour of the huge spaceship. It is here that we are shown Allen's intense love of hardware and gadgets. Although the camera moves all around the top deck, down the elevator, and then around the lower deck, there are no people at all to distract us from this paean to technology; just the invisible cameraman and the disembodied narrator.

With the tour over, the camera cuts to a panel on the front of a console. The panel opens and a reclining acceleration couch emerges with a man wearing an Air Force officer's uniform. He walks over to the large robot in the middle of the floor and fiddles with its innards. Just as he finishes, a security guard catches him and places him under arrest since no unauthorized

©20th Century Fox Television

personnel are allowed on board this close to takeoff. The officer identifies himself as Col. Zachary Smith, then proceeds to karate-chop the guard and toss him down the trash chute leading outside. With the guard out of the way, Smith pulls a small transmitter out of his pocket and calls "Aeolus 14 Umbra" to report that the Robot has been programmed to destroy the *Jupiter 2* eight hours into the mission.

Moving back inside Alpha Control, we see the last of the Robinson party, the young boy, getting his preflight physical, which involves sitting in a chair with a huge plasma arcs radiating behind it. An operator standing at a console, with his face darkened, twists the controls violently. "Do I pass?" asks the boy. "You'll do," says a familiar sounding voice. Stepping into the light we see that it's Col. Smith, who we now find out is the flight surgeon.

The voice of Alpha Control says it's now T-33 minutes and time for the Robinson party to board. As they walk aboard, the ship is swarming with technicians doing last minute checkouts. The Robinsons (and thus we) are introduced to their pilot, Major Don West.[22] West is really only the backup since the ship will be flown completely on automatic guidance while all the crewmembers are frozen. "If you wake up and find me driving, you'll know you're in trouble!" he quips in a nice bit of foreshadowing.

---

[19] *Starlog* #248, pg. 31.

[20] Research has turned up no significance to this date.

[21] Remember that this was during the heyday of Paul Erlich's *Population Bomb* and other doomsday predictions.

---

[22] Doesn't seem odd that they are only now meeting the pilot? After all, this is a one-way trip and they are supposed to be spending the rest of their lives with him.

The Saucer Fleet

Also back on board is Dr. Smith who, despite being arrested earlier for being "unauthorized personnel," is now allowed in to do the final medical check as they get into their freezing tubes. But rather than doing that, he sneaks down to the lower deck where a technician is finishing up the final checkout of the Robot. Seeing that he (the Robot) is secured and powered off, Smith lags behind as they leave so that he can power the Robot back up. Meanwhile, upstairs it's T-2 minutes and time for the crew to get into the freezing tubes. (Wow! Talk about waiting until literally the last minute!) They enter one at a time as ordered by Alpha control. This is actually the first time that we, the viewers, have had the family members each identified by name. Interestingly, Maureen is identified as *Dr*. Maureen Robinson, something that was left over from the pilot where she is introduced as being a "Doctor of Biochemistry from the New Mexico College of Space Medicine." Unfortunately, her degree and educated status was never referred to again in the series.

We are less than one minute to liftoff when John finally enters his tube and the freezing process begins with a strange glow from "the genius of advanced technology," according to the breathless narrator. Smith emerges from the lower deck

©20th Century Fox Television

and races to the hatch, just as it closes trapping him aboard. He tries in desperation to open the hatch, smash the front viewport, and use the radio, as Alpha Control counts down the last seconds to liftoff. He finally races back to the lower deck and gets into his hideaway acceleration couch (the one we first saw him in) just in the nick of time.

The ship lifts off in the best saucer fashion. No smoke, fire and attendant commotion of a chemical rocket launch; just a pulsating blue glow[23] as it lifts smoothly into the sky.

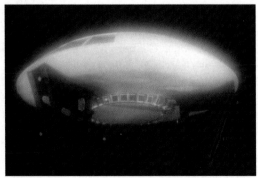

©20th Century Fox Television

Everything goes fine at first. Alpha Control announces that the ship has passed through "Maximum Dynamic Pressure,"[24] an amazingly sophisticated snippet of dialog, although the scriptwriters probably didn't understand what that really was, since stars and galaxies are already visible out the main viewport as they call it out and Max Q happens quite low in the atmosphere.

From outside, we see huge meteors start to slam the ship. The lunar tracking station reports that the "flight pattern is going negative" due to 200 pounds excess weight on board, and that they're heading right into a meteor swarm.

©20th Century Fox Television

With panels bursting into flame all around him (an Irwin Allen trademark), Smith revives West so that he can steer the ship out of the swarm. After he does so, West revives the Robinsons so that they can decide whether to continue or return to Earth. For Smith there's no question; he wants to go back. West reports that the guidance system was badly damaged by the meteors, and that he has no idea where they are. While they debate whether or not to go back, the ship's clock nears 8 hours after liftoff. Smith sees this, and in a minor panic, remembers that the Robot will start destroying things in a few minutes. He heads down to the lower deck just as West shuts off the artificial gravity in order to repair the guidance system. The connection between the a-grav and the guidance systems is not made

---

[23] At least we presume it's blue. The show was black & white for the first season, but all atomic engines glow with blue Cerenkov radiation, right?

[24] Dynamic pressure (called "Q" by aeronautical engineers) is the force of the air pressing on any vehicle as it travels through the atmosphere. It is caused by the vehicle's own motion and is a combination of the density of the air and the speed of the vehicle. Think of sticking your hand out the window of a moving car. If you're not moving, there's no force on your hand, even though it's surrounded by air. In the case of a vehicle which leaves the atmosphere, it starts out at zero as the vehicle sits on the pad (lots of air but no speed), rises rapidly to a maximum point (Max Q) as speed builds, then tapers off as the air thins out until it's finally back to zero when the craft is out of the atmosphere (lots of speed but no air). For a tragic example of how significant it is, consider the two Space Shuttles that have been lost. Both *Challenger* and *Columbia* were lost right at the point of Max Q (*Challenger* on the way up, and *Columbia* on the way down).

©20th Century Fox Television

Will and Penny go for an air-swim while Don repairs the Inertial Guidance System

clear, but it provides an excuse for some pretty good zero-g sight gags, including watching Smith trying to power down the Robot while floating around.

West turns the gravity back on, and Smith barely gets the Robot deactivated in time. Just then, Will walks in and asks what he's doing. Smith deflects this with a bit of medical doubletalk about a virus and tells Will to stay down there "in quarantine" while he goes back up to the command deck. Ever curious, Will reactivates the robot, which comes alive shouting, "Destroy!" while waiving its arms about.

Meanwhile upstairs, the debate on whether to return to Earth continues. Maureen sides with Smith and says they should return. She actually gets angry with John when he suggests they continue in the face of these "new developments" (what happened to the "pioneering resourcefulness" she was supposedly selected for?).

©20th Century Fox Television

Just then the robot comes up the elevator and starts smashing panels with more flames and explosions. The ship goes into "hyperdrive" as the crew struggles to remove the robot's power pack.

Robot Rampage. Above, a panel in the aux control room gets smoked. Left, the Robot and Dr. Smith's ultimately cozy relationship gets off to a shaky start.

In our last shot of Alpha Control, a spokesman wearing heavy black framed glasses (of the type Irwin Allen himself wore) reports to the president that "the spacecraft is beyond the limits of our galaxy, and is presumed hopelessly lost in space." (Beyond the galaxy? It looks like other stars weren't out of the range of our technology after all!)

Back on the ship, John decides to go EVA to repair the NGS (Navigational Guidance System) scanner on the roof so that they can find their position. He floats outside the ship and has a gas gun for maneuvering, almost exactly like the one Ed White used during the first US spacewalk on *Gemini IV* four months earlier (June 1965). Even though he's initially floating, and has the gas gun to maneuver, he needs to climb up the side of the ship to get to

©20th Century Fox Television

the roof. He keeps loosing his footing and falling back down (falling? Maybe it's time for Don to turn off the a-grav again). Eventually his tether rope breaks, and he starts floating away. Someone has to go rescue him, and since West is the only one who can fly the spaceship, Maureen suits up and goes out with a "rocket gun" rope launcher. She shoots the line and John strains to reach it.

To Be Continued

NEXT WEEK!

Same Time, Same Channel

Episode 2, "The Derelict," opens, of course, with John still trying to grab the rope from Maureen. Naturally, he catches it on the first try. Don detects a comet coming their way and tells them that they'll cook from the intense heat when it passes if they don't get back inside.[25] John decides to fix the antenna anyway, and they barely get back inside before perishing. While that drama is unfolding outside, we see Smith down below reprogramming the Robot to respond only to his voice.

There are a few homey, domestic scenes, including John keeping the ship's log (on paper!) with a voiceover eerily similar to Kirk's "Captain's Log" entries of a year later. Before we can get too complacent, though, they pick up a signal "unlike anything we've ever heard before." They soon

©20th Century Fox Television

---

[25] Yes, we know that comets are really intensely cold, not hot, and that they only shine and have a tail when illuminated by a nearby star, but this isn't a science documentary

come upon the episode's title character, a gigantic organic looking spacecraft more than 15 times the size of the *Jupiter 2*. As they do a fly-around, the jaws at one end suddenly open and they're drawn in by a tractor beam. Once inside, they seem to

©20th Century Fox Television

have no trouble hovering and maneuvering slowly until finding a suitable landing spot. We get our first view of the landing gear as the legs unfold slowly from the underside, and the ship gently sets down.

This whole time Smith seems convinced that this is a ship from his "employers." When they first encountered the signal he clandestinely tried to contact it with his little hand radio, only to be rewarded with an earful of intense static. Undeterred, he volunteers to join John and Don as they prepare to go outside and investigate, and even leads them down the stairs. He is remarkably condescending as the other two marvel at the stupendous size and advanced technology of the vessel.

There is much wandering around, and they eventually split up with John and Don finding the control center while Smith looks for his compatriots. Will, precocious as he is, decides to let the women fend for themselves and sneaks out to join the men. He eventually finds Smith and the two of them encounter the series' first monster, a strange, multi-eyed blob that shoots lightning bolts between its pseudo pods. This finally convinces Smith that it's not his peoples' ship. Will tries to communicate with the creature, but Smith, upset at having been wrong, zaps it with a laser pistol.

Meanwhile, John and Don have discovered the navigation library of the control center. They're just getting into it when everything starts coming apart after Smith zaps the alien. They high tail it back to the *J2,* which they find, inexplicably, is now free to leave (well, they have to cut the doors open with a laser, but the tractor beam that drew them in has, for some reason, been turned off).

Once they're back in free space, John

©20th Century Fox Television

announces that he detected a nearby planet when searching the alien navigation library. As they head for it, the narrator breathlessly announces that this could be the end of the Robinson family if the planet should not prove habitable (not really much of a cliffhanger).

To Be Continued
NEXT WEEK!
Same Time, Same Channel

Episode 3, "Island in the Sky" picks up with the *Jupiter 2* closing in on the planet. John and Don are discussing sending the Robot down to do an environmental analysis prior to landing to make sure the planet is habitable (a remarkably good idea!). Smith, of course, wants to keep it on board as his personal henchman, so he convinces John that the Robot is still malfunctioning. John decides to go down himself using some weird looking arm-mounted rockets called "para-jets."

©20th Century Fox Television

The rockets fail as he descends (due to Smith's meddling) and his last radio call says he's at 10,000 feet and falling. Interestingly, the shot of him tumbling and struggling with the rockets shows the entire planet disc in the background. Perhaps they meant to say 10,000 miles. In any case, the production notes for this episode called for a parachutist to be filmed falling to Earth to double as John falling to the planet. This could have been an amazing shot (sort of like the Bell Rocket Belt footage used later) but apparently time or budget didn't allow it.

Don tells Maureen that he has a fix on where John should have landed, so he plots a course down. Smith, though, wants to head back to Earth, and uses the Robot to threaten Don if he doesn't turn the spaceship around and head back to Earth. Don does the ol' lurch-roll with the ship to throw Smith off balance and regains the upper hand. He stuffs Smith into a freezing tube and turns it on which keeps him out of the way for the moment. He sends Maureen and the kids down below to strap into more of the acceleration couches, which have mysteriously appeared, while he hops into another of the freezing tubes.[26]

After a thrilling final approach and crash-landing, the family gets out of the couches and put out the control panel fires while Judy defrosts Don. They leave Smith frozen for the time being. Will, imitating Smith's voice, sends the Robot out to do what it was designed for, environmental surveys. It gives the area a quick once over and reports that conditions are chilly, but livable. Don breaks out the Chariot so they can go looking for John (where the heck is that thing stored, anyway? Many

©20th Century Fox Television

fans speculate that, like the tractor in *Forbidden Planet*, it is stored disassembled[27]). They eventually discover him in a pit lined with some sort of bare tree that has lightning running along its branches.

While rescuing John, the party notices the Robot leaving to head back to the ship but they are too busy to find out why. Smith had programmed it to check on his (Smith's) safety periodically, so when the Robot returns to the ship and discovers him in the freezing tube, it revives him. Before the others return, Smith directs the Robot to start killing off the members of the party whenever it finds them alone, except for West, who he needs to fly the ship.

---

[26] There's a logical disconnect here. They are following a highly dangerous and non-standard landing trajectory. Why would the pilot abandon the helm to the autopilot and hop into a freezer tube? If the tubes increase your chances of survival, why doesn't the whole family get into them? Sending them below decks makes even less sense considering they are going to be belly landing at high speed and the lower decks are the most likely to sustain damage. Why do they have to crash land at all? The hovering capability and landing gear seemed to be working fine in the last episode. The answer is that they wanted to use film from the pilot episode that looks out the front viewport as the ship nears the ground. In that version, the ship crashes (after having been damaged in the meteor storm) with the crew still in hibernation, and the footage shows West in a freezing tube on the left edge of the frame (photo, page 253) so they needed an excuse to get him into it. That still doesn't explain sending everyone else below, though.

[27] As we'll see later in Jon's Archeologists' Report, a hatch was actually located on the underside of the spaceship in both the large-scale miniature and the full-scale mockup. The edge of this ramp, in the closed position, can be seen behind John Robinson's head in the episode "Visit to a Hostile Planet" (this is the head shot used in the third season opening credits so it can be seen every episode). Of course, that still doesn't explain how they got the Chariot out of the hatch when said hatch was buried in the dirt under the ship. Irwin Allen got some use out of the ramp in the pilot film for his never-produced series, *The Man from the 25th Century*, when James Darren drives his convertible out of the saucer and down the Chariot ramp.

The Chariot has problems on the way back after running into some electric sagebrush, so they walk the last mile or so. They send the Robot out to do a complete survey of the surrounding area, after which there are more homey scenes with dinner and another log entry. Will sneaks off after dark to work on the Chariot, and naturally the Robot comes rolling along. Seeing this member of the family alone, it remembers its programming, fires up the lightning bolts between his grippers and advances on the boy shouting, "Destroy!"

©20th Century Fox Television

To Be Continued
NEXT WEEK!
Same Time, Same Channel

Episode 4, "There were Giants in the Earth" opens by having Will stop the Robot by imitating Smith's voice again. He holds it at bay by playing a verbal game of chess to "test his circuits" until help (his parents and Don) arrives.

The next day, John declares that they have to make the camp self sufficient in order to survive until they're rescued. This is followed by scenes of John and Don setting up the force field projector for defense, Penny and Judy planting a hydroponic garden, and Will fixing the spaceship's control equipment. The common thread through all of this is Dr. Smith wandering through; avoiding work while explaining to everyone how hard he is working. Except for planting a few pea seeds in the native dirt, he doesn't do anything. Later that night, they test the force field. Don tosses a couple of rocks at it (which explode), followed by a blast from a laser pistol. Finally, in an act that shows either remarkable faith or incredible stupidity (the author thinks the latter); he turns and aims the pistol straight at the family and pulls the trigger! Sometimes you can have too much faith in technology. Even stranger, John Robinson, who is ever vigilant for the slightest threat to his family, doesn't even recognize the possible danger, and simply compliments Don on a good job.

©20th Century Fox Television

The Saucer Fleet

The next morning the men go off to work on the Chariot, while the women stay home to cook and wash clothes (hmmm…). We see that the seeds Smith planted the day before have grown gigantic. John and Don fix the Chariot and start driving back. They hit some "rough spots" that weren't there on the way out. When the camera pulls back, we see that they are huge footprints. After getting back and seeing the giant pea pods, John does a soil analysis and manages to determine how the life cycle of all the life forms on the planet work from just from a few quick tests and a gander through a microscope. Bottom line is that they form a parasitic relationship with a host and grow gigantic.

Will is outside fiddling with the Robot. They hear a roaring noise and the Robot leaves to investigate. When it returns (hours later) it is babbling, "Humanoids, 16 meters tall, does not compute!"

Vegetables planted in the native soil under controlled conditions grow huge, but some seem to be dying of frost John and Don head up the mountain to check the weather station and discover the temperature is dropping rapidly. In two days it will fall to -150°F (-100°C) overnight. As they turn to leave, they encounter a giant Cyclops, which is what made the footprints that the Chariot drove through earlier. Will spots them with the "radio telescope" he has just repaired (*radio telescope?*). He grabs a laser pistol and dashes off to rescue them. He drops the giant with a single blast from the pistol, but after they all leave, the giant gets up apparently unharmed (not even a burn mark).

©20th Century Fox Television

John makes the command decision that they will have to head south to avoid freezing. Maureen asks why they just can't stay here in the spaceship with its automatic heat and insulation. John says that the spaceship is inadequate and that they must head south in the flimsy, un-insulated, Plexiglas-bodied Chariot.[28] Smith, though, seems to share Don's faith in technology and refuses to leave, additionally insisting that the Robot stay with him (thus allowing the use of footage from the pilot, which has neither Smith, nor the Robot, in the Chariot).

Just before they leave, Penny goes looking for her pet "Bloop"[29] who has wandered off. John goes looking for her using the jetpack (which works a whole lot better than the "para-jets" of the last episode). The music under this scene is not John Williams, but a Bernard Herrmann piece used directly from the pilot.

©20th Century Fox Television

Almost immediately after leaving they run into the Giant again, which Don dispatches easily and apparently permanently this time, with a laser rifle. They travel on until dark when they make camp. It's another family picnic with Will playing "Greensleeves" on the guitar while Don works on the Chariot. There certainly isn't any sign that the temperature is heading to more than 100 degrees below zero!

In the next scene, they're on the road again, still in the dark, when massive lightning bolts start hitting all around them, very reminiscent of Brack's Interocitor beams aimed at the escaping scientists in *This Island Earth* (see page 120). They find a convenient cave to drive into to wait it out. It's not just a cave, they discover, but an entire

©20th Century Fox Television

underground city from a long-dead civilization; complete with mummified corpses and hidden chambers with spin-around doors. This entire sequence (from the pilot) was left as originally scored with the "Spaceship at Night" Herrmann music, from *The Day the Earth Stood Still*.[30] Will and Penny get separated looking for the Bloop again, and wind up trapped in one of the chambers. Don and Judy find them, but then they all wind up trapped when an earthquake, uh, planet-quake hits! John and Maureen somehow hear them from behind the two-foot thick stone wall as the mas-

---

[28] For all his science smarts, John makes some pretty stupid decisions. Maureen is absolutely right here. Has John forgotten that the *Jupiter 2* is an interstellar spaceship? When it's away from a star, the outside temperature is -454°F (3 Kelvins), considerably colder than what he says the planet is going to get. The ship seems to have no trouble dealing with this temperature with its *nuclear powered* heating system!

[29] With apologies to those who think the Bloop was cute, it has to be one of the silliest and least effective aliens ever brought to TV. In the story, it is an indigenous life form of the planet, which they discovered in Episode 3 while out in the Chariot looking for John. In reality, it was an undisguised chimpanzee with a hat on it to make the head look taller, and with two horn-like ears sticking out the top.

[30] This music was used again later in the two-part episode "The Keeper" with Michael Rennie. In addition to being more mysterious and evocative than Williams' music, it made a great homage to the film when used in the scene where Rennie's character uses a large, stand-up communication device to contact his home world.

sive structures tumble all around them. John draws his laser pistol to start cutting them out of their stone prison…

To Be Continued

NEXT WEEK!

Same Time, Same Channel

Episode 5, "The Hungry Sea." This is the final episode based on footage from the pilot. It picks up with John finishing his laser cutting of the stone wall. They push the multi-ton stone block out with ease as the quake continues to rain down debris. The next shot is of the Chariot racing out of the cave. It's still night, but at least there aren't any more lightning bolts.

Cutting back to the ship, Smith and the Robot are chatting about the weather. The Robot says it has some "contradictory data" on the planet's orbit, which Smith dismisses with distain after looking at it, but we (the viewers) don't know what it is.

©20th Century Fox Television

Meanwhile, the Chariot has encountered a large inland sea, frozen solid. They start across, driving on the ice. Back at the ship, Smith and the Robot discuss freezing to death. Not necessary since, as they chat, the temperature starts rising again even though it's still night! Smith takes another look at the orbital data and becomes convinced it's for real, although we still don't know what's causing him such concern.

The Chariot continues across the sea, but they, too, notice the temperature rising, along with more quakes. John now says that maybe they should have stayed at the spaceship (duh!). Smith has a twinge of conscience[31] and decides that he has to warn them about this horrible thing happening with the planet's orbit (whatever it is). He calls them on the radio, but they refuse to even listen to him. No matter, the amount of "cosmic interference" gets so bad that no further radio contact is possible.

©20th Century Fox Television

---

[31] This is probably the only time in the entire series where Smith shows genuine concern for the Robinsons, without any sort of self-interest at stake.

©20th Century Fox Television

Smith decides to send the Robot in person with this dire information. We get a dramatic shot of it crossing the frozen sea before finally catching up with them the next evening during dinner. It enters their camp waiving its arms and announcing "Warning! Warning! Matter of Life-and-Death!" Don takes this as a threat and zaps him with a laser pistol before it can deliver the message.

The next morning John discovers the paper (!) computer tape hanging from the Robot with the orbital data on it. He instantly understands that the planet's orbit is highly elliptical and moves back and forth between deep cold and blazing heat as it travels around its sun. He says that not only are they no longer going to head south, but that they're turning around to head back to the *Jupiter 2* (yay!). There's not enough time to get there before perihelion (closest approach to the sun),

©20th Century Fox Television

which he figures should be in a few hours. They set up a big aluminum foil canopy (boy, they thought to bring everything!) that covers both them-

selves and about half the Chariot. The poor Robot, though, has to tough it out in the inferno. They further get under thermal insulation blankets at point of closest approach as the surrounding vegetation bursts into flame!

After the worst is past, everyone slowly recovers, starting with the men first (naturally). They pack up the Chariot, including disassembling the Robot to stow it securely since

the inland sea they drove over is now quite liquid and they don't want it bouncing around (the real reason is, of course, that the Robot isn't seen in any of the Chariot scenes since they came from the pilot).

©20th Century Fox Television

247

The Saucer Fleet

They start back, and when encountering the sea it is, as expected, completely thawed. Naturally a huge storm blows up which nearly ends everything, but our stalwart band makes it through, and once on the other side they stop to "give thanks" in a land that went directly from stark desert to lush jungle.[32] They arrive back at the ship, and start unpacking, all being right with the world…for the moment.

Incidentally, this planet's orbit would produce one of these cold-hot cycles every one of its years, which appears to be about two Earth-weeks long, but they never encountered them again the rest of the series.

The Vehicle

[Author's Note: In previous chapters all the descriptions of the design and construction of the filming models were presented in the production history section up front, allowing this section to be about the "real" ship. Since this chapter is about a TV show that went on for three years, interweaving that description in the production history was awkward and confusing so I'm placing it here. Sub headings will separate the models from the "real" ship description – JH]

The Sets and Filming Models

The Robot wasn't the only thing in *Lost in Space* that can trace its lineage to *Forbidden Planet*. The *Jupiter 2* not only shows strong influence from that picture, it also got considerable refinement from Bob Kinoshita who had worked on the film.

When Irwin Allen made the pilot, he used William Creber, with whom he was already working on *Voyage to the Bottom of the Sea*, as art director. Creber designed the *Gemini XII*[33] for the pilot, but before series production began, he had to return to *Voyage* to supervise the upgrading of the

Seaview sets for the second season (which included a major move from one sound stage to another).

In April of 1965 Irwin Allen hired Kinoshita as the new art director. In addition to designing the Robot, Kinoshita was responsible for the redesign of the *Gemini XII* into what we know as the *Jupiter 2*.[34]

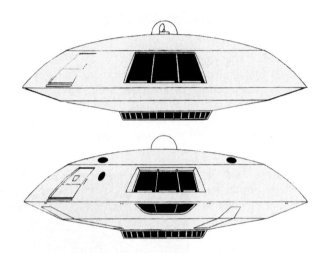

Fan website drawings modified by Jack Hagerty
A direct comparison of the *Gemini XII* (top) and *Jupiter 2* profiles. From the beltline-up they are identical in shape, but differ in the details. The *Gemini* has a door, but no window next to it. It's also lacking the two NGS scanner ports on the roof. The main viewport, however, is larger than the *Jupiter's*. The J2 obviously extends down further to accommodate the lower deck, plus it has the lower viewport and landing gear covers.

The *C57-D* interstellar cruiser from *Forbidden Planet* was the high water mark for movie spaceships up to that time. Even if he didn't work on the film, like Kinoshita did, Creber must have been a fan since he incorporated a number of the features from that ship into his *Gemini XII* design. The main one is simply the fact that it's a saucer, and not a cigar-tube rocket. Since this was a vehicle for a single family, and not the 20-man crew of an interstellar cruiser, he made it much smaller by pulling in the *C57-D*'s knife-edge disc until it became no more than a body break around the middle. The cruiser's huge top dome was flattened out to

---

[32] This scene, with them all kneeling down beside the Chariot, is where the pilot episode ended. In that story they weren't just being thankful for having successfully crossed the inland sea on their way back to the ship, they were actually now in their new home where the rest of the series was to have taken place. The final dénouement in the pilot, missing from the episode, shows two dome-headed aliens watching them from behind the trees.

[33] There is some interesting speculation regarding the *Gemini XII* name. The real Gemini program was continuously in the news when *Lost in Space* was in pre-production, so the name carried a lot of recognition vis-à-vis spacecraft. The first manned flight, *Gemini III,* was launched on 23 March 1965; roughly the same time they started filming the pilot. The number, "XII," is in Roman because that's what NASA used for the real Gemini flights, but why 12? Knowing Allen's penchant for giving trivia buffs something to chew on (e.g. the "serial number" on the side of the Space Pod in the third season was actually Allen's studio office phone number, see photo page 240), some fans speculate that it is a reference to Allen's birthday, June 12. Seeing as how his astrological sign, Gemini, coincided with the current NASA manned space program probably proved irresistible.

[34] There is more conjecture regarding the name change. The "Gemini" name was probably dropped because by the time the show premiered, NASA's Gemini program only had a year left, so if they went into a second season, the name would suddenly become historical rather than current. Also, being a cramped capsule launched atop a conventional chemical rocket, it already seemed really old-fashioned compared to the sleek atomic-powered flying saucer. But why "Jupiter" and why "2?" Again, the trivia buffs point out that in Allen's hit movie *Five Weeks in a Balloon* (1962) the title character was named *The Jupiter*. Naturally, this equally magnificent vessel would be the *Jupiter 2*.

become a curved roof (but he added back the Astrogator bubble on the top to fill in the dome's esthetic function). Down below, a sequentially rotating ring of lights in a squat, roughly cylindrical structure[35] indicates the main drive in action for both ships. Some new features added to the *Gemini XII* were a side door and a front "windshield," both of which are lacking on the *C57-D* (indicating that this spaceship was more of a motor home, while the Cruiser was more like a submarine).

While this design got Allen through the pilot, Kinoshita had to do some major upgrades for the series. The biggest addition was an entire lower deck to accommodate the Robot and, inadvertently, Dr. Smith. This area was put to good use as living quarters and laboratory space (apparently the *Gemini XII* didn't need living space since the crew made the trip frozen and didn't thaw out until they got to their destination and set up camp). For the flight deck, he carefully reworked the interior, which is where the *Forbidden Planet* influence from both designers really shows. The "flying buttress" style ribs that divide the deck into segments are straight from the *C57-D*, while the vertical "freezing tubes" for suspended animation echo the "DC" (deceleration) stations used by the big saucer's crew when passing through light speed (see page 181). This connection, though, is as much due to the ethereal light emitted by the freezing process since the DC stations had no physical tubes. Better precedents for the tubes are the "Conversion Tubes" in *This Island Earth* (see page 138).

Kinoshita's changes to the flight deck set had to be subtle since they still wanted to be able to use the footage from the pilot. Of the eight bays, only the three around the back (not really seen in the pilot) could have any substantial changes. That's where Kinoshita added the ladder and elevator between the decks and a door. Initially the door led to a storage room (as in this frame where Will is stowing a spacesuit). In the third season, however, it became an

©20th Century Fox Television

airlock leading to the Space Pod. The remaining bays only had minor changes so as not to be too noticeable. Most of the upper deck is still Creber's basic design, with Kinoshita's influence seen most notably in the door design. The lower deck is all Kinoshita's work.

©Warner Bros. (top) and 20th Century Fox Television (bottom)

The *C57-D* Astrogator (top) is more complex looking than the *Jupiter 2*'s, but the latter retains all the functionality, including the flight control panel (lower left).

The "Astrogator" in the center of the flight deck duplicates the same navigation device from *Forbidden Planet* with its gimbal-mounted model of the spaceship under a clear hemisphere. While this simple dome lacked the visual impact of the movie's complex armillary spheres, it still retained the same function of flying the ship from the small control panel on one side. This was quickly abandoned in favor of a more conventional "driver's seat." By the third episode, Don and Maureen control the ship's entry into the planet's atmosphere from the control consoles behind the main viewport (the "windshield").

On the first season set, Kinoshita had designed a single pilot's seat into the center of the master control console. It was designed in such a way that when folded down it had a smooth, blank appearance that matched the rest of the console. For the second season this was eliminated in favor of a new, instrumented center control along with a second pilot's seat so that both John and Don could sit down while driving, and the seats were never shown folded away again.

Outside, the movie influence continues with the addition of landing legs in a tripod arrangement with their integral boarding stairs. Missing, though, is the central pylon that supported most of the weight of the *C57-D*, presumably not needed since the *Jupiter 2* is much less massive. If the *Gemini XII* had a landing gear, it was never shown since it took off from its pylon launcher then crash landed on the planet, after which it was abandoned.

Other detail changes to the exterior were the reduction in size of the front "windshield" viewport and the addition of a lower deck viewport covered by shutters (usually kept closed). There are shutters for the main viewport as well, but they are almost never shown closed. Two smaller round viewports (although some fan sites show three) were also added: one next to the side hatch was added for the first season, and another around the rear of the saucer added for the second season. This second port was only shown from the inside and was never added to any of the effects miniatures. On the roof, there are three small round hatches, one on the left and one on the right for navigation instruments to deploy through, and one over the elevator for the astrodome. It's interesting to note that in the exterior shots during the

---

[35] This drive nacelle (called the "fusion core" by fans) is actually a very short cone frustum on the *J2*, and a multiple radius, straight-sided dome on the *C57-D*.

The Saucer Fleet

©20th Century Fox Television

"meteor storm" sequence of the first episode, there is neither a hatch nor a viewport on the side of the ship.

The "Real" *Jupiter 2*

Researching the actual technical specifications of the *Jupiter 2* caused a great deal of anguish. Even though the show was supposed to have a realistic feel (at least at first), Irwin Allen wasn't all that worried about scientific accuracy, just viewer impact. Dozens of consoles with flashing lights litter Alpha Control.[36] Inside the ship itself we have more twinkling lighted displays and lots of cool hardware with big curving conduits connecting odd shaped machinery. While most of the pieces of William Creber's original sets did have specific purposes in the context of the story, there's no argument that the sets were also awash in what today we call "eye candy."

As we have seen, the *Jupiter 2* mutated as the series progressed, becoming larger and carrying sub-vehicles with no place to store them. It didn't start out that way, however. Creber did an excellent job of violating the standard Hollywood rule of having an exterior far larger than the interior. In fact, the *Gemini XII* interior set was based on a 40 foot (12 m) diameter ship, so when they bumped the diameter to 48 feet for the *Jupiter 2* it gave Kinoshita extra room to design in the side airlock and the elevator, etc. The lower deck doesn't fit quite as well. The diameter is about right, but the full height deck-to-overhead (ceiling) extends too far out into what would be the tapering edges of the ship. The Polar Lights model (which we'll see in the "Modelers' Note" coming up) features an exterior and upper deck at 1:48 scale; but a lower deck at about 1:96 scale. The different scales do nothing to reconcile the model to what was seen on the series, they just help to get stuff in place that represents what we saw. The major reason for shrinking the lower deck interior becomes obvious the first

time you try to retract the landing gear. There's no place for it to retract into, otherwise!

Even establishing something as basic as the overall diameter is tricky. The official studio blueprints for the "full size" *J2* exterior set have a diameter of 43½ feet (13.3 meters).[37] Most *Space* authorities (both fans and industry professionals) consider the 4-foot (1.2 m) effects miniature to be the true representation of the ship (known as the "hero" reference that everything else is scaled from). The blueprints for the 4-foot model were drawn to an architect's scale of 1"=1' (i.e. 1/12 scale), making it representative of a full size ship 48 feet (14.6 m) in diameter.

In the case of the interior, there is no all-encompassing drawing of the entire set extant. There is plenty of fan-lit creating the "backstory" for all of the visual elements seen in the sets and the dialog heard in the scripts. This is what causes the anguish mentioned above. The credo used for all of the entries in this book is "seen on the screen," meaning that if it wasn't actually depicted in the show or film, or at least described in some official production document, then it can't be considered "real" for the purposes of documenting the ship. Fan-lit is fun, but it's useless for determining what the original creative team had in mind.

The only way to determine what's seen on the screen is to watch (or at least scan through) all 83 episodes paying careful attention to any hardware scenes.

DRIVE SYSTEMS – The *Jupiter 2* is nuclear powered, of course, but nothing as common as a straight reaction motor. Just saying it's "atomic," though, doesn't explain how the nuclear power is used to move the ship. The series gives us several different explanations, all of which, surprisingly, are not mutually exclusive.

Ion Drive - In the episode "The Cave of the Wizards", John Robinson states, "…our chief concern will be the conservation of fuel until we can switch to our ion propellant." Later in "The Deadliest of the Species" Don and John are seen repairing the spaceships "ion grid". This indicates ion propulsion as the *Jupiter 2*'s drive system, at least in deep space.

In "The Deadliest of the Species," the ion grid is shown as a square approximately 12 feet (4 meters) on a side when it's spread out on the ground for repair. Out-

©20th Century Fox Television

side of this one episode we never really see the ion drive hardware in the series. There's no indication where this would be installed in the *Jupiter 2*. For an ion drive to work,

[36] Note that video screens and keyboards were not yet part of the common concept of a computer. The technicians worked at low, wide consoles with dozens of lights and pushbuttons, playing them like virtuoso pianists. These were, in fact, real display consoles from the Burroughs 205 Data Processing System.

[37] *Lost in Space Technical Manual vol. 1*, Richard R. Messman

the grid has to be perpendicular to the direction of travel, meaning that the grid would be vertical when the ship is on the ground. A quick glance at the *Jupiter 2* shows no 144 square foot (16 m²) panels sticking up anywhere. There also needs to be a tunnel around it for the focusing magnets. So none of the required "real world" equipment or structure for this drive is shown in the series.

Magnetic Drive - In the episode "Mutiny in Space", Don gives Dr. Smith a spare propulsion unit from the *Jupiter 2* (in a wooden crate?). Later John recalls that the propulsion unit is "Electromagnetic" and constructs a device that reverses the magnetic polarity of the propulsion unit.

©20th Century Fox Television

The theory behind magnetic drive is that it uses the attracting and repelling force between electromagnets and naturally occurring magnetic fields in space. When the drive is activated the spaceship is repelled by the magnetic field of the planet you are leaving. Once in space, by controlling the polarity of the electromagnets the spaceship will be attracted to, or repelled by, the magnetic fields of planets, stars and other celestial bodies. Addressing the fuel issue, Don West says, again in "Mutiny in Space," that magnetic drive can operate almost forever on a minimum of fuel.

There are many problems with this drive concept, but again, these are "real world" objections (you can do anything you want in fiction).

- Not all planets have magnetic fields.

- Even for those that do, the fields are very weak. The Earth's magnetic field, considered strong by planetary standards, can barely move a light compass needle by itself.

- These weak fields drop off by the "inverse-square" law meaning that as your distance increases, the strength of the field drops off by the inverse of the square of the distance. This is less complicated than it sounds. Example: when you get twice as far away as your starting distance, the field is only one fourth as strong. When you are three times as far, it's only 1/9. By the time you get only 10 times the distance your propulsion strength would be only 1% (1/100) of your starting value.

- Even these rare, weak, severely diminished fields are not uniform. The star's "solar wind" distorts them into weird shapes streaming away from the star. Even worse, some planets, like Mars, have "punctuated" magnetic fields that emanate from multiple places on the planet and are all different strengths and orientations on the same planet. What a control nightmare!

Rocket Drive - So, what about the "fire all rockets" lines used throughout *Lost in Space*? Why does the *Jupiter 2* need rockets if it has ion and magnetic drive?

A system of chemical rocket motors is required for various manual maneuvers. Since there are nine push buttons on the control panel, it's reasonable to assume there are nine rocket engines, but just where they are located around the rim is kept deliberately vague. When in operation, all we see is a close-up of the *J2*'s central beltline break with a gaseous plume emerging. The rockets are probably liquid fueled using hydrogen as a fuel and oxygen as the oxidizer since those can be made

©20th Century Fox Television

locally wherever there is a source of water. We'll ignore for the moment that hydrogen/oxygen rocket motors have an almost invisible exhaust.

It is, again, no coincidence that the Space Pod also features maneuvering thrusters on the sides.[38] The hydrogen and oxygen propellants could be stored in the prominent external tanks.

Miscellaneous Drive Discrepancies - In addition to those already mentioned the biggest discrepancy is seen in the first season episode "The Raft." John and Don take the 8-foot (2.4 m) spherical "reactor core" vessel from the *Jupiter 2*'s engine room and convert it into a "space raft." We'll leave for a moment how they got this multi-ton vessel out of the ship with only hand tools, and wonder instead just where was it housed? Only the third deck

©20th Century Fox Television

"Power Core" room seen in the third season episode "The Space Creature" would be big enough to house this unit. It's too bad that the Power Core set is larger than the interior volume of the entire spaceship.

Power Systems – The ion and magnetic drive systems need to be powered by something. We're pretty much at a loss to describe it, except that we know it's "atomic."

Atomic Motors - Access to the "atomic motors" is through a hatch on the lower deck right next to the elevator. We have no idea what the actual motors look like but suspect

---

[38] It is also no coincidence because the Space Pod was based heavily on the Apollo Lunar Module that also has very prominent reaction control thrusters in four "quads" around the vehicle.

that they have to be quite compact to fit in the drive nacelle. In "The Mechanical Men," Dr. Smith is seen with a test motors as he tries to invent a liquid fuel for the spaceship. This test motor could be similar to the atomic motors that power the ship. It is very compact, approximately 12

©20th Century Fox Television

inches (30 cm) in diameter and about 18 inches (45 cm) tall. Something this size would fit easily into the drive nacelle.

Fuel - the atomic motors use the fictional element "deutronium" as the fuel (the name picked, presumably, since it sounds like a combination of "deuterium," the isotope of hydrogen used in H-bombs and fusion reactors, and "plutonium," the element used in A-bombs and fission reactors). Several episodes reference that the ship requires only six cylindrical canisters of deutronium fuel pellets. Each canister (seen in several episodes during the series) is 8 inches (20 cm) high and 2½ inches (6 cm) in diameter.

GUIDANCE & CONTROL SYSTEMS -
The art directors on Lost in Space did a good job when designing the Guidance & Control System. Nearly everything needed to guide the spaceship through space was designed into the computer area of the *Jupiter 2*. Of course, since the basic concepts for the control deck came from *Forbidden Planet*, we should say that Arthur Lonergan and his crew at MGM got it right and Creber/Kinoshita maintained those elements correctly for *Lost in Space*.

Astrogator - At the center of the *Jupiter 2* flight deck is the Astrogator (see photo, page 249). It has a *J2*-shaped model mounted under its dome. An addition for the second season was the "astrocompass," a calibrated ring around the base of the dome. The joystick on the side of the Astrogator, rather than flying the whole ship as in the beginning, is now used to manually set a course on the astrocompass. Presumably major course changes would be handled here; minor maneuvers would be done at one of the Master Control panels with the thruster rockets.

Electronic Brain - Primary guidance comes from the Electronic Brain, the spaceship's main digital computer. Physically, it is the yellow-lighted control wall located next to the airlock. The 'Brain has 54 circuit modules arranged on three 6 x 3 module frames. Each five inch (13 cm) square module has two identical back-ups, all three of which are mounted on a triangular frame. In the event of module failure, it will automatically be rotated out of service and its replacement rotated into place. A few episodes, including "The Haunted Lighthouse" show how the panels rotate for replacement.

Inertial Navigation System - Next to the Electronic Brain is the INS. This is a spherical device mounted on gimbals that seems to bounce around continuously when the ship is in flight. It's not clear how the purpose of this device is different from the Astrogator since they both seem to have gyroscopic mounts and wobble around when working. The INS is what Don had to shut off the artificial gravity to fix in "The Reluctant Stowaway." It's also referred to as the Navigational Guidance System, especially when talking about the external scanners (see below). In later episodes it was sometimes called the "Automatic Pilot."

Guidance Tape Drive - The last section of the guidance system bay has a computer tape drive. In the pilot, this is described as the guidance tape that has their course pre-programmed on it. In "The Reluctant Stowaway" Don refers to it as the "vector tapes," a more technical (and more accurate) description for their planned course,

Also in this section is the Atomic Clock. The readout is at the top just below the ship's status indicator, but the actual clock mechanism is probably the equipment below the Tape Drive.

NGS Scanners - There are three hatches located on top of the *Jupiter 2*. Two of these hatches, located on the port and starboard sides of the roof, house the NGS scanners that feed sensor data to the Navigation System. John Robinson attempts to repair one of these units during the cliffhanger ending of "The Reluctant Stowaway" and succeeds in the task at the beginning of "The Derelict".

©20th Century Fox Television

©20th Century Fox Television

Computer Bay from the first episode. The "Electronic Brain" is on the left, the Navigational Guidance System in the center and Vector Tape Drive on the upper right. Below the tape is the Atomic Clock. The overhead panels were changed occasionally throughout the series.

Elevator & Navigation Site - The *Jupiter 2*'s elevator is designed to not only transport personnel and equipment between decks, but also as an elevating platform for the Navigation Site. The site, mounted in an astrodome, is raised through a ceiling hatch for a clear view of the stars. This is the third and largest hatch on top of the *Jupiter 2*, near the back. While the dome was built for the series, it was never used in an episode. The navigation instrument platform, though, was constructed and its bottom edge can be seen above the elevator during the entire first season.

Master Control Console – This is the "driver's seat" of the *Jupiter 2*, allowing pilot control of all systems. This includes the radio transmitter, but being regular radio it can communicate with Alpha Control only at the speed of light. That means it would take 4.3 years for news of the family's safe landing at Alpha Centauri (presuming they made it) to reach home.

The master console is divided into three sections: left console, center console and right console. Each of the three consoles has a Burroughs 205 display unit on top. These three units probably relay manual inputs to the Electronic Brain.

The left and right consoles have many duplicate controls (in fact, in case of damage or emergency, almost all of the *J2*'s systems have online back-ups or can be replaced from storage). The left and right consoles each have a console cover to protect them from air-circulated dust during the long voyage.[39]

The left console has a radarscope, radio/intercom speaker, the rocket firing control to operate the nine chemical rockets located around the spaceship's perimeter, lift-off control, and a lineal acceleration indicator.[40] The radarscope uses the NGS scanners to track objects in space and effectively allows the pilot to see where the ship is going.

The lineal acceleration gauge displays the spaceship's speed. The top gauge shows the low range and the bottom gauge is for hyperdrive.

©20th Century Fox Television

The right console also contains a radarscope, radio/intercom speaker, rocket-firing control plus the lift-off rocket control and the artificial gravity control.

The center console is presumably not required for automatic space flight. During the first season, that area housed the pilot seats in a folded down position so it looked like a blank panel. At the beginning of the second season the center console was added and it remained for the rest of the series. A manual tripod control, for the landing gear is attached on the left.

After Jon's Archeologists Report, we'll be back with some final thoughts.

©20th Century Fox Television
Original Control Console from pilot episode. A bit sparse, but no one (supposedly) has to drive.

©20th Century Fox Television
Master Control Console for the second and third seasons with the (now-obscuring) pilots' chairs.

---

[39] Plus, being an Irwin Allen production, it protects the instruments from smoke and flaming debris during the not-infrequent conflagrations. Seriously, during the first season the covers helped the units match up with footage from the pilot where no controls are shown.

[40] This is not a typo. Close ups of the gauge in "The Reluctant Stowaway" shows that it really does say "Lineal" rather than "Linear." This has led some fans to speculate that "Lineal" was one of the contractors that built the *Jupiter 2*.

Archeological Report on the saucers in *Lost in Space*

By Jon Rogers

INTRODUCTION

The flying saucers in the TV series *Lost in Space* were seen around the world by more people and have had a more lasting influence than any other single flying saucer design in history. Fittingly, they too have their mysterious elements because there is much confusion to this day as to exactly what was seen during *Lost in Space's* three seasons.

This ongoing confusion stems from the fact that there were two spaceships designed for the series and both were supposed to represent one ship, the *Jupiter 2*. If you have read the previous "Vehicle section" you are aware of some of the history of how the *Gemini XII* was created and evolved into, and was presented as, the *Jupiter 2*. Additionally, the sets and models seen on TV as representing the *Jupiter 2* also evolved over the series' three seasons.

As the series progressed, the *Jupiter 2* seemed to change significantly. Portholes were added. A Pod bay, landing pod, storage room and a power deck complete with engine and machinery all appeared at different times. The interior sets were also altered. The control panels, the pilot seats, and interior fittings were all changed at various times. Another example was that the Astrogator in the *Jupiter 2* no longer rose to the top of the ship for a look out the observatory bubble as it did in the original *Gemini XII*. In some episodes it was even missing completely.

Another element that has added to the problem of identifying "what was the *Jupiter 2*" is that many different sets and models were used to create the image of this one ship. These models differed from one another and this meant that, when we looked at details of the *Jupiter 2*, there would be inevitable differences between what was seen in different scenes even in the same episode.

To help the reader understand the difficulty in capturing the essence of the "real" *Jupiter 2*, here is a list of the sets and models that were used to create the *Jupiter 2* on screen.

- Two or more 11 to 13-inch diameter flying models were made for use in some action sequences, e.g.; showing the *Jupiter 2* leaving Earth. In this scene, the model appeared to be the *Gemini XII*, however, pictures of it are inconclusive. The eventual fates of these models are unknown.

- One or more 18-inch diameter flying model was used. In one episode, "Wish Upon A Star," we see a model of this size that is clearly the *Jupiter 2*. Also, later scenes with a model against a planet background seem to use this model. No details of this model or its eventual fate are known.

- One 48-inch diameter flying model of the *Gemini XII* was made for the pilot film and sub sequentially seen in many episodes of the *Lost in Space* series as the *Jupiter 2*. This model has a different outside shape as it was intended to have no lower deck(s). This model was sold at auction and may still exist in a collection somewhere.

- Two each, 48-inch diameter flying models of the *Jupiter 2*, one with workable landing gear and one with workable Pod bay door were constructed. Both used a single working power section with rotating light patterns as needed. These models still exist and both have been restored. There is some disagreement as to the originality of the restorations as both models suffered abuse after the series was completed. Still, these are the original hero models used in *Lost in Space*.

- One 43 foot 6-inch diameter, full-sized model with movable landing gear and tractor ramp for use in external scenes of the ship was built. It was seen in several episodes at times substituting for the Internal Main Deck Set. Although it was intended to be as similar as possible, its outside shape is different than the 48-inch diameter models. This model was destroyed during Fox remodeling in the 80s.

- A *Gemini XII* Interior Set was constructed for all internal scenes. This set could be viewed from both inside and outside. It was originally intended for use only for the pilot filming but was used in beginning episodes as the *Jupiter 2*. It was either destroyed or converted into the *Jupiter 2* Flight Deck Interior Set when that set was created.

- The *Jupiter 2* Flight Deck Interior Set. This set was derived from the *Gemini XII* Interior set and was also equipped with an external skin, and personnel ramp for use in external scenes of the *Jupiter 2*. However, it had a working 'airlock, additional portholes and other differences. This set was dismantled after the series but parts of it, notably the flight control systems, have survived in private collections.

- At least one set of the *Jupiter 2's* living quarters deck was created for internal scenes needed there. This set was supposedly positioned below the main deck but was actually on another stage. This set was dismantled after the series ended.

- An additional *Jupiter 2* power room set was built for the third season. This set was shown in only one episode in the third season entitled, *Space Creature.* This set was dismantled after the series.

- A ten-foot diameter, flying model was also constructed. Evidence suggests that this model was to have all the working parts and to be the reference or "Hero" model. However, it appears that it was used for the special effects scene where the saucer crashes on an uncharted planet and was damaged, making its further use impossible. It was later discovered in crash damaged condition in the possession of a private collector. Its existence and condition are unknown today.

NOTE: all comments as to the location or survival of all the studio's props are as of this writing.

Perhaps now the reader can better appreciate the difficulty in recreating a ship with so many variables involved in its original makeup. Originally, I had planned on presenting a complete set of drawings and reports explaining all possible details of each of the models and sets used to create both the *Gemini XII* and the *Jupiter 2*. However, that effort soon outstripped the available space. Regrettably, I must leave such a thorough examination to be published at a later date, perhaps in a stand-alone article or book.

In order to capture the essence of the saucers in *Lost in Space*, I will limit myself to showing the *Gemini XII* and *Jupiter 2* as represented by each ship's 48-inch diameter hero model. For the interior, I will limit my examination to the *Gemini XII* interior set since it also served as the *Jupiter 2's* flight deck. I will indicate the major changes between the two to give the reader an understanding of the differences. I will also give a walk through description of the two spaceships highlighting their similarities and differences. However, this explanation will not be an entirely complete one. That would take another chapter the size of this one.

Gemini XII

Our examination begins with what most fans consider the "true definition" of the *Jupiter 2*. That is, the very detailed models used for the deep space flying and planetary landing scenes. We know that they made one *Gemini XII* and two *Jupiter 2* models. However, the series began with the *Gemini XII*.

The *Gemini XII* was best represented by a specially designed 48-inch (1.22 m) diameter model. It was built for the pilot episode and was seen primarily on the first several episodes of the first season. It set the style for what the *Jupiter 2* would look like throughout the series. Figure 1 shows its first appearance sitting on its launch pad.

When the *Gemini XII* launched it would be seen to glow and rise straight into the sky, just like some of the UFOs many people had seen up to that time. This was to demonstrate its advanced nature compared to rocket propelled spaceships of the day.

Besides this first introduction of the Robinson's spaceship in the series, the *Gemini XII* model would be seen in many space flight scenes. Stock footage of it

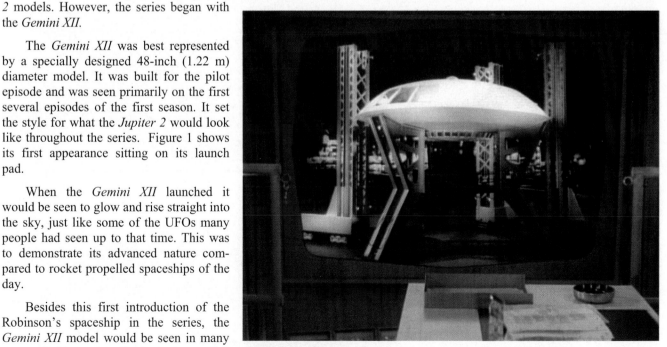

©20th Century Fox Television

Figure 1

The Saucer Fleet

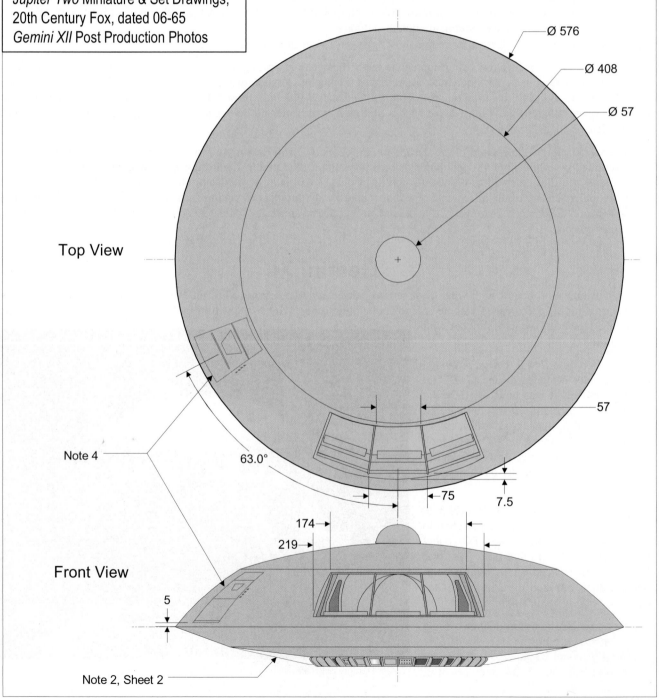

Gemini XII
48-inch 'Flying' Model
1/120 scale
Dimensions in Inches
© 2006 by Jon C. Rogers
Sheet 1

Sources:
Nowhere to Hide, Unaired Pilot-1965
The Reluctant Stowaway, Lost in Space,
1st episode-1966
Jupiter Two Miniature & Set Drawings,
20th Century Fox, dated 06-65
Gemini XII Post Production Photos

Notes:

1. Color is Silver.
2. Ship is depicted as shown in the first episode of the first season.
3. This drawing's scale is chosen to recreate the full sized ship from the 48-inch diameter model dimensions as originally intended. [1"=1']
4. The main airlock is not visible on this model. It is included here due to its requirement. Its features and location are from the *Jupiter II* model drawing.

Top View

Ø 576

Ø 408

Ø 57

57

63.0°

75

7.5

Note 4

Front View

174

219

5

Note 2, Sheet 2

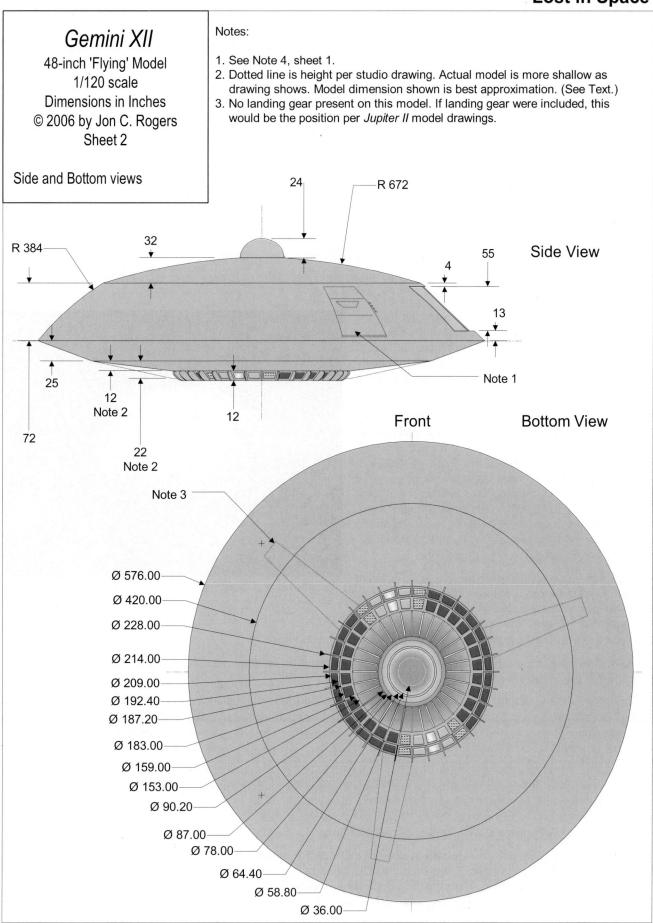

Gemini XII
48-inch 'Flying' Model
1/120 scale
Dimensions in Inches
© 2006 by Jon C. Rogers
Sheet 2

Side and Bottom views

Notes:

1. See Note 4, sheet 1.
2. Dotted line is height per studio drawing. Actual model is more shallow as drawing shows. Model dimension shown is best approximation. (See Text.)
3. No landing gear present on this model. If landing gear were included, this would be the position per *Jupiter II* model drawings.

Side View

R 672

24

32

55

R 384

4

13

Note 1

25

12
Note 2

12

72

22
Note 2

Front   Bottom View

Note 3

Ø 576.00
Ø 420.00
Ø 228.00
Ø 214.00
Ø 209.00
Ø 192.40
Ø 187.20
Ø 183.00
Ø 159.00
Ø 153.00
Ø 90.20
Ø 87.00
Ø 78.00
Ø 64.40
Ø 58.80
Ø 36.00

The Saucer Fleet

would be reused at various times during the entire series. Its use alongside the closely similar *Jupiter 2* 48-inch hero model created the confusion about the exact size and shape of the *Jupiter 2* that exists to this day. It also means that comparisons of the *Gemini XII* to the *Jupiter 2* begin almost immediately after the first episode was shown.

Recall that the *Gemini XII* was designed for a different purpose than the one it was put to when it appeared on screen as the *Jupiter 2*. Its original purpose was to carry the family Robinson *in suspended animation* to Alpha Centauri. Also the original plot called for it to be shown flying only during the first episode, then to be crashed and abandoned. Because of this, much of the deck space and equipment added later (including even the landing gear!) was originally not needed or included in its design.

However when it came time to add more equipment, the original shape was a constraint. The *Jupiter 2* still had to look like the *Gemini XII* or the pilot footage would have to be redone. Given that, the modifications to add equipment had to be subtle.

When we compare the outside envelope of the *Gemini XII* and the *Jupiter 2,* the differences are a thickening below the centerline, a new airlock, portholes, and a reduction in size of the main view window.

Of the three major changes, the portholes and the reduction in size of the front view port are the most obvious. The thickening of the lower hull is the most ambiguous.

As you can see in Figure 1, the lower hull of the *Gemini XII* is comprised of two sections. Each has a different outer angle. The one closest to the centerline has the least angle while the bottom section is the most acutely angled. It is the height of these sections that is most in doubt.

Without being able to measure the original model, the closest we can come to an accurate measurement is based on the original drawings and photos of the actual model. The problem is that these sources disagree as to the height of the lowest section. The drawing states the height to be about 22 inches (56 cm) while photos indicate it was more on the order of 12 to 16 inches (30 to 40 cm) high. Due to these ambiguities I have included both measurements in the data drawing. I leave it to a future historian with access to the original model to clear up this point.

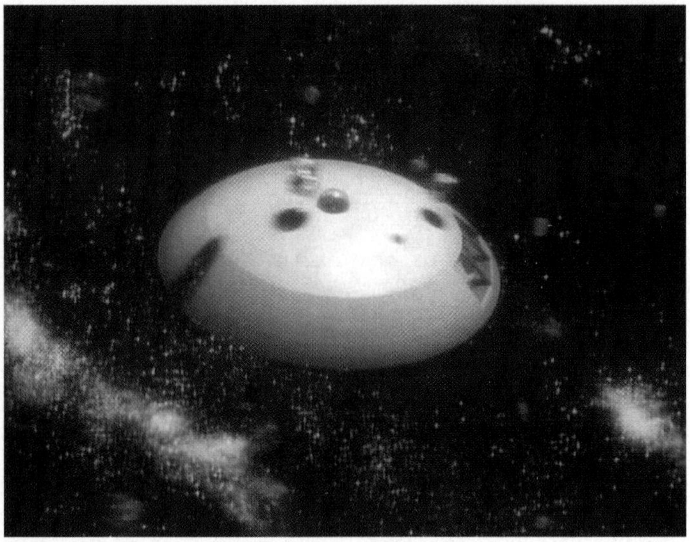

The main view port, besides being a slightly larger dimension on the *Gemini XII*, apparently did not have the protective shutters that the *Jupiter 2* had to close its main view port. The *Gemini XII's* main view port is always seen open--even during the meteor shower that knocks it off course! (Figure 2)

Figure 2

Also, close examinations of pictures of the model indicate that there is insufficient space for sliding view port shutters to be installed as was done in the *Jupiter 2*.

One point of confusion that needs to be resolved is the size of the navigation dome on top of the ship. Comparison of the original video of the *Gemini XII* model and pictures taken of it during its sale in the 1980s show domes of differing size. The post series photos show a dome of significantly larger size than original. The original dome was about 4 ¾ inches in diameter [representing almost a 5 foot dome] and the one shown during its sale is about 5 ¾ inches in diameter [representing a 6 foot dome]. Aside from the photographic evidence, little is known about when-or why the dome was changed.

There were other differences that were not visible to viewers at the time. One item was the power section with the rotating lights that propelled the saucer through space. It was a different size on the two ships. On the *Gemini XII* it was scaled to be about 18 feet (5.5 m) in diameter (excluding the fins) while on the *Jupiter 2*, it was about 15 feet (4.6 m) in diameter. What happened to the earlier power section after the series is unknown.

One significant difference likely missed by most of the audience was that the Gemini XII model had no airlock. If you look closely at Figure 2, you can see that none exists. The item that was most often noticed to be missing was the 2 foot (60 cm) diameter port hole that was right next to the airlock on the *Jupiter 2*. Still, the lack of an airlock on a spaceship is such a critical omission that I have included the one used on the *Jupiter 2* into the *Gemini XII's* data drawing as speculation.

Jupiter 2

Now we'll examine what many consider to be the most definitive version of the *Jupiter 2*, the 48-inch diameter hero model. Actually there were two 48-inch diameter models. Intended to appear identical, one was specifically designed to have working landing gear for the landing sequences, and the second featured a special bay and remote controlled door that opened to launch the landing pod, a new piece of hardware that was developed for the third season. In order to define what the real *Jupiter 2* should look like, it is necessary to combine the features on both of these models.

Of course there were inevitable other differences between the two models (none of these slight changes were to be seen by the viewer). One example was that the model that had the pod bay and functional doors did not have any landing legs. Not just non-functional legs--*none*.

This economy was possible because the landing legs were not fully recessed into the body. Even when retracted a part of them was external. The footpad was recessed behind a sliding door, but the outer skin of the underside of the landing leg was left outside the skin of the saucer. (Figure 3 Landing Leg Model)

This made it possible to duplicate the external appearance of the landing gear with thin similarly shaped pieces of flat material and that sufficed for the pod bay version of the saucer.

These flat pieces were attached in two places, probably by screws. (Figure 4-Pod Bay Model). Since the landing leg and stairwells were very intricate and therefore expensive to build, it was considered sufficient that they had one model with completely functional landing gear and the thin pieces of material sufficed to look as though the pod bay model had them also.

There were also differences between the real models and the intended design as noted in the 20th Century Fox blueprints. One is the item of stairs in the landing legs. The blueprint indicated that only one of the three landing legs would have stairs built into it. I have indicated on my drawing which landing leg this is. However, there is evidence to conclude that, when the model was actually built, at least 2 landing legs had stairs built into them. I have been unable to locate any scene or photo in which it can be shown that all three legs had stairs in them, but it is quite possible they did.

Also, other features were not built into the models. The atomic engine elevator which was to lower on four

©20th Century Fox Television
Figure 3 — Landing Leg Model

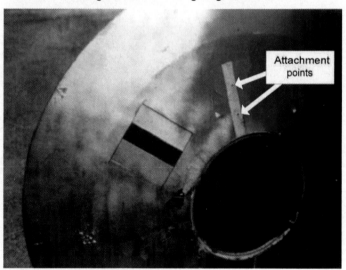
Attachment points
Figure 4 — Pod Bay Model

cables was designed but not actually built into either model. It may have been incorporated in the full-scale model but it was never seen in use during the series, so that cannot be verified.

It is unknown as to whether the Chariot ramp was incorporated in either of the 48-inch models. I have found no evidence to indicate that the ramp was included in anything but the full sized 43-foot exterior mockup.

It has been stated by fans that, although the lower half of the saucer had a functional view port; it was never used during the series. This is not entirely accurate. A complete functioning lower view port, complete with sliding protective shield doors, was designed and built into both 48-inch models. These shields could be extended or retracted into hidden pockets at the sides of each viewport. During the series, there were several occasions when the view port was shown open and one occasion we see Mrs. Robinson actually closing the view port manually in preparation for a crash landing. Of course these scenes were within the interior sets of the *Jupiter 2*. Although I have not found a screen shot where a model's lower view port is shown open, clearly the lower view port and its protective shields were designed to be used.

Jupiter 2
48-inch 'Flying' Model
1/120 scale
Dimensions in Inches
© 2006 by Jon C. Rogers
Sheet 1

Sources:
Jupiter Two Miniature & Set Drawings,
20th Century Fox, dated 06-65
Lost in Space,1966-69, 20th Century Fox

Notes:

1. Color is Silver.
2. Drawing includes all changes made in the three years of the series.
3. Viewport shields. One open, one shown partially closed.
4. Only landing leg with stairs per drawing. See text.
5. Atomic Engine Elevator. Lower with cables on each corner.
6. Chariot Ramp. Lower with cables.
7. Pod Bay. 3rd Season only. Half scale. See text.
8. Escape Hatch. Slide open. See text.

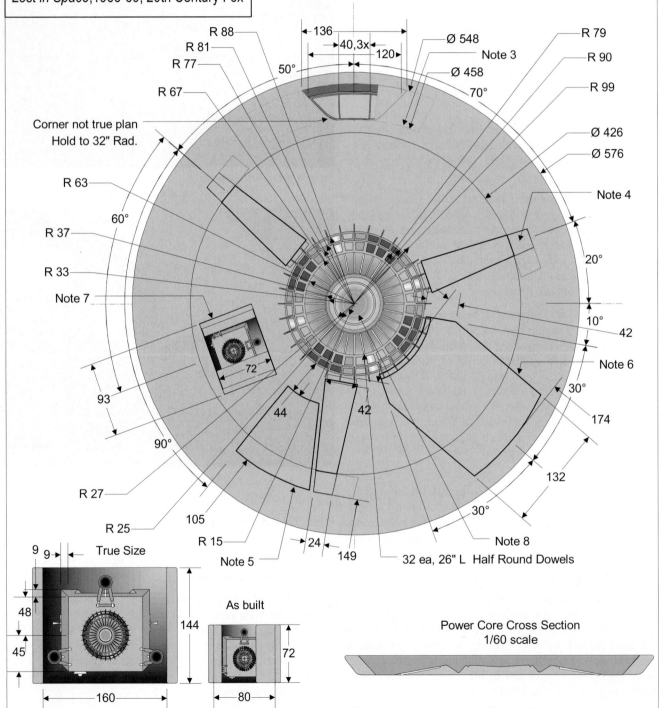

Bottom VIew

True Size

As built

Power Core Cross Section
1/60 scale

Jupiter 2
48-inch 'Flying' Model
1/120 scale
Dimensions in Inches
© 2006 by Jon C. Rogers
Sheet 2

Front and Side Views

Notes:

1. Astrogator in raised position, shown rotating during flight.
2. Viewport pockets for sliding shields. Both Upper and lower viewport had them and were operated during series.
3. Distance from Centerline to ground is a close approximate. Saucer was seen to compress its landing gear upon landing.
4. Door and Ramp for Chariot added. Shown in down position.
5. Atomic Engine Access port shown lowered on cables as planned.
6. Pod door shown partially open.
7. Airlock and Porthole was barely visible and not functional on model.

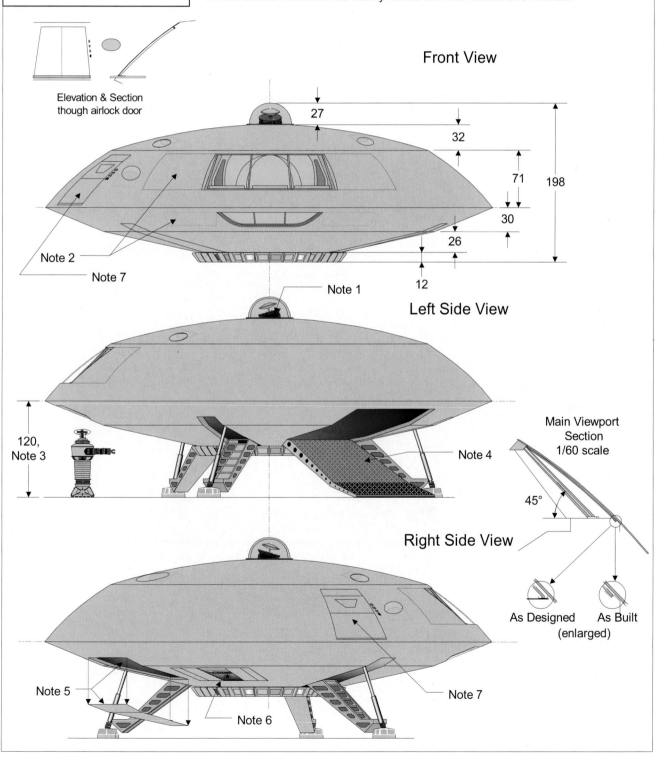

Elevation & Section though airlock door

Front View

27
32
71
198
30
26
12

Note 2
Note 7
Note 1

Left Side View

120,
Note 3

Note 4

Main Viewport
Section
1/60 scale

45°

Right Side View

As Designed   As Built
(enlarged)

Note 5
Note 6
Note 7

The Saucer Fleet

Notes:

1. Original studio drawing of viewport is ambiguous. Drawing states it has a 12ft. width (dotted line) but viewport is drawn with 14.5 ft width and appears to have been built that way. Unable to verify exact dimensions.
2. Viewport pockets for sliding shields. Drawing displays one open and one partially closed.
3. Cantilever platform and sliding pocket door as in studio plans. Neither feature was included in these models. Door has no surface details. External open-close control also not shown in plans.
4. After series photos of one model shows a base measuring 60" dia. x 1" h under the dome.

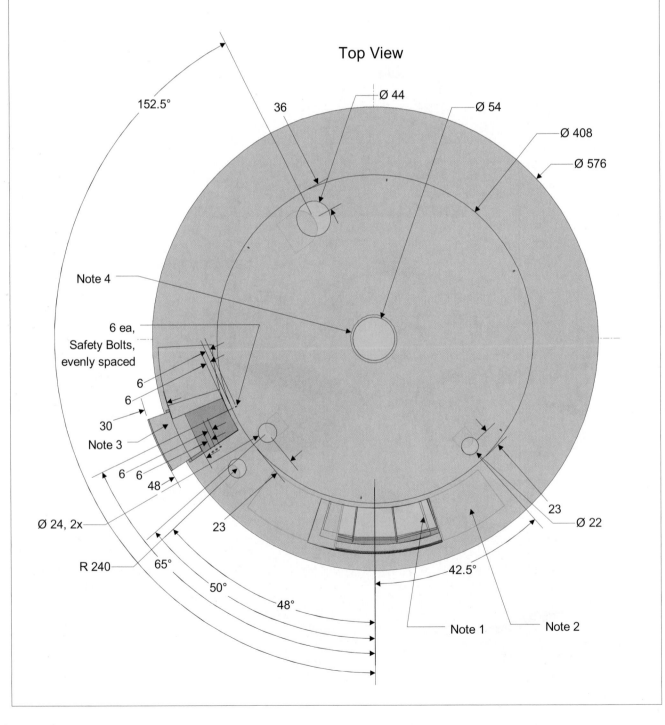

Top View

There were also some slight dimension changes between the blueprints and the actual models on the outside of the power core in the center of the bottom of the saucer. The models have longer, slightly more pointed fins than originally called for in the design drawing that has survived.

However, the largest dimensional change to any feature on the bottom of the saucer has to be to the pod bay and its doors. Several scenes show the pod being launched from its bay. Also a full-scale model of the pod was made and used in several episodes. Accurate drawings of it are available. This makes it relatively easy to get the actual size of the pod and relate it to the *Jupiter 2* (Figure 5).

In doing this, I discovered that the pod bay as shown on the 48-inch model in Figure 5 is actually being shown at half of its real size. If they had scaled the pod bay to incorporate the full-sized pod, it would not have fit into the *Jupiter 2* at all. So when you see the pod being launched from the *Jupiter 2*, it is actually a half sized model of it that is coming out below the ship. Please see Sheet 1, Note 7 of my data drawing for its location and further details.

©20th Century Fox Television

Figure 5 — Pod Launch

Another feature that was not incorporated was an escape hatch in the center of the power core at the very bottom of the ship. After reviewing the details of the completed power core it is apparent that this feature was too difficult to incorporate. The power core was filled with a myriad of lights and wires and a sliding door would have difficult, if not impossible, to include. However, it is important to note that the escape hatch is clearly indicated on both studio drawings of the 48-inch and full-sized models of the ship. It was clearly intended to be there in case one of the plots in the series required it. Please see Sheet 1, Note 8 of my data drawing for its location and further details.

One puzzling item on the blueprint was the width of the upper view port. This was one instance where the blueprint was ambiguous.

The blueprint was drawn to scale and as such the view port was scaled to be approximately 14 1/2 feet (4.4 m) wide at its widest point. However there is an additional dimension on the window that clearly states that it is to be 12 feet wide (3.7 m). Careful review of the series and pictures of both models indicate that the models were built to the larger dimension. Why this anomaly exists is puzzling.

The airlock was another item that was not included in the actual models per the studio drawing. Although the airlock and its sliding outer door and extended platform are clearly included in the blueprints, there were never more than rudimentary etchings made in the outside surface of the models. Neither 48-inch model had a functioning airlock.

A slight difference in the uppermost dome was also discovered. In photographs taken of the models after the series ended, the model appears to have a larger dome on it than it had during the time it was making the series. In addition, it later showed up with a slightly larger base between the dome and the top of the saucer (See Data Drawing sheet 3, Note 4). What the reason for the addition of this small detail is also unknown as is the reason for the later change in the diameter of the dome.

One of the things that have long puzzled fans of the series was that when the 48-inch saucer model was seen flying through space it contained a rotating object within the uppermost dome. What exactly this object was has been the subject of great conjecture. However, in scrutinizing the moving pictures of the *Jupiter 2* in its flights I discovered what it was intended to be.

Originally, in the pilot episode, the ship was to be on automatic pilot throughout its entire voyage while the passengers remained in frozen stasis until the end of the voyage. Therefore, it was necessary for the ship's computer to have Astrogator input during its flight through space. Both the *Gemini XII* and *Jupiter 2* had an Astrogator with its own dome in the center of the main flight deck that was designed to rise up to the top of the ship and look out the dome in the top of the ship during flight.

In addition, the Astrogator dome revolved and the miniature saucer in its center, rotated rapidly during its operation. This was shown during the first episode and several other times during the series. That is clearly what this detail was meant to indicate--the Astrogator in operation during the ship's space flight. If you look closely in Figure 9 you can see the Astrogator in its operating position inside the topmost dome in the *Gemini XII* interior set. Close examination of the ship in flight revealed that the power core lights and Astrogator bubble object to be in synchronization. The 48-inch model used reflecting light from the power core to simulate this feature.

I believe this clears up the mystery of what was supposed to be inside the dome during the *Jupiter 2's* flight.

The one feature that proved nearly impossible to document, though, was the maneuvering jets. The maneuvering jets were added after the first season. They were never on the original drawings of the ship. Although these jets were clearly seen in operation during several episodes, they were always shown in a close-up in such a way that it was impossible to see any details of them or identify their location on the outside of the ship. Figure 6 is one of the better close-ups of one of the jets in action.

As you can see in Figure 6, they are not round jets but rather some sort of slot in the underside of the ship. There were supposed to be more than one jet (possibly four would be needed?) with one of them on the side directly opposite of the front view port. However, little real data exists about them. The pictures of them give no detail and, because they were not included in any surviving blueprints, I was not able to include them in my data drawing. Their exact location and shape will have to remain another of the ship's mysteries.

©20th Century Fox Television

Figure 6

Another item that came to my attention was that the front main view port shield was built slightly differently than it had been designed. Originally it was intended to have a small track for the shield to fit into on a horizontal surface. However, review of the models as built show that the recess was significantly larger than originally intended and that the area was cut away below the outside skin. See sheet 2 of the data drawing for further details.

Also, it has been revealed that there was significant detailing inside the model that included miniature replicas of the crewmembers, details of the control consoles and the walls. Taking this all into perspective, these models have to be some of the most intricately detailed and expensive props built during this time period for any show on TV. Whatever minor differences these models may have had the overall impression they made was excellent. They are a very impressive piece of work and they became, to many people, prime examples of what an actual flying saucer spaceship would look like.

Jupiter 2 Interior Sets, Introduction

This brings us to the "tour-de-force" of the *Lost in Space* series, the interior sets of the *Jupiter 2*. It was mainly these sets, and their relation to the models used for exterior flying shots, that gave such convincing realism to this flying saucer spaceship.

In review, there were four interior sets that made up the saucer during the series from conception to finish of the broadcast. First there was the *Gemini XII* interior that was originally intended to be the only spaceship set in the series. This set was modified to become the *Jupiter 2* flight deck. Although almost identical in plan to the *Gemini XII*, the modifications were so significant that we could consider this an entirely new set.[1] Next, a second deck for living quarters was added to the *Jupiter 2*. And finally, in the third season, a third deck or power core area was featured in one episode.

Additionally because the main flight deck set also served as an external set in both space and at any landing site, we can compare them to the full size exterior mockup.

---

[1] One modification was to increase the set radius from 20 feet (6.0 m) to 23 feet 8 inches (7.2 m) to allow for the new airlock. It was done so cleverly few people noticed.

It must be admitted that it is possible that there were more than one internal set of the *Jupiter 2* flight deck. The studio plans call for a complete, movable set with portions of it made wild.[2] These panels could easily have been used in other partial or full sets. However, the main evidence comes from the first season episode, "The Derelict," where two distinctly different external sets are seen. We will review this evidence shortly.

There were observable detail differences of the interior set between different episodes, e.g.; the Astrogator was replaced by a table and chairs at times and re-appeared later in place of them. It is known that the production changed stages between the second and third seasons. It is not known whether there was more than one stage for the interior of the spaceship set up at the same time but given the variations seen and the urgencies of producing a weekly TV show, it is very possible.

Much equipment and information about the *Jupiter 2* flight deck set does survive at this writing. The entire flight deck control console, various set pieces, panels and such are in the hands of private collectors. The original Robot (and many duplicates) also survives, as does many of the external pieces, weapons etc.

For a spaceship archeologist or anyone trying to understand the reality of what made up this widely known flying saucer, it is fortunate that some of the original studio plans for the internal decks are still in existence, having been passed down from that generation.

Available, yes, but are they in good condition? Some yes, others no. Still, it was by examining these plans and also scenes from the original series that we were able to discover what the *Jupiter 2* was intended to be like inside.

The *Gemini XII* Interior and *Jupiter 2* Flight Decks

Let us begin by examining the original, and most consistent interior set, the flight deck of the *Gemini XII*. I will try to compare it to that of the *Jupiter 2* as I progress.

The first thing I noticed from the original Fox blueprints for the *Gemini XII* interior set was something very unusual. It was designed around a 40 foot (12 m) diameter.

Considering the full scale mockup's diameter of 43 ½ feet (13.3 m) and the 48-inch flying model's scaled diameter of 48 feet (15 m), the interior set was *smaller than its exterior*! Although it could have easily fit within the ship's outside envelope, it would have needed to be a larger diameter to have fit properly. Although not noticeable by the TV audience, for Hollywood it was *very* unusual.[3]

It was also quite clear that they always intended to have the interior set used as part of the scene where the ship is shown to have crash-landed on a strange planet. It was to be finished on the outside, making it very clearly the ship you see at the crash site (See Figures 8, 9 and *Gemini XII* Interior Set Data Drawing sheet 1). In this way they used the interior set(s) for both interior and exterior scenes.

Another major discovery was to resolve the question of whether there were two interior sets used in the first season. In observing the *Jupiter 2* crash site in various outside scenes in different first season episodes, the ship seemed to have two different outline shapes. This raised the possibility of two sets being used on two different stages.

In one scene (Figure 7) the saucer was shown outside with a profile that had a sharp pointed nose. In another scene the saucer seems to have had a "foundation" under

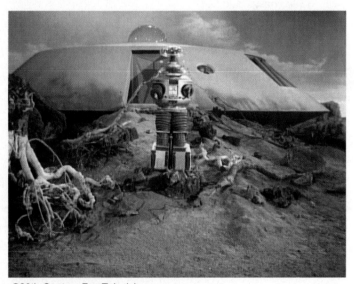

©20th Century Fox Television

Figure 7

---

[2] The term "wild" refers to independently removable and sometimes interchangeable background panels. They are removed to allow access for the camera and crew during shooting and replaced when needed as background. They can also be used on other stages to make additional sets.

[3] The *Jupiter 2* interior set's increased diameter of 47 ft 4 inches (14.4 m) was still with in the Hero model's scaled diameter.

Gemini XII
Interior Set
1/120 scale
Dimensions in Inches
© 2006 by Jon C. Rogers
Sheet 1, External View
Sources:
No Place to Hide,
  Unaired Pilot CBS (1965)
GeminiXII
  Interior Set Drawing,
  20th Century Fox (12/1964)

Notes:

Note 2
Note 3

1. Color is Silver.
2. Ship is pictured as it was
   intended to be shown at its
   crash site in the Pilot episode.
   (For vertical Dimensions, see sheet 2)
3. This is the damaged Engine area planned
   to show why the ship was no longer
   flyable. (Unfinished shot.)
4. This section (Engine area) had ceiling but
   was open for filming crew access.
   (See sheet 2 and text)
5. Airlock does not have sliding door.
   Closed with replacable panel.
6. Viewport has no protective shields.
   Always open.

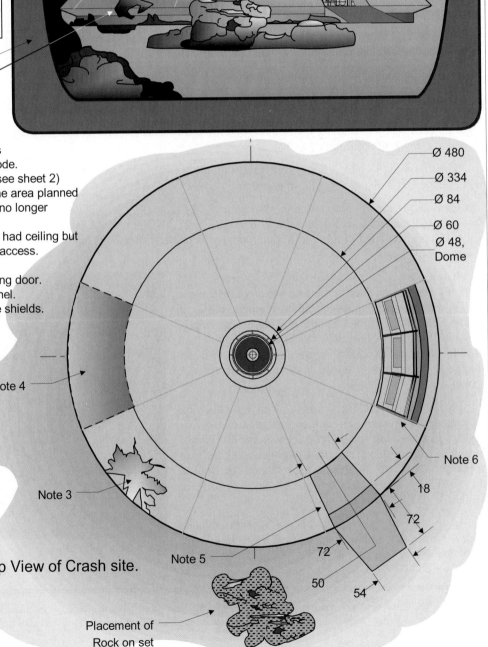

1960 style TV

Ø 480
Ø 334
Ø 84
Ø 60
Ø 48, Dome

Note 4

Note 3

Note 6

18

72

Note 5

72

50

54

Top View of Crash site.

Placement of
Rock on set

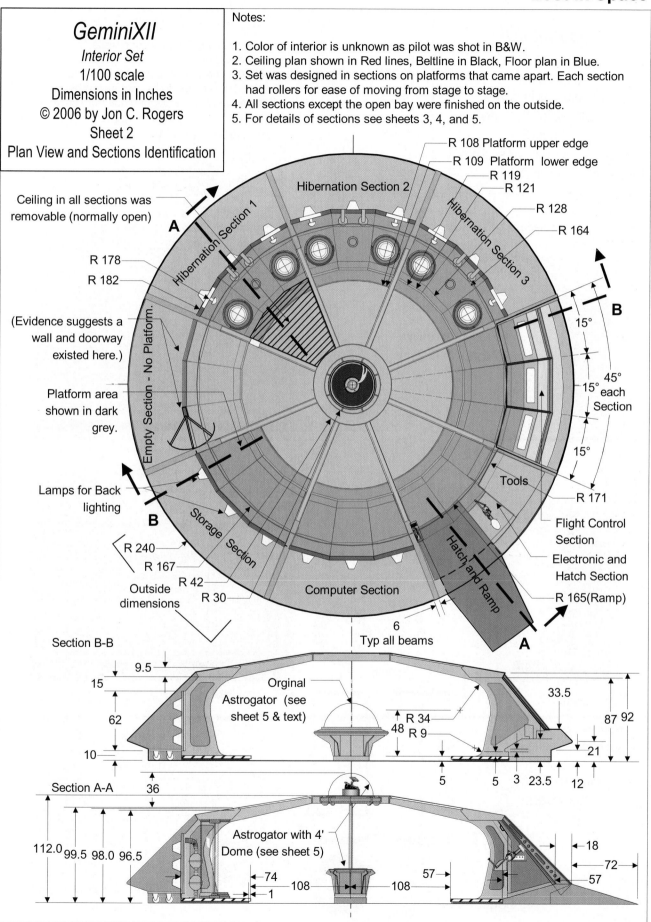

GeminiXII
Interior Set
1/100 scale
Dimensions in Inches
© 2006 by Jon C. Rogers
Sheet 2
Plan View and Sections Identification

Notes:

1. Color of interior is unknown as pilot was shot in B&W.
2. Ceiling plan shown in Red lines, Beltline in Black, Floor plan in Blue.
3. Set was designed in sections on platforms that came apart. Each section had rollers for ease of moving from stage to stage.
4. All sections except the open bay were finished on the outside.
5. For details of sections see sheets 3, 4, and 5.

R 108 Platform upper edge
R 109 Platform lower edge
R 119
R 121
R 128
R 164

Ceiling in all sections was removable (normally open)

Hibernation Section 2

Hibernation Section 1

Hibernation Section 3

R 178
R 182

(Evidence suggests a wall and doorway existed here.)

Empty Section - No Platform.

Platform area shown in dark grey.

Lamps for Back lighting

15°
15° each Section
45°
15°

Tools

R 171

Flight Control Section

Electronic and Hatch Section

Storage Section

R 240
R 167
R 42
R 30

Outside dimensions

Computer Section

Hatch and Ramp

R 165 (Ramp)

6
Typ all beams

Section B-B

9.5
15
62
10

Orginal Astrogator (see sheet 5 & text)

R 34
R 9
48

33.5
87 92
21

5   5   3   23.5   12

36

Section A-A

112.0 99.5 98.0 96.5

Astrogator with 4' Dome (see sheet 5)

74
1
108
108
57

18
72
57

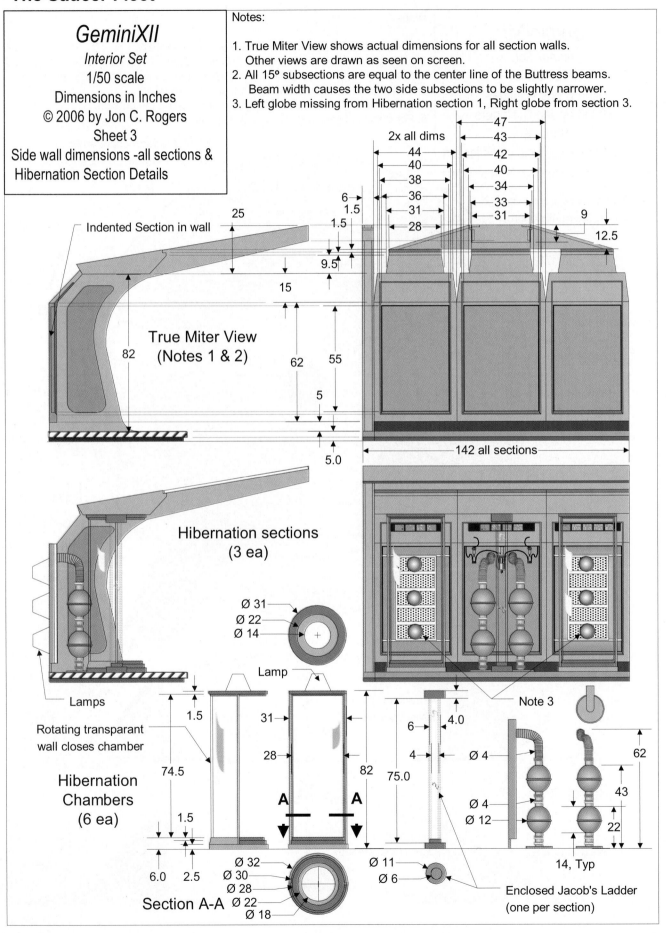

GeminiXII
Interior Set
1/50 scale
Dimensions in Inches
© 2006 by Jon C. Rogers
Sheet 3
Side wall dimensions -all sections &
Hibernation Section Details

Notes:

1. True Miter View shows actual dimensions for all section walls.
   Other views are drawn as seen on screen.
2. All 15° subsections are equal to the center line of the Buttress beams.
   Beam width causes the two side subsections to be slightly narrower.
3. Left globe missing from Hibernation section 1, Right globe from section 3.

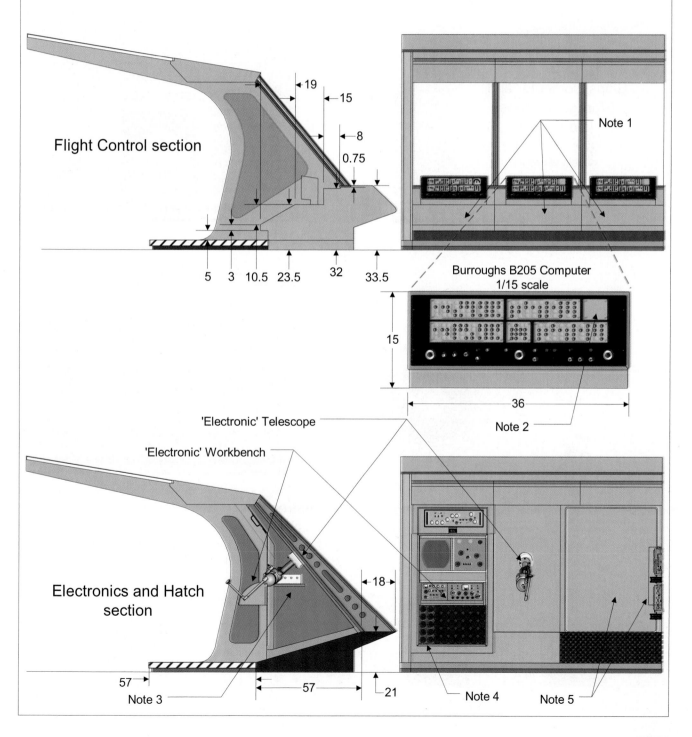

GeminiXII
Interior Set
1/50 scale
Dimensions in Inches
© 2006 by Jon C. Rogers
Sheet 4
Flight Control Section and
Electronics and Hatch Section details

Notes:

1. The Gemini had no other Control or Radar systems in these consoles. They served only as bases for the computers.
2. Of the three Burroughs Computers used as props, only the left one had a meter in this position.
3. The original blueprints called for Equipment storage in this area.
4. The 4" diameter rolls stored here per the blueprints, only the ones in black can be confirmed in the movie and set stills. The others are speculated. They would be needed to fill out the space.
5. The hatch had no working door. For the flight, it was sealed by a panel finished on the inside only. On the planet it was always shown open. The purpose of the equipment on the wall is unknown.

Flight Control section

19 · 15 · 8 · 0.75 · 5 · 3 · 10.5 · 23.5 · 32 · 33.5

Note 1

Burroughs B205 Computer
1/15 scale

15 · 36

Note 2

'Electronic' Telescope

'Electronic' Workbench

Electronics and Hatch section

18 · 57 · 57 · 21

Note 3 · Note 4 · Note 5

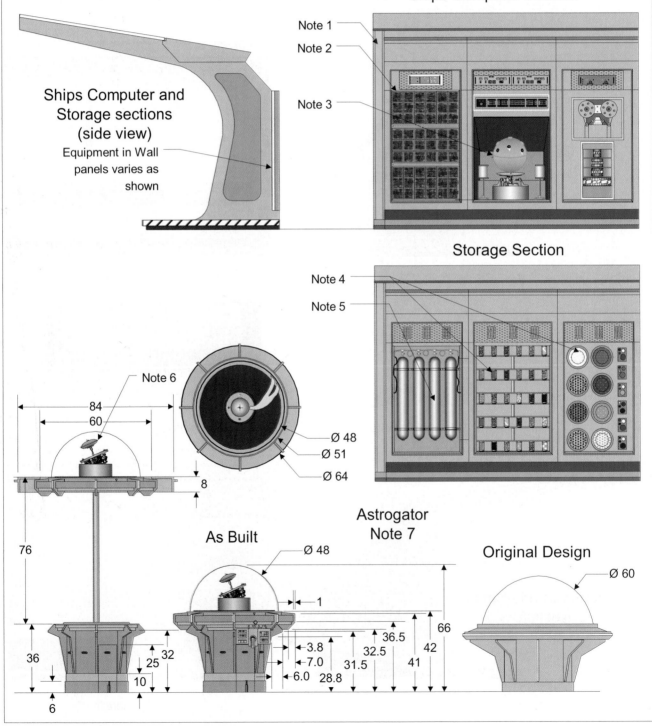

GeminiXII
Interior Set
1/50 scale
Dimensions in Inches
© 2006 by Jon C. Rogers
Sheet 5
Ship Computer Section, Storage
Section, and Astrogator Details

Notes:

1. All buttress beams are internally lit and flash on and off at random intervals.
2. Ships Computer section is backlit by rows at various times.
3. This part of the autopilot is set into the panel about 18 inches. The ball is lit internally and pivots above the stand. The two lights come on randomly.
4. This section is for food storage.
5. This section is for gas storage. It is seldom seen in the pilot.
6. The Astrogator saucer rotates at a high rpm as well as tilting in all directions.
7. The Astrogator was redesigned before the series debut. The original design is per the ships blueprints dated 12/64. (See text)

Ships Computer Section

Note 1
Note 2
Note 3

Ships Computer and
Storage sections
(side view)
Equipment in Wall
panels varies as
shown

Storage Section

Note 4
Note 5

Note 6

84
60

Ø 48
Ø 51
Ø 64

8

76

36
25
32
10

6

As Built

Astrogator
Note 7

Ø 48

1

3.8
7.0
6.0

36.5
32.5
31.5
28.8

66
42
41

Original Design

Ø 60

it that goes straight down to the ground (see Data Drawing sheet 1 and Figure 8). These two had to be different objects. Were they two different internal sets?

The first time we see the *Jupiter 2* in "Island in the Sky," it has an acutely angled underside. You can see it clearly when the environmental robot is sent outside to test the atmosphere and take a soil sample. See Figure 7. I noted that in this scene the airlock stayed closed preventing you from seeing inside the ship.

After review, I am sure that this is not the interior set but rather the full size 43-foot mockup that was made for the series. This model was also used in other scenes in episodes where they needed to show the ship in full profile or the Robinsons entering the ship via the stairs on the landing legs, such as in the first season, "The Derelict" and the third season "Journey to a Hostile Planet."

In most other scenes of the crash site, beginning with "There were Giants in the Earth," the saucer was full size, its airlock was open and the full detailing of the Main Deck Set was visible inside. In all the scenes where this was true, the saucer had a "foundation" under the rim of its beltline that went straight down to the ground (Figure 8).

©20th Century Fox Television

Figure 8

Other production stills show that this ship is the Main Deck set complete with outside "skins" so it could be viewed as part of the landing area set. Further evidence came when I reviewed the main Deck Set plans and discovered that the Main Deck Internal set had exactly the profile of the ship shown in Figure 8.

The drawings of the full size 43-foot mockup showed that it was designed to be dismantled so it could be moved from stage to stage. The outer sections of this saucer were actually removable and the center section had a dolly for transporting it.

This meant they could use the full size mockup inside one of Fox's sound stages to reproduce the crash site where viewing inside the ship wasn't necessary and outside where views of the ship in the open were needed. This also explains why the ship had two different profiles in two different episodes in the first season.

However at times other differences in the *Jupiter 2* Man Deck set were apparent. There are several scenes of the saucer where it had a taller front view port, airlock door and lacked details like a port hole next to the airlock.

These differences came during the episode, "There were Giants in the Earth." They stemmed from Fox using the original *Gemini XII* Interior set footage from the pilot episode at the same time they showed the set as it was modified to become the *Jupiter 2* Flight Deck.

©20th Century Fox Television

Figure 9

You can see how they modified the original interior set of the *Gemini XII* to become the *Jupiter 2* Flight Deck set for the series by comparing Figure 8 (*Jupiter 2*) to Figure 9 (*Gemini XII*). Since they used scenes from the pilot episode in the first five, first season episodes, we get to see both sets mixed in together several times.

In "There were Giants in the Earth," all three variations are visible. The external full scale mockup is shown in the title sequence. The *Gemini XII* interior set is seen as in Figure 9, and the changed *Jupiter 2* Flight Deck set as shown in Figure 8. To the long time fan or sharp viewer, the differences are enough to be confusing. However, at the time, most people in the audience probably did not notice the switch between differing models and sets.

The Saucer Fleet

The *Gemini XII* saucer's interior set design had already established the theme for the series. It was one circular deck divided into eight distinct sections. Eight dramatic, lighted, curved, flying-buttress-styled beams connecting the floor to ceiling. Each of these eight sections was equal in size.

In the *Gemini XII* interior set, seven of the eight sections were fully detailed although only three (the "Hibernation Sections") were designed to be fully used. The rest of the ship was largely for background. One section (the "engine" area) was not even detailed at all. Instead it was left open for easy use by the stage crew. In one rare, behind scene photograph a wall and doorway can be made out in this section.

Also the raised platform floor that circled the saucer's main deck was left out of this area so there would be a flat floor for the camera dollies. When the set was modified into the *Jupiter 2*, this raised platform was removed completely, as it was an impediment for the Robots movement about the deck.

There were two particular features of the *Gemini XII* interior set which gave away its "only for the first episode" nature. The first one was its complete lack of an airlock. The second was a scene that was planned to show its damaged engine area at the crash site.

The hatch of the *Gemini XII* had no working mechanism and no integral chamber with inner door to form an airlock. It was never meant to be shown in operation. It was only shown closed during the space flight and when on the planet, it was always open. While this is acceptable for a "one trip ship," it would be a major problem if the ship were needed to fly to more than one planet as the series developed.

It was also originally planned to show the ship badly damaged at the crash site as part of the "Space Family Robinson" pilot theme. Surviving plans of the scene show a damaged area specially fabricated for the engine area of the ship. This indicates it was to be of major importance to the story. It is not known why the shot was not actually completed and included in the pilot. However, since original evidence survives of this planned scene, we have included it in the data drawing (see *Gemini XII* Interior Set Data drawing sheet 1).

Both *Gemini XII* and *Jupiter 2* interior sets had three nearly identical hibernation stations. Each station had two freezing tubes and between them, two sets of spherical reservoirs with their 4-inch diameter pipes separated by an enclosed Jacob's ladder[4] over six feet tall (see sheet 3).

Behind this equipment are three sections of lighted wall. Two sections had three silver hemispheres on a panel with a control panel at the top. Each of these sections was removable. The lighting was varied as to when they were on or off

There were a couple subtle differences noted. The control panels vary from station to station in the number and pattern of lights in them and behind Penny Robinson's and Don West's tubes there are only two globes (see Figure 10 and Data Drawing sheet 3). Other than these items the three stations are the same.

In the flight deck section there are three computer consoles[5] mounted on top of what looks to be three blank control panels (Figure 10). In the *Gemini XII* these panels are too low to be really useful as control panels but they would be raised and modified to add radar consoles and controls when the interior was redesigned into the *Jupiter 2*.

In the center section of the control panel was a collapsible pilot's seat that would fold away out of sight when not used. This only existed for the first season. For the second season, a large radar screen occupied the place where the seat had been stored. By the third season, the

©20th Century Fox Television

Figure 10

---

[4] A Jacob's Ladder is two wires in a narrow "V", charged with high voltage so that an electrical arc will start at the bottom and crawl up between them, disappearing at the top. It has been a Hollywood tradition in every mad scientist's laboratory movie scene since *Frankenstein*. However, since the voltage is quite high and can give a person a nasty shock, these were sensibly enclosed in insulating, clear plastic tubes.

[5] These were not just props, but then-current, state-of-the-art Burroughs B205 computer consoles. Why all the flashing lights? Because they handled data, *one bit at a time* and these were the readouts. They had to be manually translated from binary to digital numbers or characters!

single pilot's seat had become double pilots' chairs. Figure 11 shows both the center radar and dual pilots' chairs.

The next section clockwise from the control area has what amounts to a workbench for electronic repair, electronic telescope and the entryway hatch and ramp. Above the bench is what could be a television monitor for communication. In the *Gemini XII* they were never used. In the *Jupiter 2*, the detail of the panel was redesigned. The TV monitor remained and the President made his speech on it just before launch. The communications set took the place of the electronic bench and there was a locker behind the entire console. Also room was made for the porthole access between the electronics console and the new airlock (Figure 11)

Next, to the right of the entry hatch (later the airlock) was the ship's computer and storage sections. In the pilot these sections were not given more than a few cameo shots and did not contribute to the plot. They did in the series. When the

©20th Century Fox Television

Figure 11

interior is modified into the *Jupiter 2* we see the *Gemini XII's* computer panels were moved forward one and a half feet and the storage section was moved downstairs to become one wall of the Galley in the living area. In its place was the ladder access to the Living quarters deck and a doorway that led to another compartment.[6]

One of the largest changes made during the development of the *Gemini XII* before its showing in the pilot was to the Astrogator. The Astrogator is shown in the original blueprints as being wider, shorter and having a 5-foot (1.5 m) diameter dome (see Data Drawing, sheet 5). By the time the Astrogator was actually constructed, it was taller, narrower, with a 4-foot (1.2 m) dome.

In both cases the Astrogator was to be capable of raising its dome up over six feet so it could look out the raised bubble on the top of the ship. Presumably, this is so it could get a fix on its position by sighting known star systems as it navigated the ship. One result of this change in domes was that the hole in the ceiling of the interior set was left at the original 5-foot diameter, which made for a loose fit!

Actually, there was no other dome in the ships ceiling. When we see the Astrogator in the raised position (as in Figure 9) it is only the Astrogator dome we are seeing. However, since there were no close-ups of this area in the series, *having a full time hole in the roof of the spaceship* made little difference.

As it was, the *Gemini XII - Jupiter 2* interior set was the most complicated prop made for the new series. It was quite detailed, and dimensionally close to what it should be in reality. With everything taken into consideration, it was the most accurate, rational representation of the intended spaceship on the series, a true high watermark. However, for all the effort in making this wonderful set, the Robinsons were supposed to abandon it after the first episode and venture off into the unknown spaces of their new "Island in the sky." But as it turned out, the set was to have a new lease on life and become something quite different than what it started out to be. With the change in plot, the interior was modified to become the *Jupiter 2* Flight Deck and would be seen through out the entire series.

Jupiter 2 Living Quarters

The change that defined the *Jupiter 2* from the *Gemini XII* more than all others was the inclusion of the second deck living quarters. This area added much to the believability of the *Jupiter 2* as a viable spaceship that could actually support seven people during their sojourn throughout outer space.[7]

---

[6] This compartment was at first just an equipment storage locker but later in the series would perform other purposes.

[7] The data drawings of the Living Quarters and Power deck are reserved for publication at another time due to the amount of space and time they would require.

The Saucer Fleet

This area was designed to reflect the design of the flight deck above it. It was divided into eight segments like the flight deck. However, the angled overhead beams protruding from a lowered ceiling were not the curved "flying buttress" styled support beams as before (Figure 12).

The living quarters contained three cabins capable of sleeping up to two persons each, one observation station complete with pilots chair, view port and controls, a fully equipped laboratory, a galley area with food storage and eating area, the elevator, ladder and engine room access area, and last but not least a small laundry and bathroom.

This last (a bathroom) has been something of a conspicuous absence from many Sci-Fi spaceship designs. It is interesting that, although it was not shown in use during the series, it was a detail included in the plans and is evident by carefully viewing scenes that it was built and was incorporated into the ship.

Next to the bath and laundry on either side were the cabins with folding door entries. Most of the time these cabins were shown with only one bed. However, the beds dropped down from a wall section and two were provided for in the studio drawings.

The observation station was forward directly under the pilot section of the flight deck. It was implied that it held auxiliary flight controls. The consoles on each side of the pilots chair contained acceleration couches which would roll out as required.

The laboratory section was the right of the observation station (Figure 12). Its consoles had roll out acceleration couches also. The laboratory and the observation sections were used frequently in many episodes.

To the right of the laboratory was the galley area and the ladder, elevator and engine room access area. A "magnetic lock station" was provided in the center of the room for the newest member of the crew, the "environmental control robot,"

There was also a narrow hallway between the laboratory and the galley that was never used or explained. Its existence is seen clearly in both studio blueprints and various scenes in the series. It remains one of the mysterious features of the *Jupiter 2*.

The most prominent feature was the galley area. This area was the social hub of the saucer. It featured the food storage area from the *Gemini XII* (seen in Figure 14 on the right). It had processing equipment on the left and back wall.

However, its most dramatic feature was a centrally located table that jutted out of the wall. This table had a centrally located conveyer belt that delivered food along the center of the table (Figure 14). The food appeared out from a small opening in the wall. How the meal was prepared is just another of the ship's mysteries.

Figure 12

Figure 13

Figure 14

274

Over the course of three seasons, changes were made to the living quarters (as were all sets) as required by the particular episodes storyline. The ships detail continuity was second to a good plotline. Still, the fundamental quality of the sets carried the show without allowing the myriad changes to destroy the illusion of reality.

And, admittedly, some continuity of the ship's details was observed. For example, great pains were taken to make it appear that the living quarters deck was indeed right below the flight deck. In reality, the two were entirely separate. The flight deck was on one stage and the living quarters were on a different stage in another area of the studio.

Another detail was that the *Jupiter 2's* living quarters deck, together with all its areas, was properly proportioned to have fit within the envelope of the saucer if allowances were made for some moderate height restriction. Indeed, the deck's ceiling was designed to be lower than standard to emphasize this restriction and add realism to the set.

This was an unusual break from Hollywood tradition. Furthermore, the consistency of the diameter and spacing of details with that of the flight deck, added more realism to the lower deck. There were detail problems (that have given later *Jupiter 2* model makers fits) but they were not apparent from the TV viewer's point of view and so it may be said that the set worked quite well during the series.

However, there were several areas of the ship's interior where accuracy and relation to the size of the ship was discarded for the sake of the story line. That was the addition of the power deck, the storage areas for supplies, the Chariot, and later in the series, the Space Pod.

Jupiter 2 Power Deck, Chariot, and Space Pod

Of all the decks, the Power deck was a return to the previous Hollywood tradition of "no relation to the outside size of the ship."

The Power deck was supposed to be below the living quarters and accessible only through the special door next to the elevator marked "Danger-Radiation." It was shown only in the third season episode, "Space Creature."

In this episode, Will Robinson must enter the Power deck to look for other missing members of his family. He opens the marked door, which reveals a small entry room, and then another door (also marked Danger Radiation) on its left wall. He then proceeds through a storage area with many boxes of supplies and spare equipment to arrive at a third door (also marked Danger Radiation). He enters into a room beyond this third door and we next see him descending into the Power Deck via a ladder on the far side of the room (Figure 15). It is here that the final scenes in this episode are played out.

I would like to point out that the storage room that Will enters is near the edge of the rim of the saucer. In areas like that, there is less height than toward its center. Therefore, the storage room is much taller than it could have been to fit within the saucer. Also, why would anyone ever put a storage room with vital equipment *in a Dangerous Radiation zone?* But that is just the beginning. There are also several real problems with this one episode's power room scene.

First, although Will has been repeatedly warned about the existence dangerous radiation, *he never puts on any radiation protection.* If this area is "hot," he is dead!

Second, if this "engine room" is "hot" what are all the control panels doing here? (Figure 15) This area should be run via remote control as in all other dangerous radiation areas.

Third, the power room as shown is about eight feet (2.4 m) high and fifteen feet (4.6 m) in diameter. This room is both too high and wide to have been within in the envelope of the *Jupiter 2's* design by any stretch of the imagination. By all calculations, Will is somewhere very near the power ring of the ship --an area not over a foot and a half (46 cm) high.

The power room was not the only problem in ongoing modifications to the *Jupiter 2.* Other areas added to the ship also violated the original design concept. One of the original problems was where to put the Chariot.

©20th Century Fox Television

Figure 15

The Saucer Fleet

Since the Chariot was included in the pilot episode where the Robinsons crashed, it was originally assumed to have been stored in a dismantled condition. Its size is such that there is literally no space for it within the ship. However, there is no evidence or dialog that was offered in the series to indicate that the Chariot was stored disassembled. It is a viewer assumption that has no basis other than the nagging question, "How did they get that thing inside that little ship?"

Another problem with this assumption is that there was a ramp designed and built into the bottom of the ship for the Chariot to drive out from its storage area. Although never seen in use, it was in the plans and included in the full scale mockup version of the ship. If the Chariot was stored inside this ramp area, it would have taken up a large part of the living quarters and possibly even intruded on the flight deck.

The Space Pod is another problematical addition. When it was introduced it was shown to be positioned in the ship off of the flight deck in a corridor to the right of the elevator.

In Figure 16, Don West is shown facing the area where the Pod is stored. He is already outside the normal flight deck area, about to shut the door that accesses it. The space he is in already has a restricted height if the original design means anything. The room is already too high.

Still he closes the door and enters the Pod. Since detailed drawings of the Pod still exist it is possible to say that the full size Pod was over 9 feet high by 9 ½ feet wide by 8 ½ feet deep (2.7 x 2.9 x 2.6 m). It isn't possible to fit such an object in the envelope of a saucer where its height is at a minimum.

Finally, as far as the Pod is concerned, I have discovered and shown on Sheet 1 of the Jupiter 2 Data drawing, that the model of the Pod was reduced by one half before they tried to add it to the 48-inch hero model. The Pod was just too large to have fit within the *Jupiter 2*.

©20th Century Fox Television

Figure 16

Still these elements were desired for dramatic (or whatever) reasons and so they were incorporated anyway into what had been a reasonably believable interstellar spaceship design.

The storage areas for supplies, the power room, the Chariot and the Space Pod were added extras that would not have fit within the envelope of the *Jupiter 2* by themselves, let alone all together. The power room, Chariot and Pod were the greatest offenders. Although, during the pilot episode, "Nowhere to Hide," with the lack of the Space Pod, Power Deck, and if we assumed that the Chariot was shipped disassembled we could claim that the *Gemini XII* was entirely consistent--something not seen in the movies since *Destination Moon* and perhaps, *Forbidden Planet*.

Quickspec: *Gemini XII*

Vehicle Morphology..............Saucer

Year 1965

Medium….....................TV Series

Designer.......................W. Creber

Diameter (exterior).…....48 ft (14.6 m)

Diameter (interior)..…..40 ft (12.19 m)

Height ....…14.75 ft (4.5 m)

Quickspec: *Jupiter 2*

Vehicle Morphology…..…........Saucer

Year 1965-68

Medium…..........…..........TV Series

Designer........................R. Kinoshita

Diameter (exterior)….......48 ft (14.6 m)

Diameter (interior).….47.3 ft (14.42 m)

Height (gear up)...........16.5 ft (5.03 m)

Height (gear down)…20.83 ft (6.35 m)

With the modifications added to make the *Jupiter 2*, however, we can only point out that, while the diameters of the first and second decks are consistent, the height was a problem. With the additions of a power deck, Chariot and Space Pod the ship is pushed out of the universe of reality and into the realm of Hollywood.

So the *Jupiter 2* began life as a very interesting and fairly convincing presentation of the concept of an interstellar flying saucer spaceship. It had sets and models possessing a large amount of detail, consistency and realism--a vision that greatly appealed to many science fiction fans. Over time, due to the demands of the studios producers who were not worried about its authenticity, the *Jupiter 2* was gradually modified into an unbelievable fantasy.

Modelers' Note

With the popularity of this show as great as it still is, it's no surprise that there are several wonderful models of the *Jupiter 2* to choose from. Lunar Models has a highly detailed 16.5" (42 cm) diameter styrene kit, but it's quite pricey if you include the optional resin-cast interior.

The *Jupiter 2* touches down on a strange planet! To the right are a view of the backside and main hatch detail.

Polar Lights, has a much more affordable 12" (30 cm) diameter kit in all styrene that includes the interior. One of the most spectacular build of a Polar Lights *J2* is by Steve Payne of Marietta, Georgia. Payne, who took over a year to build his enhanced version, included not only the parts you see in the show, but much of the hardware in the periphery around the rim. LED's and fiber optics bring all of the displays and living areas to life, right down to actual freezer tube glows and the twinkling lights above them! All photos © Steve Payne.

Far left: With the top shell removed we can see the fiber optics and other behind-the-scenes magic. The floor around the Astrogator is cutaway to reveal the lower deck.

Near left: Removing the top deck reveals the highly detailed lower deck. The elevator is at the top in this view. The red spheres are the auxiliary rocket motors.

Lower deck details. Left to right: Sleeping quarters, the elevator and Power Core entrance, and galley / landing leg exit.

Epilog

Lost in Space is, by far, Irwin Allen's most popular and long lasting show in terms of fan devotion. His other Sci-Fi hit shows, such as the earlier *Voyage to the Bottom of the Sea*, and the later *Time Tunnel* and *Land of the Giants*, were popular in their day, but today are not much more than footnotes to his career. *Lost in Space* is second only to *Star Trek* in the size of its fan base for mid-60's space-theme shows, and is probably first in the intensity of fan devotion. Of course, while *Star Trek* went on to eventually spawn four (as of this writing) later television series, *Lost in Space* had only its three seasons (the same as the original Star Trek series) on CBS; but that was enough to generate its fierce fan loyalty.

Organized fandom for the show can even point to a definitive starting point. Three years after the series was cancelled, the November 1971 issue of Forrey Ackerman's *Famous Monsters of Filmland* (issue #87) contained a letter by Ron Sapp, the first "publicized" *Lost in Space* fan. He included his home address so that fans could contact him and organize into a formal group. One fan that contacted him was Bob Richards who created the first fanzine devoted to the show, *The LIS Review*. From that sprang multiple organized fan clubs and fanzines (fan-magazines).

Today the fan base is massive. A brief web search reveals hundreds of sites either wholly, or partially devoted to the show or its cast members (with a sizable number of them devoted to just the Robot). It has become such a social icon in the Sci-Fi community, that you can evoke a response with little or no explanation. For example, when Joel Hodgson, creator of the Sci-Fi satire series *Mystery Science Theater 3000*, needed a name for his character, who had been shot into space against his will (and thus "lost"); he picked "Joel Robinson."

In a later interview,[41] June Lockhart (Maureen Robinson) relates how the Sci-Fi "cons" (conventions) with a *Lost in Space* theme are always well attended, and that the cons celebrating the 25th and 30th anniversary of the show both had attendance in excess of 30,000 fans.

Other parallels with *Star Trek* are that the TV show was followed by an animated series (voiced by some of the original cast members) and comic books. During the '70s, both shows attempted to produce big screen versions with the original cast, and while *Star Trek* eventually made it (in 1979, and is about to release film #11), the *Space* scripts were all stillborn. With Science Fiction films being so popular in the mid '80s, it looked like the Robinsons might make

it to the screen after all, with a script penned by Billy Mumy (Will Robinson) where the family and crew finally making it back to Earth after 15 years. Unfortunately, Guy Williams (John Robinson) died in 1989 and Irwin Allen in 1991, which shattered all hope of a film using the original cast and producer.

Lost in Space did eventually make it to the big screen in an effects-laden extravaganza in 1998,[42] but, while thrilling, it was a major disappointment to the fans. Both director Stephen Hopkins and screenwriter Akiva Goldsman were quite overt in their intent on making the film much "darker and more serious" than the TV show,[43] while apparently not understanding that the show was a success because it was just the opposite.

The first half hour of the film managed to work in a nice tribute to the TV series with appearances by June Lockhart as Will's teacher (making a holographic phone call to her movie self), while Mark Goddard (Don West) plays the general who gives his latter incarnation the assignment to pilot the *Jupiter 2*. Angela Cartwright (Penny) and Marta Kristen (Judy) don't get to play off of their doppelgangers, but are reporters at the pre-launch press conference interviewing John Robinson. Jonathan Harris was offered the role of Dr. Smith's superior who gives him the assignment to sabotage the Robinson mission, but he turned it down cold:

> They did offer me an innocuous six-line bit that they laughingly referred to as a cameo, and I told them exactly where to shove it. The director was a very nice man and he called from London and we had a very nice chat. He asked me what he could do or say to convince me to be in the movie. I said "There is nothing you can do and there is less you can say." I felt, and I still feel, rather proprietary about Smith. I created him and I feel I own him.[44]

That was the last chance to get Harris in the role as he passed away on 3 November 2002. Love him or hate him (many people did both), Dr. Smith was the heart of the show. Without him, it's hard to imagine it lasting three seasons.

Billy Mumy was offered the role of the grown-up Will in the movie's time-travel climax, but he used the more diplomatic "scheduling conflicts" when refusing. The only member of the original cast to perform in his original role was the invisible Dick Tufeld reprising his part as the voice of the Robot.

The original *Jupiter 2* makes a cameo as well when the ship blasts off from the top of its launch tower (inexplicably a thousand feet tall), but after leaving the atmosphere it turns out to be only an aeroshell that is blown off into space

---

[41] TV special *Lost in Space Forever*, 1998.

[42] The premier was 3 April 1998. It would have been much more significant for the fans if they could have brought it out six months earlier in October 1997, preferably on the 16th…

[43] *Starlog* #249, April 1998, pg. 33, and #250, May 1998, pg. 44.

[44] *Starlog* #248, pg. 27.

©New Line Cinema

The plot, which is not germane to our discussion, incorporates a lot of real concepts, such as gravity slingshots, etc., into the dialog. But the awkward and incorrect way that it's done (doing a gravity-assist by going **through** the planet?) shows that the scriptwriters really didn't have a clue what they were talking about. By comparison, this film makes *Rocketship X-M*[45] look like a NASA mission proposal.

One final note. The film's $70 million budget was exactly 100 times that of the TV pilot. That the "most expensive pilot in TV history" in the '60s cost only 1% of a moderate Sci-Fi film in the '90s shows that "space" might not be the only thing we've lost.

revealing the all-new, completely modern *J2* which zooms off, leaving thousands of new pieces of space junk behind in orbit. Of course, it's not really the same ship at all. The original *Jupiter 2* was, as described previously, a motor home of a spaceship, practical and functional with all of the amenities and equipment necessary for a successful multi-year voyage, but still quite compact. This new ship is a stupendously gigantic vehicle, a flying concert hall with 40-foot (12-meter) ceilings and huge sweeping galleries full of nothing but air. No explanation is ever given as to why the ship has to be so gargantuan, or why a family of five plus a pilot needs millions of cubic feet to live in.

©New Line Cinema

It would probably be a lot easier to put the overhead controls down where the pilot actually sits.

---

[45] See *Spaceship Handbook* for the history of this film.

The INVADERS

By Jon Rogers

Your are David Vincent. It is after midnight. You are alone on a country road, asleep in your car. Strange lights and eerie sounds awaken you. In a nearby field, you see an alien flying saucer descend and land. Luckily, you get away. Now you must warn the people of Earth that the invasion has begun! But the ugly truth is: *you have no proof!*

You soon discover that no one wants to believe you. You cannot easily tell friend from foe. And…*"they"* are after you!

You are alone in a battle against a cleverly disguised, technically advanced race bent on humanity's destruction. Posing as humans, *they* betray you. *They* kill your best friend. *They* hound you. *They* cause you to lose your dreams, your reputation, your job, your home, your family and your girlfriend. Your life has become a living nightmare.[1]

©CBS Paramount Network Television

Figure 1

You are now in the world of *The Invaders.*

What makes a nightmare truly frightening? Is it the Danger of the situation? Its reality? The timing of when it occurs? All of the above?

The Quinn Martin production of *The Invaders* that aired over the ABC network beginning on 10 January 1967 had all those elements going for it. As a result, it was a truly frightening TV series, perhaps the most nightmarish of all TV shows of the decade.

Consider its timing with regard to society and the UFO phenomenon.

Between 1965 and 1967 there had been a resurgence in the number of UFO sightings throughout America. According to Project Blue Book, the number of sightings in 1966 exceeded all previous years except 1952.

By the 1960s there had been several widely publicized cases of people who claimed to have had contact with, or even been abducted by, aliens in flying saucers. This had begun with George Adamski in 1953, who claimed to have met with aliens from the planet Venus and Saturn. In many cases, the public greeted their stories with disbelief, just like David Vincent's.

Also, America had already gone through the Senator McCarthy-driven Communist witch-hunts and the extremely paranoid times he created. We were in the depths of the Cold War. The Vietnam War was frustrating us overseas. Anti-war riots and civil rights "demonstrations" were tearing up our domestic peace. Russia was ahead of us in space and there were counter-culture Hippies in the streets. Everyone was worried about who could be trusted—and who couldn't. Never before had American Society been assaulted on so many fronts.

At the same time, the ABC Television Network was also in trouble.

The network was having a terrible time during 1966-67 because its line-up [of shows], including *The Green Hornet, The Milton Berle Show,* and *The Monroes* were bombs, and old-standbys such as *Combat, Batman,* and *The Fugitive* were fading. The network needed a hit.[2]

It was at this moment that Quinn Martin Productions stepped in to help.

Quinn Martin, who later built a reputation as a crime drama producer…already had the *"The Fugitive," "The FBI,"* and *"The Untouchables"* (which he worked on for *Desilu*) to his credit. His name promised quality to the networks.[3]

Quinn Martin had little trouble selling ABC the idea of a new prime time TV series called *The Invaders.*

---

[1] *Unseen Invaders,* Mark Phillips', *Starlog # 206.*

[2] *A story of watching The Invaders,* Mark Phillips, http://www.theinvaders.co.uk/, used with permission.

[3] *Unseen Invaders* by Mark Phillips *Starlog # 206.*

The original concept for the series was the work of writer-director Larry Cohen. Cohen had envisioned the program as a twice-weekly half hour serial. He had also decided to use the atmosphere created by the McCarthy era—where Communist secret agents were believed to have infiltrated American society—as the main theme for the series. During that time, there had been plenty of suspicion and fear upon which to base a dramatic series. Cohen worked up twenty-two half hour story outlines and went to pitch the idea to Quinn Martin.

Martin liked the concept but wanted to change the format to a one-hour show. He and Cohen also disagreed as to who should have control over the series' production. This resulted in Cohen leaving his basic idea and story outlines with Martin, then quitting QM to begin directing feature films.

Figure 2 - Quinn Martin

Quinn Martin (Figure 2) hired Anthony Spinner to direct the series. Like his other productions, *The Invaders* would be first class, with top-notch actors, writers, sets, special effects, and guests.

At the time, QM (as the Quinn Martin production company was called) had several series in production, so they had a regular group of good actors to pull from.[4] Well known stars like Diane Baker, Anne Frances, Peter Graves, Arthur Hill, Jack Lord, Roddy McDowell, Burgess Meredith, Jack Warden, Fritz Weaver, James Whitmore, and William Windom would appear on the show. Interesting newcomers like Karen Black, Gene Hackman, and Peggy Lipton, would have featured guest roles, while other newcomers like Louis Gossett Jr. and Diana Muldaur would get bit parts in the series. Many of these actors would become stars later in their careers.

From the beginning *The Invaders* would use a different approach and focus on a different audience than the current two most popular Science Fiction TV Series. Of these two, *Lost in Space* was a family drama and *Star Trek* appealed to more hard core SF enthusiasts. They were set in either the distant future or another solar system. *The Invaders* would be set in the here and now and would aim at an adult audience, especially those who might be turned off by the imaginary settings of the other two Sci-Fi TV shows.

By doing so, *The Invaders* gained a measure of realism that the others lacked. This also allowed it to pick up on current events that added still more realism to the show. This realism increased the emotional impact of its main thesis—that Aliens have landed, are hidden among us, and are working toward our destruction.

Adding to that impact were the excellent special effects used in the series. While not of the type seen in the multi-mega-budget movies of today, they were on par with the best available in movies at the time. Of course, they would not need as many effects scenes as the other shows would. Still, showing a dying alien vanishing in a ball of fire, leaving nothing behind for poor David Vincent to claim as evidence, was very effective.

Even though the special effects were well done, they were used as little as possible. Director Spinner said that this was mostly done due to budgetary reasons, "We tried to invent effects that were not only plausible but that we could afford."

> One of the series' most interesting effects was the flying saucer. The saucer was particularly striking in scenes where aliens milled around the landed craft. "There was no full–scale saucer mock-up," says Spinner. "All we had were the four landing struts. The rest was matted in by the effects team."[5]

Spinner created a number of elements that became associated with the series. Besides the introductory prolog about the aliens coming from a dying planet, he also created the idea that the aliens glowed and then disappeared when they died on Earth. Spinner also pointed out that the 22 original plots that Larry Cohen had created were not used because, being written for a half hour serial, they wouldn't work for a one hour show.[6]

In fact, for the writers, the theme and plots that it required became the series most challenging aspect. The theme itself was well suited to the requirements of TV as it was a tough uphill battle for David Vincent to prevent aliens, who were almost indistinguishable from real humans, from taking over the planet.

However, in order to continue the theme from week to week required several forgone conclusions. One was that David Vincent would never amass enough evidence to reveal their presence and allow the human race to retaliate against them since that would end the series. The stories always had to end with the idea that, while David Vincent

---

[4] QM Productions would have one or more shows in prime time on ABC for 21 straight years, from 1959 to 1980, a still unbeaten record of success. Martin would sell the company in 1978 while it was on top with the condition that he not compete with the new owners for 5 years. Quinn Martin died in 1987, at age 65 after producing a career total of 17 TV series, 20 made-for-TV movies, and one feature film.

[5] *Unseen Invaders*, Mark Phillips *Starlog* # 206.

[6] Cohen claimed in a later interview (*Starlog*, 1977) that his stories *were* used. However, this has never been proven.

The Saucer Fleet

might be able to win a battle, he would never win the war. Furthermore, for their part, the aliens would never be able to kill David as that would also end the series.

This was a tough problem for screenwriters to keep plausible. After all, the aliens were supposed to be far superior to humans in technology and had many gadgets to prove it. Why then was it that they couldn't eliminate just one pesky human who endangered their entire project?

Walking the tightrope between these two extremes eventually produced an unintentional effect on the audience—frustration at David's lack of progress.

In the meantime, many inventive plots did manage to produce the desired effect—fear and a good deal of paranoia. The aliens were supposed to appear threatening, and for the most part, they were. Even the technique of having very few things distinguish an alien from a human gave the audience reason to fear almost every stranger David met. In the context of the tumultuous times of the '60s, this was very effective.

One thing that *could* identify an Alien was a straight and immovable little finger. What made this minor gimmick so scary was that it was picked up and mimicked by many people who watched the series. Suddenly, this *little* item started popping up *all over the country!* In the

©CBS Paramount Network Television
Figure 3 - The famous, frightening little finger

entertainment industry, Hollywood trade magazines reported that various actors and actresses were seen at important cocktail parties sporting stiff, straight little fingers. On other TV shows, anyone—even Newscasters—might suddenly display a stiff, extended little finger.

During an episode of the rival Sci-Fi show *Lost in Space,* all of the Robinson ladies—Penny, Judy and even mother Maureen played by Angela Cartwright, Marta Kristen, and June Lockhart, respectively—were seen to have stiff, extended little fingers. Were they Invaders too? If even the Robinsons, lost on some distant planet, could be "infected" was anybody safe? Rumors spread. Seemingly, *The Invaders* were *everywhere!*[7]

The author recalls seeing the series in the '60s, and personally observing how disconcerting this little "gimmick" became. One might go to work or school some morning and discover a co-worker displaying a rigid, ex-

tended pinky. Normally, a joke like this would be taken as "all in good fun." But this was the already paranoid '60s featuring war overseas, riots in the streets, space being penetrated further every day, and mounting reports of UFOs. No American could feel secure. These factors, combined with David Vincent's serious (and continuingly fruitless) struggles against an enemy cleverly hidden among us, made the unexpected discovery of a stiff, straight little finger on someone *near you* a very unsettling experience.

As for timing, the introduction of *The Invaders* as a mid-season replacement on 10 January, 1967 was perfectly in sync with the paranoia of the time. This was the year China announced that they had a thermonuclear bomb. Dozens died in six days of race riots in Newark, NJ and more died in eight days of riots in Detroit. Russia made their first soft landing on the Moon, and about two weeks after the show premiered, three Apollo astronauts died on the pad at Cape Canaveral. Anti-war protestors were marching everywhere. The military was bombing Hanoi, US casualties were mounting, and the Viet Cong were cooking up something called "the TET Offensive." The last thing Americans needed was another threat!

The cumulative effect was that *The Invaders* became one of the most watched shows on National TV. It beat out its two direct competitors—CBS's previously unbeatable *Red Skelton Show* and NBC's *Occasional Wife*—and landed among the top 20 shows in the country. Against the backdrop of strife in real life everywhere, the show's presentation of "enemies in our midst" made for compelling watching.

When the first season was over, *The Invaders* had proven more popular than either of the other two science fiction series on TV. It finished with a Nielson rating of 20.5, slightly ahead of *Lost in Space's* 19.5 and well ahead of *Star Trek's* 17.9. Millions of Americans had been "*Invaded.*"

However, as it turned out, this first season would be its high point. Anthony Spinner, the original writer-producer, had left saying, "Quinn, it's a good series and it's well made, but the show has no future."[8] His replacement was David Rintels who must have agreed with him. When *The Invaders* returned in fall for the 1967-1968 season, he had changed the show's format.

Mr. Rintels' new direction for the series was to move away from the shock and horror that had permeated the first season and to tackle some of the current issues of the day.

> I felt that by using this format, I could write about things I cared about. I wanted viewers to feel and think.' Vietnam, racism, nuclear war, drug abuse, and the other themes were touched upon [in the second season]. Rubber suited monsters need not apply. The Invaders [would be] the most adult science fiction series ever made.[9]

In the second season, the show would do just that. It

---

[7] This gesture should be ranked together with the split finger Vulcan "Live Long and Prosper" sign as the most widely recognized alien gesture in entertainment history.

[8] *Unseen Invaders*, Mark Phillips *Starlog* # 206.

[9] *A story of watching The Invaders*, Mark Phillips.

would touch on such contemporary problems as the Vietnam War, racism, Watts and other riots. It even mentioned the current football star Johnny Unitas.

While many established viewers were unhappy about the new direction that the second season took, Roy Thinnes, the star of the series (Figure 4) was very happy. Like many people, he had become frustrated with the standard ending to each episode. "I would come home and say to my wife, 'it's depressing, David Vincent is going to lose another battle.'"[10] With this new direction he felt that, because the series was leaning more toward drama than science fiction, he would get to show a more complex, sensitive nature in the character of David Vincent.

©CBS Paramount Network Television
Figure 4 - Roy Thinnes as David Vincent

One of the series' writers, John W. Bloch, agreed that it was a chance to explore more of the wide variety of issues present in 1960's society.

> It was done during a time of disillusionment with the government. We had recently had the purges [McCarthyism/blacklisting in the '50s] where the entertainment industry had been really hurt by people who testified against others. Some of that was in *The Invaders*. We also wanted to do programs that spoke to the concerns of the 1960s.[11]

According to Bloch, the aliens were not evil, just threatening. He drew the parallel between them and the explorers of the 15th through the 18th centuries. He said, "To me, the aliens were pioneers from space. We were the Indians they found."[12]

However, with this new direction, the show's ratings began to slip. This gave the producers something new to

worry about. Even though David Vincent now had people who believed in him, the ratings were not recovering.

The fault could not be blamed on the show's quality. The writers, actors, and directors were some of the best. They worked hard to make each show believable and they did an excellent job. The show had many loyal, highly vocal fans.

Still, according to the Nielsen ratings, the show went into a decline during the 1967-68 season and finished with an average of 14.5. ABC made the decision to cancel the show in March. Even though the ratings began to improve in April, it was too late.

After its cancellation, *The Invaders* did not die, but went into syndication where it did very well. The series had several successful runs in the 1980s and even became a cult hit in France. It has been shown almost continuously in one area of the world or another since 1969.

The first time *The Invaders* would resurface was when QM used the basic plot of *The Invaders'* pilot in one of its *Tales of the Unexpected* episodes. The episode was called *The Nomads*. It was about a man (David Birney), who witnesses a flying saucer landing, tells a local woman (Lynne Marta) what he had seen, and tries to stop the invasion single-handedly. However, the woman is an alien in disguise who betrays him, the hero ends up being declared insane, and the invasion succeeds in the end.

The next bit of *The Invaders* came, strangely enough, during a famous comedy show in the late 1970's. In 1979, in one episode of *Mork and Mindy* where Raquel Welch appeared as a guest star playing an Alien Queen, they used a clip from *The Invaders* to show her space ship landing. Sadly, these effects shots were not credited.

Next, QM made an attempt to revive *the Invaders*. In a 1980 TV film called *The Aliens are Coming* starring Tom Mason, Melinda Fee, Max Gail and John Milford, Quinn Martin retold the original pilot of *The Invaders*. It did not have much of a budget for special effects and did not get another series started as was hoped.

In 1995, there was a remake of *The Invaders* as a four-hour mini-series. It starred Scott Bakula as Nolan Wood, Elizabeth Pena as Ellen Garza, Delane Matthews as Amanda Thayer, and Richard Thomas as Jerry Thayer. Even Roy Thinnes made a cameo appearance as an elder David Vincent. It too failed to restart another series and was edited into a long, made-for-TV style movie.

As we proceed into the 21st Century and look back at the series, we can better understand the many reasons for its popularity. It was well made, with excellent actors, writers, and special effects. It was well timed, bringing stories of compelling drama during a tumultuous era. And perhaps more subtly, it tapped into people's interest in UFOs. After all, it was the first TV show to address the UFO phenomenon as seen by people on Earth.

---

[10] Ibid

[11] From *"Unseen Invaders"* by Mark Phillips *Starlog* # 206

[12] Remembering the early explorers' treatment of the Indians does not make one feel any better about his approach!

The Saucer Fleet

The possibility that UFO aliens could be already here and invisibly mixed in amongst us was the series' defining characteristic and the source of its power to command our attention.

And of course the first question to always ask when dealing with aliens in UFOs or flying saucers is, "Are they for us, or again' us?" People's response to this question has been evenly mixed over time. But here was a show that presented the valid viewpoint that they are definitely against us. And it did so in a realistic, adult fashion.

The Stories

There were 43 stories in *The Invaders* TV series. Most of them are told entirely from David Vincent's viewpoint. Although the plots sometimes varied widely from the basic theme shown in the pilot film, the series was always perceived as a continuation of David Vincent's battle to bring the alien invasion to light.

There were many stylistic similarities among all the shows in the series. They would all start with a recap of David's seeing the alien saucer land that one dark night. They would all use the same title styles, with the words "The Invaders" being revealed by a series of spotlights expanding in a pulsing fashion on a black screen. Then the screen would be torn apart to reveal the guest stars and the preview scene. The several segments of the story would be clearly marked by the titles, "Act I, Act II, Act III, Act IV" and an "Epilog" ending the show. Each act was contained between the commercial breaks. This was a style shared with other QM shows made during this period. It was more formal than other TV series then being shown.

At the time, these stylistic features were very effective in heightening the suspense of the program.[13] Today these features, along with the over saturated colors typical of every color TV show of the era, clearly date the series.[14] In viewing the series today, it must be kept in mind that these characteristics marked *The Invaders* as a premium show during its time.

Since I do not wish to recount all 43 shows, the following synopsis will concentrate mainly on the pilot, which set the theme for the whole series.

It is the evening of 10 January 1967. You tune your TV to your local ABC station and this is what you see (Figure 5):

---

[13] Rival SF series, Irwin Allen's *Lost in Space,* used a similar titling feature, i.e.: "to be continued…" in the old "playbill" font at the end of each episode.

[14] When *The Invaders* debuted, color TV's were very expensive and color programming was only in its third full season on ABC. Color was intentionally broadcast at a very intense level to help color TV sales. It made the contrast striking when compared side -by-side to the much more economical B&W sets in the store.

It's a lonely night on a country road. A pair of headlights peers out of the mist. A single car splashes across a shallow puddle as it comes toward you. Eerie music floats in the background. A hidden announcer intones:

"How does a nightmare begin?"

The camera closes in on the driver. He appears to be a square jawed man in a light business suit. His tie is loose and his shirt collar is open. He looks tired. The announcer goes on:

Figure 5 - *The Invaders*. Alien beings from a dying planet. Their destination: The Earth. Their purpose: To make it *their* world.

"For David Vincent, Architect, returning home from a business trip, it began at a few minutes past four on a lost Tuesday morning…looking for a shortcut…that he never found."

The camera does a close up on the car's headlights and then drops to show the front wheels rolling over a "Road Closed" sign laying flat in the dirt road.

The music drops dramatically. The driver peers out into the mist as the car bumps over the uneven surface. Dimly in the headlights a sign appears on the roadside. The sign reads "Buds DINER." The announcer continues:

"It began with a welcoming sign that gave hope of black coffee."

The car pulls past the sign, goes a little further and then pulls into a wide driveway next to an old building. The man looks out through the windshield. The headlights illuminate the building. It is deserted and ramshackle. Abandoned. Dejectedly the driver pulls up in front of the diner and stops the car.

"It began with a closed deserted diner, and a man too long without sleep to continue his journey."

The driver shuts off the engine, slides down in the front seat of the car, and lays his head against the seat back. He has decided to catch some sleep in his car. After a few movements to get more comfortable, he slides into a deep sleep. (Figure 6)

Figure 6 - Sleeping in front of Bud's Diner

The camera pulls away from his sleeping form to show the car sitting in the driveway in front of the deserted diner.

"In the weeks to come, David Vincent would go back to how it all began. Many times!"

We see a close up of his sleeping head, resting on the seat back. From far off a whirring sound rises, thrumming, whistling, and growing louder. It is joined by other strange noises. Suddenly David's face is caught in a bright glowing light that slowly pulses on and off.

Figure 7 - David Awakes to the amazing sight

The camera closes in on his sleeping face as his eyes start to flutter. The sound and the pulsing light are waking him up. Slowly he lifts his head and looks toward the light. He blinks in amazement. (Figure 7)

There, hovering over a nearby field is a flying saucer. Slowly it is coming down. It is landing. Astonished, he watches as it lands. David Vincent has just witnessed a flying saucer land on Earth. (Figure 8)

The music comes back, jarringly. The scene of David's astonished face is torn apart to reveal a black background. Slowly, lights pierce the darkness to reveal letters that spell, *The Invaders.*

The announcer states that it is "A QM Production." Again the screen is ripped, this time to show other faces. "Starring Roy Thinnes," the announcer intones, "with guest stars Diane Baker, J.D. Cannon, James Daly, John Milton." Tonight's Episode:

Figure 8 - The Alien Saucer Lands on Earth

Figure 9 - He points to his truck

Beachhead

ACT 1

A car is seen pulling up in front of the Sheriff's office. A man gets out and rushes inside.

Inside, David Vincent is telling Lt. Ben Holman (J.D. Cannon) about his experience. Lt. Holman brings in the man who just arrived. The man introduces himself as David's partner, Alan Landers (James Daly). Alan confirms that David was on a business trip and then asks David, "What's this about some wild story about you having seen a spaceship?"

Lt. Holman, clearly unconvinced by David's story, suggests that he is fatigued, that he should leave his car at the office, and have Mr. Landers drive him home. David gets upset at Lt. Holman's lack of concern and demands that he investigate immediately.

It takes some time and argument but finally—and reluctantly—Lt. Holman and Alan agree to accompany David out to the site where he saw the saucer land.

When they arrive at the diner, David is surprised—and disturbed—to see that the name of the Diner has been changed. The sign now reads, "Kelly's Diner." Furthermore, when they search the area, they can find no saucer and no evidence that one had been there. This convinces Lt. Holman that David has been seeing things. Alan is also concerned about David's belief that he saw something.

While they are searching, they hear several shots. Running over the hill, David comes upon a young man shooting at flying birds.[15] A young woman is with him.

Before he can question him, Lt. Holman arrives and asks if the couple had seen anything strange the previous night. The young man claims that they have just gotten married and they have been camping there on their honeymoon for the past two days. But when he goes to point out where his camper is parked, David notices he points with the small finger on his left hand—as though the finger was stiff and would not bend. (Figure 9)

The man says their name is Brandon, but when he shows Lt. Holman his driver's license, David observes the same thing about the small finger on his right hand. He becomes suspicious about this fact but cannot interest either Holman or Landers in it.

Lt. Holman now considers David's story to be thoroughly discredited and that David must be a "nut." He refuses to investigate further. David is frustrated and upset. He decides to continue the investigation on his own.

Later that night, David drives back to the young couple's campsite to question them further. As he drives up to the camper, he is greeted by the young woman carrying a bag and the rifle. Her husband is hurriedly packing the camper to leave. When David asks her about changing the diner's sign, her husband approaches and orders her back into the truck.

He tells David to go away and turns to leave. David notices that he has gloves on and grabs the man's arm saying, "I want to see those hands! Take off those gloves!" The man refuses and tries to leave. When David wrestles with him, he turns and hits David, knocking him down. The two men struggle for a while until finally the young man knocks David down hard. While David is down, the young man picks up a big rock and struggles toward him. Just as he lifts the rock, the palms of his gloves begin to glow brightly. (Figure 10) He drops the rock, looks at his gloves in horror and backs away toward the camper. His wife shouts at him to hurry. David stares on in disbelief.

He climbs into the truck while David struggles to get up. The young woman guns the engine. Unable to move out of the

---

[15] No one seems to notice that he is using a lever-action open sight 30-30 rifle, a gun that is very unsuited for flying birds. Strange.

Figure 10 - His hands start to Glow

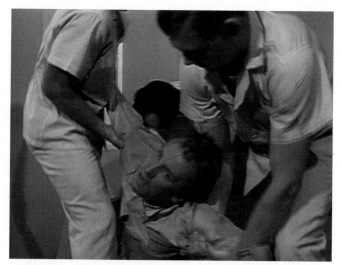

©CBS Paramount Network Television
Figure 11 - David struggles wildly

way, David watches in horror as the camper comes charging toward him. The right headlight is the last thing he sees.

David slowly regains consciousness. An overhead light comes into focus. He is in a hospital room, lying in a bed and a nurse is leaning over him. He asks the nurse, "What happened?" The nurse replies, "Do you remember anything?"

David mumbles something about the truck and the nurse cuts him off saying, "Don't try to remember anymore. Here, take these," and offers David some pills and water. When David asks what they are, she says they are to help him sleep and if he doesn't take them, she'll have to give them to him intravenously. Weakly, he protests.

When he raises his arm to push the pills away, he notices that the identification band on his wrist says "Arthur Gordon." He immediately becomes suspicious—and frightened.

"What are you trying to do to me? What is it?" he mutters. Then, as the nurse pushes the pills at him again, he becomes frantic. "You're one of them, aren't you!" he shouts and slaps the pills away.

Weakly he struggles to get up and staggers out of the room. Just outside in the hallway, he is grabbed by a male intern who wrestles him back into the room and forces him back into bed, shouting and screaming.

As the male intern holds him down, the nurse prepares an injection and approaches him menacingly. David struggles wildly but he is too weak. Then, as the nurse starts to give him the injection, his eyes go wild and, with strength born of fear, David manages to throw off the intern and run out of the room.

David staggers down the hallway, followed slowly by the nurse and intern. But he can find no exit and two more male interns approach him from the next corridor. Desperately he picks up a small bench, throws it through a window and tries to climb out just as the three interns grab him.

They wrestle David to the floor with him shouting, "Help me! For God's sake, help me!" At that moment Lt. Holman and his friend Alan Landers come rushing up. As he begins to black out, David realizes the Police Lieutenant and his friend have put him here. Then he loses consciousness. (Figure 11)

ACT II

As the scene opens, David is being wheeled into a courtyard in a wheel chair by a male intern. He asks if David is okay and then leaves him in the sunshine and fresh air. A moment later Alan Landers arrives. He apologizes for the misunderstanding and explains that he put David in the hospital under an assumed name because he was afraid that newspapers would get wind of David's sighting and twist the story.

David tries to convince Alan that the two people camping out were "some kind of alien beings" but Alan won't hear of it. He tells David to wait until he picks him up that evening before he will discuss it. He gives David some magazines and leaves.

As David dejectedly wheels himself away, a little old lady is watching him closely. Slowly, she takes off her glasses and wipes them –with some difficulty—as her little fingers will not bend.

Alan picks up David that evening as promised. And surprisingly he has done a little checking on his own. He has found out where the young couple lives. Although he does not believe David, he is clearly trying to calm him and show some sympathy for the anguish David is going through. He repeatedly asks David to "drop the whole thing." Finally he tells David that the couple lives in a little town named Kinney. At the end of the evening, David heads off to his hotel room to get some much-needed rest.

That night, he is awakened by the smell of smoke. He opens his eyes to discover his hotel room door is open and the room is ablaze, blocking his exit. (Figure 12) He picks up a rug and tries to beat the flames out. Failing that, he turns, picks up a large candelabra stand and as he turns

©CBS Paramount Network Television
Figure 12 - Fire in David's Apartment

back, he sees a little old lady peering into his room from the hallway. Just then, part of the ceiling collapses. He ducks. When he looks back, she is gone.

Using the heavy iron stand, he breaks open the door to the balcony, rushes out, and jumps. It is only one story to the ground and he is unhurt. A fireman rushes up to help him. He says it looks like David is the last one in the building. David tells him about the old lady, but when they search the building they find no one.

Figure 13 - David discovers alien machines

ACT III

David drives into Kinney. He looks around and notices that the town looks empty. The streets are deserted. Driving to what seems to be the center of town, he parks his car and walks over to a hotel. In the yard beside the hotel he meets Kathy Adams (Diane Baker). She explains that she runs the hotel, but it is closed now that only twelve people live in Kinney. It seems the town is being bought up by a Mr. Kogan for investment purposes. David asks about the young couple named Brandon, but Kathy doesn't remember them.

She says he can go inside and check the hotel register. As he does, Sheriff Carver (John Milford) enters the room. Carver mentions that the town was doing well until the state shut down the hydroelectric plant. Since then everyone is selling out to Mr. Kogan. David asks Kathy if this includes her. She says she sold out when her husband died and is just staying on as caretaker until the new owner takes over.

As they leave, the sheriff makes a phone call to Lt. Holman. He leaves a message for Lt. Holman that "He was right. That psycho did show up here. I'll keep an eye on him."

After David says good bye to Kathy, he leaves his car parked in town and goes towards the hydroelectric plant on foot. He hides under a small bridge until after he sees Sheriff Carver's car pass by and drive up to the front gate of the

plant. He watches as the sheriff checks the locks on the gate and then leaves.

David then runs up to the gate and manages to squeeze through into the compound. Finding the plant door barred by another locked iron gate, he smashes the lock with a heavy metal pulley he finds on the ground and steps inside the building.

Inside, he finds strange machinery, not at all like a power generating station. As he enters he sets off a hidden, silent sensor. Unknown to him, miles away an alarm sounds, "Kinney station to Bakersfield Control, Red Alert, Red Alert." In a mysterious workshop, several men stop their activities, get into a big truck marked "Kogan Enterprises" and drive off toward Kinney.

Meanwhile, David discovers some strange pedestals. They have transparent tubes that lower down over them when he puts his hand close to the crystal at their center.[16] (Figure 13) Quickly he rushes out of the plant.

In town, Sheriff Carver is looking for him but David manages to avoid him and goes into the only café in town that is still open. There he phones Alan Landers and pleads with Alan to come to Kinney right away.

When Alan says he doesn't know if he can get away just then, David says, "They're here in Kinney! They may be all over this planet. Now I need one witness, just one!" David adds that, if he doesn't get there soon, his proof will be gone. Reluctantly, Alan agrees to come right away. David tells Alan to meet him at the hotel.

ACT IV

David discovers Sheriff Carver is looking for him. David tries to avoid the Sheriff by staying in the café. After a while Kathy comes in and, desperate for someone to talk to about all this, David starts telling Kathy about what he has seen. (After all, there is nothing abnormal about her hands.)

During their conversation he asks Kathy to telephone her aunt Sarah at the hotel and tell Alan to meet him at the café and to be sure not to let Sheriff Carver find out. Kathy phones her aunt from the café but, when her aunt answers the phone at the hotel, we discover it is the same old lady who was at the hospital and David's hotel room fire.

Meanwhile, the old man running the café closes up and lets David and Kathy continue to wait inside. While they are there, alone, Sheriff Carver comes along and rattles the door. Quickly, they both hide from the Sheriff.

At this moment Alan drives into town and pulls up to the hotel. He is confronted by Aunt Sarah. She asks if he

---

[16] David will later find out that these are rejuvenating stations that the Aliens must use regularly to maintain their human form.

Figure 14 - David discovers Kathy is an Alien

is looking for David and when he says yes, she tells Alan that David said for him to meet David at the power plant. Alan parks his car at the hotel and walks to the power station alone.

In the café, Kathy is trying to persuade David that his friend won't come. She says that nobody wants to know the truth; that it was the same way with her husband. She goes and turns on the jukebox. When David protests that he is listening for Alan, she said she did it to wake him up. She tells him, "Stop listening for cars that never come, looking for faces you can never find, fingers… We're not all like that David!"

Suddenly, he realizes that she, too, is an alien. She pleads with him, "Don't go! Don't fight us! *You can't stop it! It's going to happen*!"

Horrified, David backs away from Kathy. He realizes he has told everything to an alien he thought was a friend. (Figure 14) In a panic he rushes out of the café.

He runs toward the hotel. There he sees Alan's car. But before he can leave Sheriff Carver finds him. When David tries to get away to go after Alan, the sheriff puts him in handcuffs and throws him in jail.

Meanwhile, Alan is searching the power plant, looking for David. He turns a corner and sees two men removing the pedestals that David had seen earlier. He realizes that David was right. (Figure 15a)

He tries to leave, but as he does, other men in coveralls silently surround him. (Figure 15b) They drive him back to the area where there is one remaining pedestal and force him into it. A high-pitched sound is heard as the transparent cylinder lowers itself down over Alan's head. He *screams*! (Figure 15c)

EPILOG

[Note: This "Epilog" as previously mentioned, is a trailer to the story that QM often used to wrap up an episode in their various series. The Epilog to this chapter is in its usual place at the end.]

It's the next morning and a stretcher is taking Alan's body out to the ambulance. In handcuffs, David is escorted

Figure 15 - Alan discovers the machinery, is trapped by aliens and killed

289

Figure 16 - David has lost his best friend

out of jail and looks sadly down on his friend. (Figure 16)

Lt. Holman drives up, gets out, shakes hands with Sheriff Carver and, looking down at Alan's body, tells David that, "Well, he'd be alive today if he hadn't wanted to believe your crazy dream. Furthermore," he says, "Sheriff Carver checked out the power station. Except for him, there was nothing." David admits that there wouldn't be anything now.

"The Coroner's report says it was heart attack." Lt. Holman tells David, "what ever you think, that's the way it was." When the Sheriff asks Lt. Holman what he wants done with David, he replies, "Let him go." As the amazed Sheriff takes off the cuffs, he tells David not to come back. Dejectedly, David says there isn't any reason to anymore. (Figure 17)

David appeals one last time to Lt. Holman for help in finding an answer to all this. Lt. Holman replies that, for his

Figure 17 - No more reason, now

own good he should "let it end here." David tells him that he wishes he could, and walks to his car.

As David Vincent is driving out of Kinney, aliens in human form, that he cannot expose, are watching him.

And as he leaves, the announcer's voice is heard saying,

"How does a nightmare end? Not here in the forgotten town of Kinney. Perhaps at Bakersfield? Perhaps at some undiscovered beachhead in another state? ...or another continent?"

Perhaps, for David Vincent, it will never end.

The Vehicle

According to the research done by Mark Phillips, the saucer in *The Invaders* was only going to be used for the pilot episode. Quinn Martin didn't see any reason, with the format they had established, why it would be needed in any ongoing episode.

But two things worked against Martin. The marketing group at ABC realized the value of the saucer to the series. They probably pointed out (correctly) that it wouldn't be very convincing to have a superior alien race here on Earth without ever showing their means of getting back and forth. Besides, the saucer created a lot of interest in the series and ensured that it was considered a science fiction show and not just a remake of QM's earlier production, *The Fugitive.*

The second thing was that the writers kept working the saucer into the script. Show after show they had it as an element in the plot. As a result, the saucer became quite prominent in the series.

When audiences first looked at the saucer during the series debut, many must have felt that there was something very familiar about this spaceship. Indeed, this feeling may have helped people accept the saucer and that, in turn, helped people buy into the series.

I will examine why this saucer would have given the audience the feeling that they had seen it before in the "Design History" section. First, let's take a look at the design itself.

The Design

It's not clear who, exactly, designed the alien's saucer. The art director for the pilot episode (who is usually responsible for prop design) was James D. Vance, but it was built by the effects team led by Darrell Anderson (the son of Howard Anderson whose company produced the special effects).

After viewing a number of scenes and behind the scenes production stills, several interesting facts emerge about the saucer they used.

This flying saucer was more of a "hat" design, taller than others in this book. As viewed from the outside, its flat top and convex sloped sides gave it a slight similarity to a sun hat with a wide brim. However, there were many differences.

The saucer was mostly seen from the external side view, giving little information about what it looked like from above, below, or inside.

However, from the side, the first thing one would notice was the ring of brightly glowing objects that spun counter clockwise near its top. At first you would be tempted to consider these objects to be windows with light coming out from inside the ship. But then you would have to ask yourself, "Why would anyone want to spin the view ports like that?

From an operational point of view, unless the interior was rotating counter clockwise too, it would make no sense. And if the interior of the saucer were rotating like that, it would make *a lot* of trouble for the designer.[17] The more likely probability is that they were energy panels, not viewports, and that their spinning had something to do with the ship's power source or propulsion unit.

As usual for TV/movie props, they never built a complete saucer. Whenever we see all of it in the show, it is actually a process shot by the Howard Anderson Co. (Figure 18) Even the sequences where we see the saucer landed on the ground and people going in and out of it would be composite shots. The lower landing gear portion of the saucer was a full-size set and everything else above it was a matted miniature.

Figure 18 - The Saucer

This helps explain some of the inconsistencies about its appearance. For example, it kept changing its color slightly throughout the series. In the initial landing sequence, it is shown to be light silver blue. Later, during one of the close ups of the landing gear, it is seen to be metallic silver. Then

---

in one of the long range views of it, it again appears light blue and glowing.

To add to the confusion, a color photo taken behind the scenes of just the landing gear set show them to be a light blue; in other pictures they seem silver. With no way to resolve this conflict, we've listed both colors on the data drawing.

The saucer had five hemispherical domes on the lower side of the ship. It also had five landing legs that (presumably) rotated down to form a platform for the main body. On the end of each leg were pegs that extended independently to the ground to hold the main body of the saucer at level on any uneven surface.

Built into one of these legs were two ladders so crew and passengers could ascend up into the saucer. The ladder's lower half would extend to the ground by rotating down, out of the recess in the landing legs made by the upper half of the ladder. When they were closed, they folded together to make a smooth surface on the leg. (Figure 19)

Figure 19 - The Landing legs and ladders

The existence of two ladders would require two hatchways in the floor of the saucer's internal structure. Only one hatchway is ever seen in use during the series, but the other one is presumed to have been there.

The final feature on the lower side of the saucer was a central, circular, lens-like structure that glowed whenever the saucer hovered or flew. Presumably this is the main propulsion system. No information is known about its functions or makeup but it appears very similar to the ones on the saucers from *Forbidden Planet* and *Lost in Space,* whose power units were also glowing structures, centrally located on the underside.

The saucer's interior was also mysterious. There were only three episodes where David Vincent (and the TV viewers) went inside the saucer: *Dark Outpost, The Innocent,* and *Saucer.* In the *Dark Outpost* episode, David would only see a small room with no details at all. In *The Innocent* and

---

[17] Not to mention the passengers!

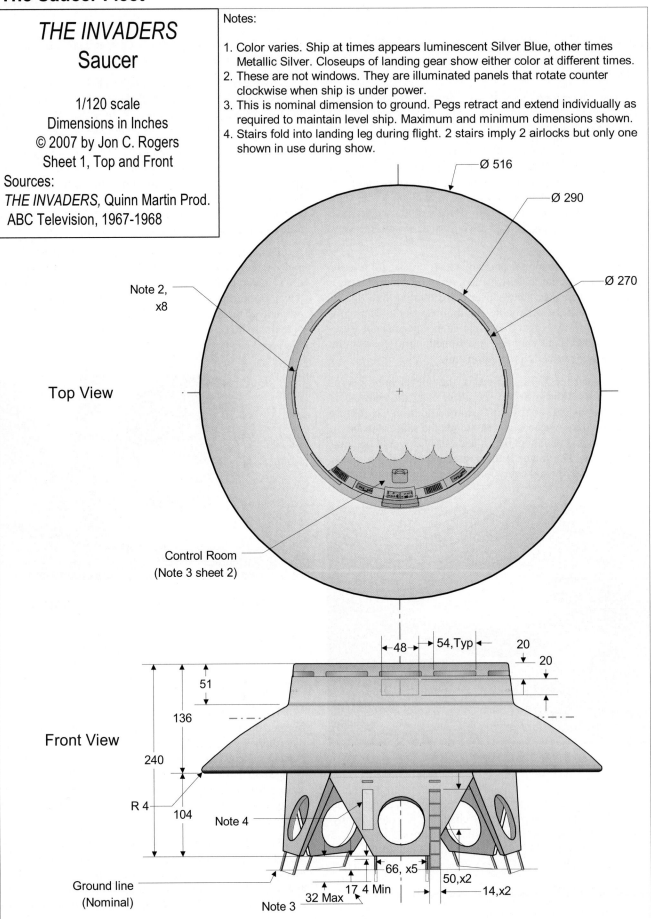

THE INVADERS
Saucer

1/120 scale
Dimensions in Inches
© 2007 by Jon C. Rogers
Sheet 1, Top and Front
Sources:
THE INVADERS, Quinn Martin Prod.
ABC Television, 1967-1968

Notes:

1. Color varies. Ship at times appears luminescent Silver Blue, other times Metallic Silver. Closeups of landing gear show either color at different times.
2. These are not windows. They are illuminated panels that rotate counter clockwise when ship is under power.
3. This is nominal dimension to ground. Pegs retract and extend individually as required to maintain level ship. Maximum and minimum dimensions shown.
4. Stairs fold into landing leg during flight. 2 stairs imply 2 airlocks but only one shown in use during show.

Ø 516

Ø 290

Ø 270

Note 2, x8

Top View

Control Room
(Note 3 sheet 2)

48   54, Typ   20   20

51

136

Front View

240

R 4

104

Note 4

Ground line
(Nominal)

Note 3   32 Max   17 4 Min   66, x5   50,x2   14,x2

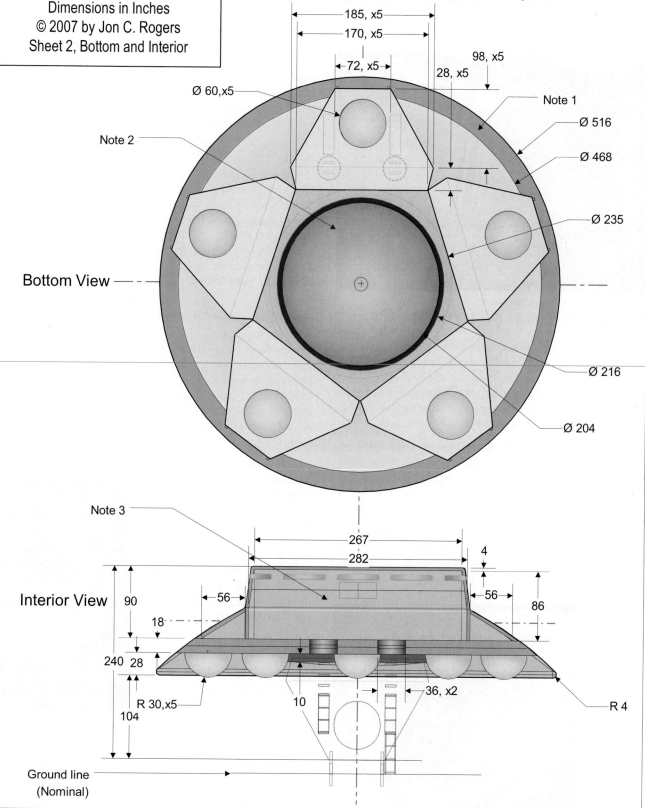

THE INVADERS
Saucer

1/120 scale
Dimensions in Inches
© 2007 by Jon C. Rogers
Sheet 2, Bottom and Interior

Notes:

1. Color of outer ring is undetermined. Best estimate is a dark blue.
   Inside of ring is same as ship outer skin. Hemispheres glow a light Blue.
2. Center Drive glows combination of Red, Orange, or White when in operation.
3. Interior layout of ship is not certain. See Text. Only the Pilot compartment
   is seen with any detail. Sheet 1 shows approximate layout.

185, x5
170, x5
72, x5
98, x5
28, x5
Ø 60,x5
Note 1
Note 2
Ø 516
Ø 468
Ø 235
Bottom View
Ø 216
Ø 204

Note 3
267
282
4
Interior View
56
56
90
86
18
240 28
R 30,x5
10
36, x2
R 4
104

Ground line
(Nominal)

The Saucer Fleet

Saucer episodes, he was only in the control room of the saucer.

In the control room we see the main view port, some controls, and seating for one pilot. (Figure 20) The position of the room, its relative size, and its shape are not clear. There are sliding doors to another room but we do not see much of its details or features.

Figure 20 - The Control Room

Analyzing the shape and size of the saucer, the internal volume would allow a circular area between 22 and 23½ feet (6.7 and 7.2 m) in diameter, and about 7' 2" (2.2 m) high. There would be additional triangular storage space of about 4½ feet (1.4 m) in width outside that main area. This space could also house machinery and/or equipment.

All this implies that the saucer was not a ship for long voyages. Perhaps the aliens used hibernation or other means, but there was certainly no room for living quarters during the voyage.

The only other area available inside the saucer was a very thick floor that the hatchway passed through. The hatch itself slid into this space. This section was a full 18 inches (46 cm) thick and 33½ feet (10.2 m) in diameter. It could have held quite a lot of machinery. (Figure 21)

Figure 21

The Hatch and Floor Thickness

Altogether, the design of the ship was quite straightforward. It was very much in keeping with saucers seen in the '50s. Like those earlier saucers, it was not a design that was intended for travel through interstellar, much less intergalactic space.

Still, this was sufficient for the series because in the days before actual space travel, interplanetary travel was also considered "too far away." The fact that the script called them "Aliens from a dying planet..." and the saucer was "a craft from another galaxy..." did not disturb audiences at the time because both planets and galaxies were in outer space. And anything in outer space was just too far away to comprehend.

Design History

If we go back to look at the saucer designs that were common in motion pictures and other entertainment media before *The Invaders*, we would conclude that this saucer design did not come from any of them. It does not resemble the saucers in *The Day the Earth Stood Still, Earth vs. the Flying Saucers, Forbidden Planet, This Island Earth, Twin Earths,* or *Lost in Space*. This design's inspiration must have come from somewhere else.

We need to look outside the entertainment industry for other saucer designs that influenced the public at the time. After the books and articles declaring that UFOs were flying saucers from outer space came out and the first big movies featuring alien flying saucers appeared, a new phenomenon emerged in society. Formerly, people had only claimed to have seen UFOs or flying saucers; now some people actually claimed to have been *in* a flying saucer and to have actually *met* the aliens who flew them. This new group became known as "Contactees."

The first of these was George Adamski. He claimed to have met a flying saucer, traveling alien from Venus in the desert outside of Desert Center, California on 20 November 1952. He told this story to reporters in 1953 and, in October of that year, published his first widely read book, *Flying Saucers Have Landed*. It detailed his experiences with the friendly aliens and contained photos of the saucers he allegedly rode in. Adamski also independently published the photos contained in his book in the March 1954 issue of *Mechanics Today*.

Figure 22

George Adamski became a public figure and the center of controversy after these public declarations. Many considered his statements and pictures to be nothing but crude hoaxes. Others firmly believed in him, while most people merely wondered. Regardless of one's belief, Adamski became very popular in the media. He was interviewed on network TV and gave many guest lectures. For a while, his stories and saucers were well known throughout America.

Adamski's popularity and the bizarre nature of his claimed experiences spawned a whole new cultural phenomenon. Nameriew contactees quickly sprung up, writing similar books about meeting similar kind aliens (termed "Space Brothers") and gathering followers into what became a cult movement. Space Brother conventions were held annually where their universal message (peace and a warning that Earthlings should "get their act together") was delivered to the true believers. The movement still survives to this day.

Quickspec: The Invaders Saucer

Vehicle Morphology	Saucer
Year	1967-68
Medium	Television Series
Designer	Vance/Anderson
Diameter	43 ft (13.1 m)
Height	11.3 ft (3.45 m)

Modelers' Note

Aurora released a 1/72 scale styrene kit of *The Invaders* saucer in 1968, and then did a re-release in 1975. In 1977 Aurora was bought by Monogram who reissued it in 1979, 1996 (after the 1995 *Invaders* TV movie), and once more in 2003. There were some changes to the box on each production run from major to minor. The biggest trend was watching the series fade from memory. The original release put the series logo prominently on the cover for maximum association, but by the first re-release (with the series seven years past) they'd done all-new cover art, re-christened it a "Flying Saucer" and reduced the series to a tiny mention below the name. The new art also changed the underside hemispheres from the series-correct blue to red.

1968

The 1979 Monogram release reverted to "U.F.O." (now with periods) and kept the series title. The 1996 release (not shown) had an identical cover except that the series name was removed. The most recent release, 2003, removed all of the international language flags to make room for a larger "U.F.O." title.

1975

The kit itself was always molded in silver-grey plastic except for the underside hemispheres. Originally they were the same silver-grey but were molded in red and then clear in the later releases. The dimensions of the saucer remained the same as the same molds were used.

The Invaders Saucer was also released in 1989-90 by Tsukuda of Japan, but since it was the same 1/72 scale silver-gray plastic, some modelers suspect that it may have still been produced by Monogram and repackaged by Tsukuda..

1979

It's important to realize that, while the outside envelope of the model is fairly accurate, the interior is largely imaginary since we didn't see much of it in the series. The control room is properly done but the other rooms are put together from set pieces in other places in the series, such as the rejuvenation chambers. Also, the placement of the rooms in an "X" pattern is complete speculation.

A wonderfully precise build of the kit was done by ace modeler Allen Ury. The images below are from his "Fantastic Plastic" website (http://www.fantastic-plastic) and are used with permission.

2003

All ©Allen Ury

The Saucer Fleet

George Adamski died in 1965, just a few years before the Russian *Venera* probes landed on Venus where Adamski stated many of his saucer-riding "Alien Brothers" came from. They proved that Venus could not possibly support life.[18] Still, during his lifetime, which ended just prior to *The Invaders* being on TV, his saucer designs were very familiar to many throughout the country.

When comparing the saucers in Adamski's book with the saucer in *The Invaders*, the relationship is obvious. The only major external differences are the number of globes underneath the saucer, the lack of portholes in its side, and a flat top instead of a rounded one as in Adamski's saucers.

Further evidence indicates that Adamski's saucers were derived from the description of saucers given in Frank Scully's 1950 book, *The Truth about the Flying Saucers*. In that book, Scully clearly described the saucers as being "equipped with landing gear which had steel-looking balls instead of wheels."

There is further evidence that Adamski knew Frank Scully and Silas Newton [the source of Scully's story] personally.[19] This would mean that the saucer in *The Invaders* could trace its heritage directly back to one of the earliest claimed flying saucer shapes. In spite of the fact that the Scully/Newton and Adamski saucers were *complete hoaxes,* they still made a significant impression on many people interested in the UFO phenomenon. And they were familiar shapes to almost all Americans of the time.

Seeing *The Invaders* come to earth in a saucer very similar to the one seen in Adamski's book (and subsequentially in media all over the world) would have made many TV viewers feel that they "had seen that saucer somewhere before" even if they didn't recognize its shape right away.

Which is just what the producers would have wanted.

Epilog

With the flying saucer in *The Invaders* series we have come full circle. We are back to the style of the saucers from the original post war UFO/flying saucer scare, the first books claiming that all UFOs were flying saucers from outer space, and the first contactees who claimed to have met the aliens flying them.

It had originally started as a genuine fear of something in the sky, something real, but strange and unknown. But the fear had evolved.

UFOs were, at first, feared to be rockets or missiles of unknown origin and therefore dangerous. Then the fear became the focus of tricksters and advertisers. When UFOs could not be identified, it was suggested that they came from some region still unexplored by humans, somewhere like outer space. Because that idea was so preposterous, it took some time to persuade the public. But with magazines, newspapers, comics, and movies all promoting the idea, after a while, it became "possible."

With that amount of acceptance, movies began exploring what the intentions of those proposed aliens in those flying saucers from outer space might be. The movies presented them as everything from benevolent to malevolent.

And now, finally, a TV series about UFOs had come out. *The Invaders* had gone back to the roots of the phenomenon. It was now presenting one of the original shapes associated with flying saucers. It was also re-creating the original emotion that their appearance in the sky above America had created.   *Fear.*

---

[18] At least not life as we know it. The surface temperature is nearly 900°F (470°C) and the pressure is the same as being more than 3,000 ft (925 m) down in the ocean. Additionally, that hellish atmosphere supports sulfuric acid clouds with acid concentrations about the same as water in the clouds of Earth's atmosphere.

[19] Both are referenced in Adamski's book and Newton is quoted as an "expert."

Afterword
What we've learned from *The Saucer Fleet*

By Jon Rogers

With *The Invaders* flying saucer we have come full circle.

We are back to the style of the first saucer promoted as a "real extraterrestrial-flying-saucer-spaceship."

We are back to the original emotion produced by the first mass UFO sightings---*fear*.

But although *The Invaders* paid homage to the flying saucer spaceship's birth, it was, by no means, the end of flying saucers. By the time *The Invaders* appeared on TV, the number of people who believed that extraterrestrials were visiting Earth in flying saucers had multiplied greatly. In fact, they had become an international community, a subculture of their own. That subculture still exists today.

The number of "flying saucer spaceships" also continued to grow in the entertainment industry. By this time, the idea that UFOs were flying saucers flown by extraterrestrials (the "extraterrestrial hypotheses") had gained sufficient notoriety and popularity that it had become profitable basis for entertainment. Few people realized how the idea had gained so much acceptability. Lets take a moment to skim over what we've just documented, and you will see just how and when the idea was "sold" to the public.

For the very first time right after WWII, the public had a new chronic fear in peacetime. That fear was generated by the unexpected Cold War and the very real threat of atomic bombs raining down from the sky. The news media's immediate reaction to the first UFO sightings in 1947 was a reflection of that period's Cold War fears. The UFOs were greatly feared because they might be enemy rockets or nuclear bombers coming to destroy us. The UFO sightings were quickly capitalized on by pranksters and advertisers who, among other things, played on that fear. This fear (of the first UFOs called "flying saucers") was also reflected in comic books and the first movie serial *Bruce Gentry-Daredevil of the Skies*. It was assumed they were flying bombs sent by some malevolent villain. The first serious suggestion that UFOs were from Space was more than two years later in an article by Donald Kehoe in December 1949.

Kehoe refused to admit that he didn't know what they were and that he could find no evidence of their origin. Therefore, lacking any positive supporting evidence, he stated they were "interplanetary" because he believed it to be the "only remaining possibility." Although sensational, the majority of the population thought Kehoe's opinion ridiculous. Released at the same time, the first movie, *The Flying Saucer,* presented saucers (UFOs) as the invention of a "lone scientist using a new principal" to create an advanced flying machine which could A-bomb us with impunity.

However, within a year Frank Scully's best selling book (based on the fraudulent statements of two con artists) expanded on Kehoe's idea in the minds of millions of readers. It claimed that the government was lying, that it had several crashed interplanetary saucers and that it was studying them in secret. Naturally, this scandalous, but widely read, book reinforced the extraterrestrial hypotheses, and started many people wondering. Then, *The Flying Disc Man From Mars* came out with both flying saucers *and* a villain from another planet. However, it was just a serial. The majority of the public still wasn't buying it. According to a Gallup poll in 1950, most people still thought the UFOs were secret weapons from…somewhere.

But then big Hollywood movies started promoting the idea. First, the truly frightening movie, *The THING!* appeared. It showed us a crashed saucer carrying an alien monster that wanted to eat us. Then came *The Day the Earth Stood Still* with a flying saucer flown by an alien who warned us that if we extended our violence out into space, they would destroy us. Soon, *TWIN EARTHS* advanced a similar idea in hundreds of newspapers all over the world. It showed us that flying saucers were spaceships from our sister planet behind the Sun. They were piloted by a technically superior race of women who were also willing and able to "kick our butts".

Less than a year later, the movie *War of the Worlds* showed us an evil group of Martians in their manta-ray saucers, bent on destroying us and taking our Earth for themselves. If we wanted war, they'd give us *War!* Two years later, in 1955, *This Island Earth* reversed the idea and suggested that, maybe the Aliens were in their own war and needed our help! Next came *Forbidden Planet* proposing that, in the far distant future, we would have our own flying saucers and explore interstellar space ourselves. The movie projectors had hardly cooled down on this new revelation when the aliens were back with a vengeance in *Earth vs. the Flying Saucers* to conquer and enslave us.

The Saucer Fleet

This all happened before Sputnik changed our outlook on space travel forever. It wouldn't be until the early 1960s that we could enjoy our own *Flying Saucer Ride* in Disneyland and also the TV exploits of a future "Space Family Robinson" as they got *Lost in Space* in their very own flying saucer, the *Jupiter 2*. But then, in 1967, the flying saucer aliens came back. *The Invaders* arrived to infiltrate, spy, kill and conquer us. Once again we were afraid. The circle was now complete.

These were the major entertainment events involving flying saucers, in the order they occurred during the first twenty years after UFOs were sighted in America, and as we have described in *The Saucer Fleet*. Observe that the last time any movie or other entertainment media proposed the idea that flying saucers were *not* extraterrestrial was in the beginning, in the 1950 movie, *The Flying Saucer*. And that movie came out just before Frank Scully and Donald Kehoe's sensationalizing books. After those books were published, flying saucers in entertainment were all spaceships from outer space. Just consider the movies *Close Encounters of the Third Kind* (1977), *The Flight of the Navigator* (1986), *Independence Day* (1996), *Mars Attacks* (1996), or *Men in Black* (1997) to see that it's still true. For over fifty years now, the only flying saucers in movies or on TV have either been spaceships flown by aliens or by humans in the far distant future!

So many people today believe that UFOs are alien flying saucers from outer space because they've grown up hearing it all their lives. In fact, the entertainment industry has been telling that to more than two generations now. Add to this the hundreds of books arguing for the extraterrestrial hypothesis and you can see the tremendous amount of effort that some people have expended trying to convince others of this one unproven belief.

Remember the two points I made in the first chapter: 1) "Movies can make you believe in almost anything" and, 2) "If you make a lie big enough, and keep repeating it enough times, people will begin to believe it."[1] That has already happened to us. However, now you know the truth. You know how the belief in extraterrestrial flying saucers started, how it grew, how the entertainment industry promoted it for the first twenty years and how it was firmly established in our society as an urban legend. In short, now you know how "the Saucer Fleet" came to be with us today. What happens in the future depends on you.

The truth is, UFOs are unknown. Flying saucers are products of our own imaginations.

J.C. Rogers

The Mainland, 2008

---

[1] Statement attributed to Joseph Goebbels, Nazi Minister of Propaganda, 1933-1945

Appendix

The Night The Saucer Fleet Bombed Bellingham!

By Jon Rogers

Bellingham?   Where's that?

Bellingham, Whatcom County, Washington State, USA, also known as Xwotqem in the local Lummi Indian language, is one of the more unknown and geographically isolated cities in the entire contiguous 48 United States. It sits in the farthest Northwestern most corner of the U.S. about 20 miles below the Canadian border, cut off from the rest of the country by the Cascade Mountain Range. Its more than 70,000 residents share a naturally beautiful agriculture and forested land that has more in common with the "Lower Mainland" area of British Columbia, Canada than it does the US. Even though it has little heavy industry and a lower than the state's average per capita income, Bellingham has a mild, cool climate and is ranked among the best cities in the country for its excellent air quality, community involvement, and desirability as a retirement location. Residents of this "City of Subdued Excitement" have little to complain of, other than their being behind the rest of the country when it comes to fads, the latest styles and politics. Altogether, Bellingham is a pleasant, quiet community in a far off forgotten corner of the country with little to no military significance.

Why would anybody want to bomb them? *Ever?*

Well, you discover strange things when you start digging into strange subjects. And as subjects go, the "Great Flying Saucer Scare of the summer of 1947" is one of the strangest. As I researched it, I noted that Bellingham did join the rest of the country in seeing flying saucers the day after they were first reported. But that was nothing unusual. So I focused my research on the "Saucer Panic" itself.

The general fear that the first saucer sightings caused was genuine. It is difficult for people of today to appreciate the intensity of the anxiety that the unexpected sighting of unknown unusual objects in the sky above them brought to the people of America during that early phase of the Cold War. It was a tense time and the appearance of threatening objects overhead seemed to drop the world's troubles right into the lap of individual citizens everywhere. To get the feel for the atmosphere of the times, here is one personal experience described by Eric Nesheim in the introduction of his book, *Saucer Attack!* [1]

> IT IS A SUMMER MORNING, early 1950s. A man drives his wife and two small children home from church through a quiet neighborhood in Marion Indiana. Suddenly, high overhead, he spots dozens of small, silvery discs fluttering slowly to Earth. He runs his car into the curb, leaps out, and abandoning his family, goes shouting down the street: "THEY'RE HERE! THE SAUCERS ARE LANDING! THEY'RE HERE!" When no one responds, the man starts pounding on people's doors. Still in the car, his mortified wife can only crouch down in the seat. People finally start to gather, wondering what the shouting is all about. At the same time, the discs start to hit the ground all around. Someone picks one up and says: "Hey, buddy, look. [paper] Pie plates with printin' on 'em. Musta dropped from a plane." The man who saw the saucers was my father. He won't admit it, and I don't remember it, but my mother has made it a family legend; our small contribution to the saucer craze.

Small contribution or not, this was the same kind of personal panic caused by Orson Welles' famous *War of the Worlds* broadcast in the late '30s. It was real. And it was also nationwide.

It was amazing how quickly American ingenuity reacted. While the first sightings were still "in the air," entrepreneurs and advertisers, eager to get in on the excitement, started using it to sell stuff! (Chapter 1 Figures 6 thru 9) Opportunists with new flying saucer hamburgers, toys, and fashions immediately popped up all over the country. Then some genius thought up the idea of dropping *real saucers* (paper plates) out of airplanes with advertising on them. And this new gimmick was just too good to pass up! It really got noticed!

As I researched, I found evidence of this new type of advertising campaign all over the country. It seemed spontaneous. Two recruiters used it to recruit soldiers for the Army on July 11, 1947 in Troy, New York *within two weeks of the original sighting!* Someone else used it on July 14, in Seattle, Washington to announce a local celebration, and an insurance agent used it on July 19 in Harrisburg, Iowa. This new gimmick had started a trend in advertising that would continue until 1954. For example, on May 20, 1950 in Lufkin, Texas, the V.F.W. used flying saucers to promote poppy sales. (Figure 1)

---

[1] *Saucer Attack!* Ó 1997, By Eric Nesheim, Kitchen Sink Press/General Publishing Group, Inc. Los Angeles, ISBN 1-57544-066-0

The Saucer Fleet

'FLYING SAUCERS' USED TO PROMOTE ARMY RECRUITING

Sergeant Drops Paper Disks From Plane Over Cohoes, Watervliet

Flying saucers fell on Watervliet, Cohoes and Albany yesterday afternoon but the air attack proved to be a scheme for promoting Army enlistments.

Investigation revealed the flying disks were paper plates with information attached concerning opportunities afforded youths by enlisting in the Regular Army.

T/Sgt. Joseph Brookstein, in charge of recruiting in Watervliet, disclosed he went up in a plane about 3:30 p.m. from Albany Airport and dropped the plates over Watervliet, Cohoes and Albany while flying at about 1,000 feet. Several reports came in today that the disks had been found.

Capitalizing on the current public fancy for celestial crockery, Sergeant Brookstein said approximately 300 "saucers" were dropped over the three cities. On each was pasted literature concerning the Army and its opportunities for youths today.

The single-wing Fairchild plane was borrowed from a friend and was piloted by M/Sgt. William Pelesz of Cohoes, also of the recruiting staff and. former Army pilot. Since the flight was not strictly S.O.P. as far as Army procedure was concerned, the two recruiting sergeants financed the flying saucer venture out of their own pockets.

SEARCHLIGHTS PLAY ON FLYING DISKS

First report of the so-called flying disks being seen from this area was made by Roland Peay, formerly of Seattle and Snohomish police forces who now is working relief on the Snohomish force. He reported to Police Chief Charles Adams that on Saturday at 11:30 p. m. while returning from Redmond to his home at Clearview south of this city he and his wife stopped and saw three of the circular contrivances.

Searchlights from Seattle were being plied on the disks which were of silver color and flying in a northerly direction over Snohomish, according to the report. Mr. and Mrs. Peay watched the trio for a brief while until they disappeared after they had gotten past the range of the searchlights.

Flying saucers dropped from the sky into the residence district of the city Saturday. They were paper plates advertising the Tyee celebration to be held at Marysville and were scattered from an airplane.

Genuine Flying Saucer Shower Here Today

The air over Harrisburg was full of "Flying Saucers" at 11:30 today.

There was no guess work about it, nor no "if, ands or buts."

The discus shower was real, bonafide and genuine, as all who picked up one of the "saucers" can tell you

They were dropped from a plane by Carl B. Brooks, general insurance agent for the Farm Bureau.

Service Officer John Hensley (right), Post 1836, Lufkin, Tex., and A. O. McQueen, flight instructor, display the "flying saucers" they launched over Lufkin.

The "Flying Saucer" Mystery

Figure 1 — Examples of the widespread use of real flying saucers in advertising

As I gathered more material about this advertising scheme for *The Saucer Fleet*, a strange thing happened. I began to suspect there was some kind of personal connection here. I couldn't quite put my finger on it. Then one night I had a dream, one of those fragments that come to you when you're semiconscious, before you're fully awake.

In the dream, I was a young boy, sitting on a small oval rug next to a large console radio, playing with it, trying to tune in something on the short wave band. The radio was sitting in one corner of an old fashioned, 1940's style room. The room was furnished with a large overstuffed leather chair and ottoman, a roll top desk, and a fireplace. I recognized it as the parlor in my grandparents' house in Bellingham. An old man, my Grandfather, was sitting in the chair reading his newspaper.

The old man looked over his newspaper at his grandson, "Bill?" he said.

"Yes, Granddad?" I replied.

"It says here there will be a whole fleet of flying saucers landing here in Bellingham tomorrow evening about 9 pm."

"WHAT!?" Came my shocked reply. If you ever wanted an eight-year-old boy's attention, this would certainly do it.

"Yep," he went on, "It says if you want to see them land you should be at the High School Stadium about 9 pm tomorrow night. What do you think? Do you want to go? I can take you down in the Buick."

"Wow! You BET!" my reply was enthusiastic. "Gee—Flying Saucers!!"

Then I woke up. Needless to say, this fragment of a dream really bugged me for days.

It was so familiar, like I should know it. And yet... Was it real or imagined? The connection it had to flying saucers was obvious enough. I had been researching enough to know that the whole flying saucer craze had started in Washington

State. Kenneth Arnold had sighted the first ones around Mt. Rainier just south of Seattle. But I hadn't found anything with regards to flying saucers landing in Bellingham.

Then, another night, more of the dream came to me.

The big Buick sedan slowed as it came near the football field. It was well lighted and there were other cars parked around it. People were gathering on the field like something big was in progress.

"There's one Granddad!" I had spotted a parking spot by the fence.

"I see it." He replied and slowly parked the car.

"Look, people are already here ahead of us."

Out in the middle of the field, at least 50 children and twice as many adults were milling around in random groups of five or ten people.

"Yes," he said, "But I don't think we've missed the saucers. It's not quite time yet."

"Let's go then!" I scrambled out of the car and onto the grassy field, my head bent backwards, scanning the black night sky.

"Granddad! —I think I hear something coming!!"

I could hear a steady drone coming out of the west. All heads went up and eyes peered into the heavens. Then suddenly, a searchlight turned on and stabbed a finger of light into the inky sky. Slowly, it began scanning back and forth.

"See anything yet, Bill?" my Granddad asked. "Not yet, Granddad" I replied.

An airplane appeared, and was silhouetted in the searchlight beam. The light followed it as it slowly flew over the field. The airplane flew out of the searchlight's beam and disappeared.

And then someone shouted, "Look! There they are!!"

Dozens—maybe hundreds of white discs had appeared in the sky, illuminated by the searchlight. At first they seemed bunched together but then they split apart and came floating silently down toward earth. People were running around trying to see where they would land. The night sky seemed full of them.

"Wow! Granddad!" I was jumping up and down. "Here come the Flying Saucers!!"

This one woke me up in the middle of the night! I could *see it!* The football field, the people, the searchlights, the sight of hundred of white round discs floating through the air, high above the stadium. This wasn't a dream. This was some kind of old, long forgotten memory resurfacing. I was *remembering this*.

Since I was now living close to Bellingham, I determined to visit the library there and see if I could find some kind of proof to support my memory of my own personal flying saucer story. After all, this was not the story I had been planning to tell in our book on famous fictional flying saucers.

For days, I poured through old newspaper microfilms, searching for a personal connection—what, I wasn't sure. I found the headlines that confirmed Arnold's UFO sighting. I read other stories that told of sightings which had occurred both locally and throughout the country. I sensed the people's fear toward the strange things in the sky seeping out of these old stories. Yes, the whole country had been affected by this phenomenon, even in this remote corner of the country. Of course, when I had lived around here as a child, I had been too young to take much notice of it.

However, when one isn't sure what month, or even what year something happened, there are a lot of old newspaper pages to search through. Days went by and I couldn't find any specific link in the newspapers to this event of flying saucers happening in Bellingham. When I inquired of the librarian about any other records that may have been kept on businesses in those days, I was referred to the local Whatcom County Museum of History and Art. Finding the Museum's archives located on the second floor, a very helpful and knowledgeable curator, Mr. Jeff Jewell, met me. I discovered that they had a very large collection of photographs from the 1930s to the 1970s that were taken for the local paper by the resident photographer, Jack Carver as well as another collection from an independent photographer, Frank Pitt.

The Saucer Fleet

More days followed looking through boxes of old photographs, taken with one of those old, large 5 by 7 press cameras with the big flash bulbs that you see in old movies from the 30s. There were pictures of car wrecks, officials opening businesses, parades with local beauty queens and…

Suddenly I was looking at three photos of kids standing outside an ice cream stand in the middle of the night. A powerful feeling of déjà vu swept over me. This *was IT*. This was the connection to my dream!

Figure 2 — The Grand Opening of Thompson's Ice Cream Freeze, 2 Dec. 1949
Photos by Frank M. Pitt, Whatcom Museum of History & Art

Looking at these old photos, it all came back to me now.

The multitude of glowing, white saucer shaped discs continued floating slowly down to earth. They were falling toward the groups of people. Several of them headed directly at my grandfather and me.

I ran excitedly around in circles with my hands reaching skyward.

My Granddad looked on and smiled. He watched as I reached up and plucked one of the floating discs out of the sky before it could land. And then another and another. I missed several and had to pick them up off of the ground. I went running toward him.

"Look what I got, Granddad!" In my hands were several white paper plates with writing on one side.

The Freezer[2]
Good for 1 FREE Ice Cream Cone!
Tonight only!

"Well, let's go get you one." He smiled. It would be past my bedtime but it was worth it this once. With that, he started up the Buick and we headed to the ice cream stand.

When we arrived, we could see the line of kids already milling about. (Figure 3) It was going to be a bit of a wait, but again, it would be worth it.

I still remember sitting in the rear seat of the old Buick later that night, on the way home, happily munching on my free ice cream cone.

----

[2] Although *Thompson's* was the first to open on 2 Dec., developers Glen Sornberger and Earl Stubler opened their nearly identical stand called, "*The Freezer* #1 on 31 Dec. 1949 at Dupont & "I" streets. While it is believed that they were responsible for the innovative saucer advertising campaign, these are the only pictures I could find of these two, nearly identical ice cream stands; the first ones in Whatcom County.

Today, decades later, I look back at that tense, post World War II period of the early '50s, when people were frightened and many looked up in the sky and imagined they saw flying saucers. I think how ironic it was that, simultaneously, some people were actually seeing real ones, and a few--just a few--were getting free ice cream just for catching one, like I did, the night Bellingham was bombed with a fleet of Flying Saucers!

Figure 3 — Who wants a free cone?
Photo by Frank M. Pitt, Whatcom Museum of History & Art

About the Authors

Jack Hagerty

Jack is a consulting engineer to the semiconductor and medical equipment industries, his current assignments involving both plasma etchers and plasma separators.

He founded ARA Press to publish and distribute *Spaceship Handbook*, but is more than happy to let Apogee take over those tasks for *The Saucer Fleet*. Now maybe the next one won't take six years!

After *Spaceship Handbook*, people began to get the idea that he might know something about the subject. He's been interviewed multiple times on the Sci-Fi interview webcast "Hour 25" and on the SETI Institute's "Are We Alone" program to comment on the eve of SpaceshipOne's historic first flight.

On the occasion of publishing The *Saucer Fleet*, he's been warned that he'll now have lots of "friends" in the saucer community whether he wants them or not!

Photo by Bronwyn Hagerty

Jon Rogers

Although Jon Rogers was born, raised, and now resides north of Seattle, Washington, for many years he called the SF Bay Area his home. There he was an electronics and Aerospace engineer responsible for building microwave systems for the Space Shuttle, as well as the TDRSS, IntelSat IV, Goes, and SCS series Satellite systems. He holds a degree in Industrial and Systems Engineering from SJSU, Certifications in Program Management from UC, and Manufacturing Engineering from SME.

Jon has been involved with Rockets and Aerospace his entire career. A witness to the Mercury space program, he was a Nike Fire Controller in ARADCOM, a communications technician at White Sands Missile Range, and a QA Inspector on the Apollo/LEM Comm. Antennas all before he became an engineer working on the Space Shuttle and Satellites.

Since then, he has studied the development of spaceships from concept to reality becoming a recognized spaceship archeologist. He co-authored and illustrated *The Spaceship Handbook,* as well as articles on spaceships for *Filmfax* magazine and the AIAA Houston's newsletter, *Horizons (April, 2008)*.

He was a speaker at the National Association of Rocketry Convention in Austin (NARCON 2002), the American Astronautical Society (AAS) 50th Annual National Convention (2003), and was interviewed along with co-author Jack Hagerty on *Mike Hodel's Hour 25* online Radio program at the San Jose World Science Fiction Convention (ConJose).

He is presently developing a website (rogersrocketships.com), dedicated to the history and Art of the spaceship.

About the Artist, Kristen Harber

Here I am in my natural habitat.

I've known Jack for a long time now, and I'm always glad to be included in one of his projects. The rest of the time I keep busy with my own painting, various athletics, and music. These days, while still taking care of my family, I teach fencing at a prep school in Oakland. I'm an avid reader and sometimes write things. I watch far more movies than is probably good for me, and some of them even feature *flying saucers* — K.H.

Photo by Geoffrey Kahler

About the Painting

In our cover illustration, the "saucer fleet" is approaching from deep space. The background is based on several Hubble images of emission nebulae, which Kristen interpreted using her own techniques.

Spearheaded by the *Jupiter 2* (which is already on the ground and surrounded by an aurora-like halo), the rest of the fleet is preparing to land. The ship from *Earth vs. the Flying Saucers* is next, extending its landing pylon while being protected by the green neutrino beam from the Metaluna ship above. In the foreground, the C57-D and Klaatu's ship are hovering, making their final preparations. Gliding in from the right is one of the locals. Is it friend or foe? We can only wonder how the meeting will play out. The other craft, hovering about near and far, well, we'll let you figure those out on your own.

The Saucer Fleet

Index

The Saucer Fleet

The Saucer Fleet

The Saucer Fleet

The Saucer Fleet